Wave Equations in Higher Dimensions

Shi-Hai Dong

Wave Equations in Higher Dimensions

Springer

Shi-Hai Dong
Escuela Superior de Física y Matemáticas
Unidad Profesional Adolfo López Mateos
Instituto Politécnico Nacional
Edificio 9
Mexico DF 07738
Mexico
dongsh2@yahoo.com

ISBN 978-94-007-1916-3 e-ISBN 978-94-007-1917-0
DOI 10.1007/978-94-007-1917-0
Springer Dordrecht Heidelberg London New York

Library of Congress Control Number: 2011933232

Cover design: VTeX UAB, Lithuania

Printed on acid-free paper

Springer is part of Springer Science+Business Media (www.springer.com)

This book is dedicated to my wife Guo-Hua Sun and lovely children Bo Dong and Jazmin Yue Dong Sun.

Preface

This work will introduce the wave equations in higher dimensions at an advanced level addressing students of physics, mathematics and chemistry. The aim is to put the mathematical and physical concepts and techniques like the wave equations, group theory, generalized hypervirial theorem, the Levinson theorem, exact and proper quantization rules related to the higher dimensions at the reader's disposal. For this purpose, we attempt to provide a comprehensive description of the wave equations including the non-relativistic Schrödinger equation, relativistic Dirac and Klein-Gordon equations in higher dimensions and their wide applications in quantum mechanics which complements the traditional coverage found in the existing quantum mechanics textbooks. Related to this field are the quantum mechanics and group theory. In fact, the author's driving force has been his desire to provide a comprehensive review volume that includes some new and significant results about the wave equations in higher dimensions drawn from the teaching and research experience of the author since the literature is inundated with scattered articles in this field and to pave the reader's way into this territory as rapidly as possible. We have made the effort to present the clear and understandable derivations and include the necessary mathematical steps so that the intelligent and diligent reader is able to follow the text with relative ease, in particular, when mathematically difficult material is presented. The author also embraces enthusiastically the potential of the LaTeX typesetting language to enrich the presentation of the formulas as to make the logical pattern behind the mathematics more transparent. In addition, any suggestions and criticism to improve the text are most welcome. It should be pointed out that the main effort to follow the text and master the material is left to the reader even though this book makes an effort to serve the reader as much as was possible for the author.

This book starts out in Chap. 1 with a comprehensive review for the wave equations in higher dimensions and builds on this to introduce in Chap. 2 the fundamental theory about the SO(N) group to be used in the successive Chaps. 3–5 including the non-relativistic Schrödinger equation, relativistic Dirac and Klein-Gordon equations. As important applications in non-relativistic quantum mechanics, from Chap. 6 to Chap. 12, we shall apply the theories proposed in Part II to study some important quantum systems such as the harmonic oscillator, Coulomb potential, the Levinson theorem, generalized hypervirial theorem, exact and proper quantization rules and Langer modification, the Schrödinger equation with position-dependent mass and others. We shall illustrate two important applications in relativistic Dirac and Klein-Gordon equations with the Coulomb potential in Chaps. 13 and 14. As crucial generalized applications of Dirac equation in higher dimensions, we shall study the Levinson theorem, generalized hypervirial theorem and Kaluza-Klein theory in Chaps. 15–17. Some conclusions and outlooks are given in Chap. 18. Some useful reference materials such as group theory, group representations, fundamental properties of Lie groups and Lie algebras, the angular momentum theory and the confluent hypergeometric functions are sketched in Appendices A–E.

This book is in a stage of continuing development, various chapters, e.g., on the quantum gravity, on the Kaluza-Klein theory, on the supersymmetry and string theory, on the high dimensional brane will be added to the extent that the respective topics expand. At the present stage, however, the work presented for such topics should be complete enough to serve the reader.

This book shall give the theoretical physicists and researchers a fresh outlook and new ways of handling some important and interesting quantum systems in several branches of physics. This book can be used by graduate students and young researchers in physics, especially theoretical and mathematical physics. It is also suitable for some graduate students in theoretical chemistry.

Mexico city, Mexico Shi-Hai Dong

Acknowledgements

First, I would like to thank Prof. Zhong-Qi Ma for encouragement and continuous support in my research. Prof. Ma was always there to give me some positive and helpful suggestions.

Second, I shall give special thanks to Profs. E. Ley-Koo, S.A. Cruz and L.H. Fabila Castillo, who have offered me unselfish help in my life. In particular, I would like to thank Prof. Ley-Koo for many positive and invaluable suggestions on this topic in 2004. The mild and bracing climate in Mexico city has kept me in a good spirit.

Third, I am very indebted to my collaborators Profs. Child, Lin, Lemus, Frank, Lozada-Cassou, Popov, Vezerra, Gu, Sun, Tang, Lara, Qiang, Chen, Yu, García, Peña, Wei and others. I also would like to thank Prof. Alwyn van der Merwe for greatly supporting this book proposal. In particular, I really appreciate the support from projects 20100297-SIP-IPN and 20110491-SIP-IPN, Mexico.

Fourth, I thank my previous supervisor Prof. Feng Pan for getting me interested in group theory and Lie algebras.

Last, but not least, I would like to thank my family: my wife, Guo-Hua Sun, for giving me continuous encouragement and for devoting herself to the whole family except for her own's study and job; my lovely son and daughter, Bo Dong and Jazmin Yue Dong Sun for giving me encouragement; my parents, Ji-Tang Dong and Gui-Rong Wang, for giving me life, for unconditional support and encouragement to pursue my interests, even when the interests went beyond boundaries of language and geography; my older brother and sister S.S. Dong and X.F. Dong for looking after our parents meticulously in China.

Shi-Hai Dong

About the Authors

Prof. Shi-Hai Dong got his PhD at Institute of High Energy Physics, Chinese Academy of Sciences, China in 1999. After that he joined the Physical Theoretical Chemistry Laboratory of University of Oxford as a Visiting Professor, the Department of Physics of Kansas State University as a Postdoc fellow, the National Autonomous University of Mexico as a Visiting Professor, the Mexican Institute of Petroleum as a Postdoc fellow and Distinguished Visiting Professor, the Superior School of Physics and Mathematics, National Polytechnic Institute as a full-time Professor through Program of Excellent Position and now has a permanent position of Professor at that Institute. He is the Editor of Central European Journal of Physics and Editorial Board Members of The Open Nuclear & Particle Physics Journal and Advances in High Energy Physics. He has authored more than 140 papers, one book Chapter and one book with over 1500 citations excluding self-citation. His research interests include algebraic method and quantum physics.

About the Authors

Contents

List of Figures

List of Tables

Part I
Introduction

Chapter 1
Introduction

1 Basic Review

The exact solutions of wave equations with a spherically symmetric potential have become an important subject in quantum mechanics [1–6]. It should be noticed that many works along this line have been carried out in the usual three dimensional space. However, what extra dimensions could there possibly be if we never see them? It turns out that we do not really know yet how many dimensions our world has. Nevertheless, all that our current observations tell us is that the world around us is at least $(3 + 1)$ dimensional space-time as illustrated in general relativity.

The idea of extra dimensions has a rich history, dating back at least as far as the middle of 1910s and earlier 1920s when the Nordstrom-Kaluza-Klein theory[1]—usually named as the Kaluza-Klein theory—was proposed [8–11]. This theory is a physical model that seeks to unify two fundamental forces of gravitation and electromagnetism. More precisely, the idea of additional spatial dimensions is from string theory, the only self-consistent quantum theory of gravity so far. For a consistent description of gravity, scientist needs more than $(3 + 1)$ dimensions, and the world could have up to 11 or more spatial dimensions. The reason why we do not feel these additional spatial dimensions in our life is because they are very different from the

[1]The theory was first proposed by the Finnish physicist Gunnar Nordström in 1914. Before Einstein's general relativity theory was presented, Nordström proposed a relativistic theory for gravity. He unified his gravity theory with Maxwell's electromagnetism through introducing a 5-vector gauge field where the first four components are identified with Maxwell's vector potential A^μ and the 5th component with the scalar gravity field. After that in 1919 a German mathematician Theodor Kaluza performed similar calculations but with Einstein's gravity theory and Maxwell's electromagnetism. In terms of a circular extra dimension Kaluza obtained a 4-dimensional action from a 5-dimensional one. The 4-action contained a graviton, an Abelian gauge boson identified as the photon and a scalar field that Kaluza put to be constant. The resulting equations can be separated out into further sets of equations, one of which is equivalent to Einstein's field equations, another set equivalent to Maxwell's equations for electromagnetic field and the final part an extra scalar field now termed the "radian". In 1926, it was Swedish physicist Oskar Klein who focused on the resulting higher modes of the particles and the size of the extra dimension [7].

S.-H. Dong, *Wave Equations in Higher Dimensions*,
DOI 10.1007/978-94-007-1917-0_1,

three dimensions. It is possible that our world is "pinned" to a three-dimensional so-called brane located in a higher dimensional space. We could be restrained to a usual three-dimensional world, which is in fact a part of a more complicated multi-dimensional universe.[2] Perhaps, we could feel these extra dimensions through their effect on gravity. While the forces such as the electromagnetic, weak, and strong interactions that hold our world together are constrained to the $(3+1)$ dimensions, the gravitational interaction always occupies the entire universe, thus allowing it to feel the effects of extra dimensions. Unfortunately, since gravity is a very weak force and the radius of extra dimensions is tiny and as large as 1 mm so that the gravitational interaction between them becomes very weak. Until now, no evidence for extra dimensions was found from the high-energy particle accelerators experiments, but we cannot say that they do not exist at all. The search for extra dimensions is not over yet. On the contrary, it has only just started. Scientists have been looking for the effects of extra dimensions in collisions that produce different types of particles, such as quarks and searching events where gravitons are produced in the collisions and then leave our three-dimensional world, traveling off into one of the other dimensions [12].

We have noticed that almost all works about higher dimensional wave equations addressed the generalized orbit angular momentum [13–15], in which Louck studied the harmonic oscillator potential as an exactly solvable model. In fact, such a generalization should go back to the earlier works by Appel, Fock, Bargmann, Sommerfeld *et al.* [16–19], the notes left by Bateman edited by Erdélyi in 1950s [20] and others [21]. Most of them paid more attention to the harmonic oscillator [13–15, 22, 23] than hydrogen atom [24–29]. Following Louck's work, de Broglie and his collaborators [30] proposed the generating bases as the hyperspherical harmonics to analyze the higher dimensional harmonic oscillator and molecular vibration. They considered the rotator model of elementary particles as relativistic extended structures in Minkowski space under the assumption that elementary particles are not pointlike, but are rather, extended structures in Minkowski space. Two years later, Granzow presented orthogonal polar coordinate systems in N dimensional space and showed explicit representations for total orbital angular momentum operator [31]. He also proved that the transformation from polar coordinates to Cartesian ones has a unique form $x^n = Rf^n(\theta)$, $n \in Z$; $\theta = (\theta^1, \theta^2, \dots, \theta^{N-1})$, where x^n could be interpreted as the wavefunction in quantum system. Based on the generalized orbital angular momentum theory, Bergmann and Frishman established the relation between the hydrogen atom and multidimensional harmonic oscillator by performing simple transformations on wave equations and wavefunctions [32]. Following this, Čížek and Paldus presented a relation between them for the special case of even dimensions [33]. Kostelecky, Nieto and Truax obtained a more general

[2]This is just like an insect crawling on a sheet of paper. For this insect, the universe is pretty much two-dimensional since it cannot leave the surface of that paper. As a result, the insect only knows the surface, but up and down does not make any sense as long as it has to stay on the sheet of that paper. These extra spatial dimensions, if they really exist, are thought to be curled-up, or "compactified".

mapping for arbitrary d and even D that involves a free parameter along with the corresponding mappings to the supersymmetric partners of these systems [34], in which they adopted the results about the D-dimensional oscillator with spin-orbit coupling obtained by Balantekin [35]. One decade later, Kostelecky and his collaborator Russell restudied this topic, but following the supersymmetry-based quantum defect theory [36]. Among the special cases is an injection from bound states of the three-dimensional radial Coulomb system into a three-dimensional radial isotropic oscillator where one of two systems has an analytical quantum defect. Also, they considered the issue of mapping the continuum states [36]. It should be pointed out that most of contributions about the relationship between the hydrogen atom and harmonic oscillator in D dimensions are based on the transformation of the radial equations.

Closely related to this, however, Zeng, Su and Li have made use of algebraic method, i.e., an $su(1,1)$ algebra as a bridge to establish a most general and simplest relationship between their energy levels and eigenstates [37]. Similar to this, Lévai, Kónya and Papp proposed a unified $su(1,1)$ algebraic treatment to the Coulomb and harmonic oscillator potentials in D dimensions by using Green's operator calculated from a Hilbert basis and the generalized Coulomb-Sturmian basis [38]. Except for these relations, it is noticed that there exist the degeneracies between the hydrogen atom and harmonic oscillator. For example, Shea and Aravind studied the degeneracies of the spherical well, harmonic oscillator and hydrogen atom in arbitrary dimensions from the view point of group theory [39]. In a similar way, Jafarizadeh, Kirchberg and their coauthors investigated the degeneracies of the Coulomb potential in higher dimensions d by using the irreducible representations of the group $SO(d+1)$ [40, 41]. The reason why the harmonic oscillator and hydrogen atom are taken as typically soluble models is because their study represents an interesting field of mathematical physics in itself, but more importantly results from them are essential for the description of realistic physical problems.

Obviously, there are no more essential advances on the higher dimensional wave equations in 1970s. On the contrary, the study on this field has revived and attracted much attention to many authors in 1980s, e.g., the eigenvalues of the Schrödinger equation for spherically symmetric states for various types of potentials in N dimensions by using perturbative and non-perturbative methods [42], the $1/N$ expansion technique for the Schrödinger equation [43–53], the generalized D-dimensional oscillator [54]. It should be noticed that the special case about the $1/N$ method was extended by Papp [55], who dealt with the q-deformed radial Schrödinger equation in N dimensions through the underlying $SO(N)$ group realized in Refs. [56, 57] and opened a new way to derive q-deformed $1/N$-energy formulas for arbitrary spherically symmetrical potentials such as the harmonic oscillator and the Coulomb potential.

Except for these, the higher dimensional Schrödinger equation are also concerned with the following scattered fields such as the position and momentum information entropies of the D-dimensional harmonic oscillator and hydrogen atom [58], the Fermi pseudo-potential in arbitrary dimensions [59], the uncertainty relation for Fisher information of D-dimensional single-particle systems with central potentials [60], the dimensional expansion for the Ising limit of quantum field theory [61], the scalar Casimir effect for an N-dimensional sphere [62], the multidimensional extension of a WKB improvement for the spherical quantum billiard zeta functions [63], the study of bound states in continuous D dimensions [64], the suppersymmetry and relationship between a class of singular potentials in arbitrary dimensions [65], the bound states and resonances for "sombrero" potential in arbitrary dimensions [66], the renormalization of the inverse squared potential in D dimensions [67], the generalized coherent states for the d-dimensional Coulomb problem [68], the quantum particles trapped in a position-dependent mass barrier [69, 70], the harmonic oscillator in arbitrary dimensions with minimal length uncertainty relations [71], the stable hydrogen atom in higher dimensions [72], the relation between dimension and angular momentum for radially symmetric potential in D-dimensional space [73], the D-dimensional hydrogenic systems in position and momentum spaces [74], the first-order intertwining operators and position-dependent mass Schrödinger equation in d dimensions [75], intertwined isospectral potentials in arbitrary dimensions [76], convergent iterative solutions for a sombrero-shaped potential in any space dimension and arbitrary angular momentum [77].

On the other hand, a number of contributions related to the higher dimensional Schrödinger equation have been carried out in atomic physics. For example, Hosoya investigated the hierarchical structure of the set of atomic orbital wavefunctions of D-dimensional atoms by using the set of their rectangular coordinate expressions [78]. In terms of group theory Dunn and Watson developed a formalism for the N electron D-dimensional Schwartz expansion and applied it to study the Schrödinger equation for two-electron system [79, 80]. However, their method seems rather complicated. To overcome the difficulty occurred in [79, 80], Ma and his coauthors made use of the group theory method [81] to develop a different formalism to separate the D-dimensional rotational degrees of freedom from the internal degrees of freedom. They have studied quantum three-body system [82], interdimensional degeneracies for quantum three-body and N-body systems [83, 84], the quantum four-body system [85] and the D-dimensional helium atom [86].

As illustrated above, we find that most of contributions have been made to higher dimensional Schrödinger equation. In comparison with the non-relativistic Schrödinger equation case, undoubtedly the studies of relativistic Dirac and Klein-Gordon equations in higher dimensions seem less than those in the Schrödinger equation case. Nevertheless, there are considerable works appearing in the literature. For example, Nieto dealt with the hydrogen atom in arbitrary dimensions D and particularly studied the Klein-Gordon equation case [87]. This might be the earliest contribution to the generalized Klein-Gordon equation, to our best knowledge. In fact, such a generalization can be easily achieved from the Schrödinger equation since the same Laplacian is involved for both equations. On the other hand, Joseph made a great contribution to self-adjoint ladder operators [88–90], in particular he applied this method to study the solutions of the generalized angular momentum problem. This revealed many interesting aspects of this approach to eigenvalue problems and specially its relationship to the addition of angular momentum. In that work, he obtained a complete set of irreducible unitary representations of the underlying algebra so(n) and calculated the corresponding Clebsch-Gordon coefficients

(CGCs) for the addition of spin and angular momentum in arbitrary dimensions. Without doubt, this shall provide some useful preliminaries including the spin algebra and Dirac operators to study the Dirac equation in higher dimensions [91], which was derived by using the fundamental properties of symmetry group SO(N). Unfortunately, we have not recognized his work [88–90] before our study [91]. In the middle of 1980s, Bollini and Giambiagi extended the Wess-Zumino model to higher dimensions, which led to a generalized Klein-Gordon equation [92]. In terms of the $1/N$ expansion technique, the relativistic Dirac and Klein-Gordon equations were performed [93–98]. Lin carried out the path integration of a relativistic particle on an N-dimensional sphere [99]. Recently, we have studied the Klein-Gordon equation with a Coulomb potential in N dimensions by traditional approach [100]. Others related to the Klein-Gordon equation with the Kratzer and pseudoharmonic potential potentials as well as the comparison theorems for the Klein-Gordon equation [101–103] have also been studied.

For solvable higher dimensional wave equations, since the energy levels depend on the dimension N and then bound state energy levels in different dimensions would be of interest. With this spirit we have studied the effect of dimension N on the energy levels for some interesting and important quantum systems. For example, we have dealt with the higher dimensional Klein-Gordon equation case [100], the Dirac equation with a Coulomb potential [104], the D-dimensional relativistic equations with a Coulomb plus a scalar potential [105, 106], the D-dimensional Schrödinger equation with the pseudoharmonic potential and the Coulomb plus an inverse squared potential [107, 108]. On the other hand, we have established the Levinson theorem[3] for the Schrödinger equation and Dirac equation in N dimensions [111, 112] and obtained the generalized hypervirial theorem [113, 114].

One of the reasons why the higher dimensional theories have attracted much attention to many authors is that the higher dimensional theories allow us to reduce enormous amounts of information into a concise, elegant fashion that unifies the two great theories of the 20th century: Quantum Theory and Relativity. It is evident to show that the contributions mentioned above are made within the framework of quantum theory. Consequently, it is necessary to review the development of the relativity and gravity in higher dimensional wave equations for completeness. For example, based on our recent work [91] Lin studied the Friedel sum rule, the Levinson theorem and the Atiyah-Singer index [115, 116]. Such method was also generalized to quantum modes of the scalar field on AdS_{d+1} space-time [117] as well as geometric models of the $(d+1)$-dimensional relativistic rotating oscillators [118]. More importantly, it should be noted that the generalization of the Dirac equation to higher dimensions might shed light on the solution of the Kaluza-Klein theory in higher dimensions if the extra dimensions are space-like. This theory has become a focus of attention for many particle physicists in past several decades. Its revival stems from the work on the string theory and also from

[3]It was first proposed by Levinson in 1949 [109] and reviewed by Ma [110]. The Levinson theorem establishes the relation between the number of the bound states and the phase shift of the scattering states at the zero momentum.

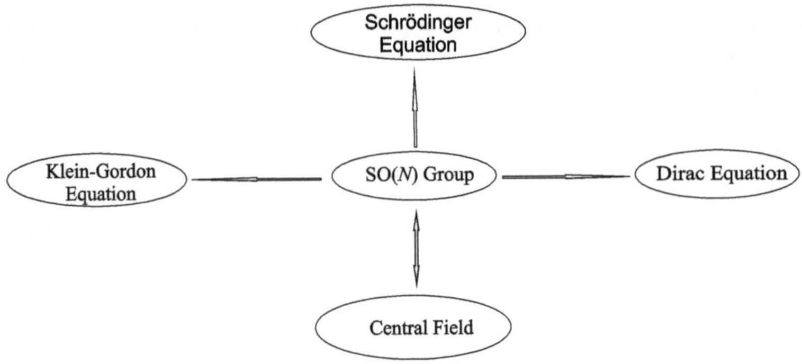

Fig. 1.1 The relations among the SO(N) group, central fields, non-relativistic and relativistic equations

the supergravity theory. Until now, the study of gravity theory and other relevant fields has become a main and interesting topic. These contributions can be summarized as follows: the brane models [119], scalar field contribution to rotating black hole entropy [120], brane cosmology [121], N-dimensional Vaidya metric with a cosmological constant in double-null coordinates [122], the spherical gravitational collapse in N dimensions [123], the motion of a dipole in a cosmic string background [124], repulsive Casimir effect from extra dimensions and Robin boundary conditions [125], extremal black hole/CFT correspondence in gauged supergravity [126], massive fermion emission from higher dimensional black holes [127], magnetic and electric black holes [128], fermion families from two layer warped extra dimensions [129], quasinormal behavior of the D-dimensional Schwarzschild black hole [130], the study of the Schrödinger-Newton equations in D dimensions [131], rotating Einstein-Maxwell-Dilaton black holes in D dimensions [132], the Kaluza-Klein theory in the limit of large number of extra dimensions [133], gauge invariance of the one-loop effective potential in $(d+1)$-dimensional Kaluza-Klein theory [134] and the multicentered solution for maximally charged dilaton black holes in arbitrary dimensions [135].

Heretofore, it should be emphasized that the symmetry group SO(N) for the symmetrically central fields plays an important role in higher dimensional wave equations. Therefore, we shall outline this group in next Chapter. The relations among those related topics are shown in Fig. 1.1.

2 Motivations and Aims

The motivations of this work are as follows. Since the literature related to this field is inundated with scattered articles on this topic we try to give a comprehensive review of the wave equations in higher dimensions and their wide-spread applications in quantum mechanics, which shall fill the gap in the existing quantum mechanics textbooks. In particular, we attempt to make use of fundamental properties of the rotational group $SO(N)$ to study the higher dimensional wave equations with symmetrically central fields. In this book, we are going to put the mathematical and physical concepts at the reader's disposal and to pave the reader's way into this territory as rapidly as possible.

Part II
Theory

Chapter 2
Special Orthogonal Group SO(N)

1 Introduction

Since the exactly solvable higher-dimensional quantum systems with certain central potentials are usually related to the real orthogonal group O(N) defined by orthogonal $n \times n$ matrices, we shall give a brief review of some basic properties of group O(N) based on the monographs and textbooks [136–140]. Before proceeding to do so, we first outline the development in order to make the reader recognize its importance in physics.

We often apply groups throughout mathematics and the sciences to capture the internal symmetry of other structures in the form of automorphism groups. It is well-known that the internal symmetry of the structure is usually related to an invariant mathematical property, and a set of transformations that preserve this kind of property together with the operation of composition of transformations form a group named a symmetry group.

It should be noted that Galois theory is the historical origin of the group concept. He used groups to describe the symmetries of the equations satisfied by the solutions of a polynomial equation. The solvable groups are thus named due to their prominent role in this theory.

The concept of the Lie group named for mathematician Sophus Lie plays a very important role in the study of differential equations and manifolds; they combine analysis and group theory and are therefore the proper objects for describing symmetries of analytical structures.

An understanding of group theory is of importance in physics. For example, groups describe the symmetries which the physical laws seem to obey. On the other hand, physicists are very interested in group representations, especially of the Lie groups, since these representations often point the way to the possible physical theories and they play an essential role in the algebraic method for solving quantum mechanics problems.

As a common knowledge, the study of the groups is always related to the corresponding algebraic method. Up to now, the algebraic method has become the subject of interest in various fields of physics. The elegant algebraic method was first

S.-H. Dong, *Wave Equations in Higher Dimensions*,
DOI 10.1007/978-94-007-1917-0_2, © Springer Science+Business Media B.V. 2011

introduced in the context of the new matrix mechanics around 1925. Since the introduction of the angular momentum in quantum mechanics, which was intimately connected with the representations of the rotation group SO(3) associated with the rotational invariance of central potentials, its importance was soon recognized and the necessary formalism was developed principally by a number of pioneering scientists including Weyl, Racah, Wigner and others [136, 141–144]. Until now, the algebraic method to treat the angular momentum theory can be found in almost all textbooks of quantum mechanics.

On the other hand, it often runs parallel to the differential equation approach due to the great scientist Schrödinger. Pauli employed algebraic method to deal with the hydrogen atom in 1926 [145] and Schrödinger also solved the same problem almost at the same time [146], but their fates were quiet different. This is because the standard differential equation approach was more accessible to the physicists than the algebraic method. As a result, the algebraic approach to determine the energy levels of the hydrogen atom was largely forgotten and the algebraic techniques went into abeyance for several decades. Until the middle of 1950s, the algebraic techniques revived with the development of theories for the elementary particles since the explicit forms of the Hamiltonian for those elementary particle systems are unknown and the physicists have to make certain assumptions on their internal symmetries. Among various attempts to solve this difficult problem, the particle physicists examined some non-compact Lie algebras and hoped that they would provide a clue to the classification of the elementary particles. Unfortunately, this hope did not materialize. Nevertheless, it is found that the Lie algebras of the compact Lie groups enable such a classification for the elementary particles [147] and the non-compact groups are relevant for the dynamic groups in atomic physics [148] and the non-classical properties of quantum optical systems involving coherent and squeezed states as well as the beam splitting and linear directional coupling devices [149–153].

It is worth pointing out that one of the reasons why the algebraic techniques were accepted very slowly and the original group theoretical and algebraic methods proposed by Pauli [145] were neglected is undoubtedly related to the abstract character and inherent complexity of group theory. Even though the proper understanding of group theory requires an intimate knowledge of the standard theory of finite groups and of the topology and manifold theory, the basic concepts of group theory are quite simple, specially when we present them in the context of physical applications. Basically, we attempt to introduce them as simple as possible so that the common reader can master the basic ideas and essence of group theory. The detailed information on group theory can be found in the textbooks [138–140, 154].

On the other hand, during the development of algebraic method, Racah algebra techniques played an important role in physics since it enables us to treat the integration over the angular coordinates of a complex many-particle system analytically and leads to the formulas expressed in terms of the generalized CGCs, Wigner n-j symbols, tensor spherical harmonics and/or rotation matrices. With the development of algebraic method in the late 1950s and early 1960s, the algebraic method proposed by Pauli was systematized and simplified greatly by using the concepts of the Lie algebras. Up to now, the algebraic method has been widely applied to various fields of physics such as nuclear physics [155], field theory and particle physics [156], atomic and molecular physics [157–160], quantum chemistry [161], solid state physics [162], quantum optics [149, 151, 163–168] and others.

2 Abstract Groups

We now give some basic definitions about the abstract groups[1] based on textbooks by Weyl, Wybourne, Miller, Ma and others [136, 137, 139, 140, 169].

Definition A **group** G is a set of elements $\{e, f, g, h, k, \ldots\}$ together with a binary operation. This binary operation named a group multiplication is subject to the following four requirements:

- **Closure**: if $f, g \in G$, then $fg \in G$ too,
- **Identity element**: there exists an identity element e in G (a unit) such that $ef = fe = f$ for any $f \in G$,
- **Inverses**: for every $f \in G$ there exists an inverse element $f^{-1} \in G$ such that $ff^{-1} = f^{-1}f = e$,
- **Associative law**: the identity $f(hk) = (fh)k$ is satisfied for all elements $f, h, k \in G$.

Subgroup: a subgroup of G is a subset $\mathcal{S} \in G$, which is itself a group under the group multiplication defined in G, i.e., $f, h \in \mathcal{S} \rightarrow f\ h \in \mathcal{S}$.

Homomorphism: a homomorphism of groups G and $\mathcal{H}$ is a mapping from a group G into a group $\mathcal{H}$, which transforms products into products, i.e., $G \rightarrow \mathcal{H}$.

Isomorphism: an isomorphism is a homomorphism which is one-to-one and "onto" [169]. From the viewpoint of the abstract group theory, isomorphic groups can be identified. In particular, isomorphic groups have identical multiplication tables.

Representation: a representation of a group G is a homomorphism of the group into the group of invertible operators on a certain (most often complex) Hilbert space V (called representation space). If the representation is to be finite-dimensional, it is sufficient to consider homomorphisms $G \rightarrow GL(n)$. The $GL(n)$ represents a general linear group of non-singular matrices of dimension n. Usually, the image of the group in this homomorphism is called a representation as well.

Irreducible representation: an irreducible representation is a representation whose representation space contains no proper subspace invariant under the operators of the representation.

Commutation relation: since a Lie algebra has an underlying vector space structure we may choose a basis set $\{L_i\}$ $(i = 1, 2, 3, \ldots, N)$ for the Lie algebra. In

[1] There exist two kinds of different meanings of the terminology "abstract group" during the first half of the 20th century. The first meaning was that of a group defined by four axioms given above, but the second one was that of a group defined by generators and commutation relations.

general, the Lie algebra can be completely defined by specifying the commutators of these basis elements:

$$[L_i, L_j] = \sum_k c_{ijk} L_k, \quad i, j, k = 1, 2, 3, \ldots, N, \tag{2.1}$$

in which the coefficients c_{ijk} and the elements L_i are the structure constants and the generators of the Lie algebra, respectively. It is worth noting that the set of operators, which commute with all elements of the Lie algebra, are called Casimir operators.

We shall constraint ourselves in the following parts to study some basic properties of the compact group SO(N) alongside the well-known compact so(n) Lie algebra of the generalized angular momentum theory since it shall be helpful in successive Chapters. We suggest the reader refer to the textbooks on group theory [136–140, 144, 154, 169] or Appendices A–C for more information.

3 Orthogonal Group SO(N)

For every positive integer N, the orthogonal group O(N) is the group of $N \times N$ orthogonal matrices A satisfying

$$AA^T = \mathbf{1}, \qquad A^* = A. \tag{2.2}$$

Because the determinant of an orthogonal matrix is either 1 or -1, and so the orthogonal group has two components. The component containing the identity **1** is the special orthogonal group SO(N). An N-dimensional real matrix contains N^2 real parameters. The column matrices of a real orthogonal matrix are normal and orthogonal to each other. There exist N real matrix constraints for the normalization and $N(N-1)/2$ real constraints for the orthogonality. Thus, the number of independent real parameters for characterizing the elements of the groups SO(N) is equal to $N^2 - [N + N(N-1)/2] = N(N-1)/2$. The group space is a doubly-connected closed region so that the SO(N) is a compact Lie group with rank $N(N-1)/2$.

4 Tensor Representations of the Orthogonal Group SO(N)

In this section we are going to study the reduction of a tensor space of the SO(N) and calculation of the orthonormal irreducible basis tensors [139, 140].

4.1 Tensors of the Orthogonal Group SO(N)

We begin by studying the tensors of the SO(N). For a given rank n of the SO(N), we know that there are N^n components with a following transform,

$$T_{c_1\cdots c_n} \xrightarrow{R} O_R T_{c_1\cdots c_n} = \sum_{d_1\cdots d_n} R_{c_1 d_1} \cdots R_{c_n d_n} T_{d_1\cdots d_n}, \quad R \in \mathrm{SO}(N). \tag{2.3}$$

It is noted that only one nonvanishing component of a basis tensor is equal to 1, i.e.,

$$(\boldsymbol{\theta}_{a_1\cdots a_n})_{b_1\cdots b_n} = \delta_{a_1b_1}\cdots\delta_{a_nb_n} = (\boldsymbol{\theta}_{a_1})_{b_1}\cdots(\boldsymbol{\theta}_{a_n})_{b_n}, \tag{2.4}$$

$$O_R(\boldsymbol{\theta}_{a_1\cdots a_n}) = \sum_{c_1\cdots c_n}(\boldsymbol{\theta}_{c_1\cdots c_n})R_{c_1a_1}\cdots R_{c_na_n}, \tag{2.5}$$

from which one may expand any tensor in such a way

$$\mathbf{T}_{b_1\cdots b_n} = \sum_{a_1\cdots a_n} T_{a_1\cdots a_n}(\boldsymbol{\theta}_{a_1\cdots a_n})_{b_1\cdots b_n}. \tag{2.6}$$

The tensor space is an invariant linear space both in the SO(N) and in the permutation group S_n. Since the SO(N) transformation commutes with the permutation so that one can reduce the tensor space in the orthogonal groups SO(N) by the projection of the Young operators, which are conveniently used to deal with the permutation group S_n.

Note that there are several important characteristics for the tensors of the SO(N) group:

- The real and imaginary parts of a tensor of the SO(N) transform independently in Eq. (2.3). As a result, we need only study their real tensors.
- There is no any difference between a covariant tensor and a contra-variant tensor for the SO(N) transformations. The contraction of a tensor can be achieved between any two indices. Therefore, before projecting a Young operator, the tensor space must be decomposed into a series of traceless tensor subspaces, which remain invariant in the SO(N).
- Denote by T the traceless tensor space of rank n. After projecting a Young operator, $\mathrm{T}^{[\lambda]}_{\mu} = \mathrm{y}^{[\lambda]}_{\mu}\mathrm{T}$ is a traceless tensor subspace with a given permutation symmetry. $\mathrm{T}^{[\lambda]}_{\mu}$ will become a null space if the summation of the numbers of boxes in the first two columns of the Young pattern[2] $[\lambda]$ is larger than the dimension N.
- If the row number m of the Young pattern $[\lambda]$ is larger than $N/2$, then the basis tensor $\mathrm{y}^{[\lambda]}_{\mu}\theta_{b_1\cdots b_m c\cdots}$ can be changed to a dual basis tensor by a totally antisymmetric tensor $\epsilon_{a_1\cdots a_N}$,

$$^*[\mathrm{y}^{[\lambda]}\theta]_{a_1\cdots a_{N-m}c\cdots} = \frac{1}{m!}\sum_{a_{N-m+1}\cdots a_N}\epsilon_{a_1\cdots a_{N-m}a_{N-m+1}\cdots a_N} \mathrm{y}^{[\lambda]}\theta_{a_N\cdots a_{N-m+1}c\cdots}, \tag{2.7}$$

whose inverse transformation is given by

---

[2]A Young pattern $[\lambda]$ has n boxes lined up on the top and on the left, where the jth row contains λ_j boxes. For instance, the Young pattern $[2, 1]$ is

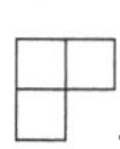

It should be noted that the number of boxes in the upper row is not less than in the lower row, and the number of boxes in the left column is not less than that in the right column. We suggest the reader refer to the permutation group S_n in Appendix A for more information.

$$
\begin{aligned}
&\frac{1}{(N-m)!}\sum_{a_{m+1}\cdots a_N}\epsilon_{b_1\cdots b_m a_{m+1}\cdots a_N}{}^*[\mathrm{y}^{[\lambda]}\theta]_{a_N\cdots a_{m+1}c\cdots} \\
&\quad=\frac{1}{m!(N-m)!}\sum_{a_1\cdots a_N}\epsilon_{b_1\cdots b_m a_{m+1}\cdots a_N}\epsilon_{a_N\cdots a_{m+1}a_m\cdots a_1}\mathrm{y}^{[\lambda]}\theta_{a_1\cdots a_m c} \\
&\quad=(-1)^{N(N-1)/2}\mathrm{y}^{[\lambda]}\theta_{b_1\cdots b_m c\cdots}.
\end{aligned}
\tag{2.8}
$$

After some algebraic manipulations, it is found that the correspondence between two sets of basis tensors is one-to-one and the difference between them is only in the arranged order. Thus, a traceless tensor subspace $\mathrm{T}_{\mu}^{[\lambda]}$ is equivalent to a traceless tensor subspace $\mathrm{T}_{\nu}^{[\lambda']}$, where the row number of the Young pattern $[\lambda']$ is $(N-m) < N/2$,

$$
[\lambda'] \simeq [\lambda], \qquad \lambda_i' = \begin{cases} \lambda_i, & i \leq (N-m), \\ 0, & i > (N-m), \end{cases} \tag{2.9}
$$

where $m \in (N/2, N]$.

- If $N = 2l$, i.e., the row number l of $[\lambda]$ is equal to $N/2$, then the Young pattern $[\lambda]$ is the same as its dual Young pattern, called the self-dual Young pattern. To remove the phase factor $(-1)^{N(N-1)/2} = (-1)^l$ appearing in Eq. (2.8), we introduce a factor $(-i)^l$ in Eq. (2.7),

$$
{}^*[\mathrm{y}^{[\lambda]}\theta]_{a_1\cdots a_l c\cdots} = \frac{(-i)^l}{l!}\sum_{a_{l+1}\cdots a_{2l}}\epsilon_{a_1\cdots a_l a_{l+1}\cdots a_{2l}}\mathrm{y}^{[\lambda]}\theta_{a_{2l}\cdots a_{l+1}c\cdots}, \tag{2.10}
$$

$$
\mathrm{y}^{[\lambda]}\theta_{a_1\cdots a_l c\cdots} = \frac{(-i)^l}{l!}\sum_{a_{l+1}\cdots a_{2l}}\epsilon_{a_1\cdots a_l a_{l+1}\cdots a_{2l}}{}^*[\mathrm{y}^{[\lambda]}\theta]_{a_{2l}\cdots a_{l+1}c\cdots}. \tag{2.11}
$$

Define

$$
\psi^{\pm}_{a_1\cdots a_l c\cdots} = \frac{1}{2}\left\{\mathrm{y}^{[\lambda]}\theta_{a_1\cdots a_l c\cdots} \pm {}^*[\mathrm{y}^{[\lambda]}\theta]_{a_1\cdots a_l c\cdots}\right\}. \tag{2.12}
$$

We observe that $\psi^{+}_{a_1\cdots a_l c\cdots}$ keeps invariant in the dual transformation so that we call it self-dual basis tensor. On the contrary, we call $\psi^{-}_{a_1\cdots a_l c\cdots}$ the anti-self-dual basis tensor because it changes its sign in dual transformation. For example, for even $N = 2l$ we may construct the self-dual and anti-self-dual basis tensors as follows:

$$
\psi^{\pm}_{1\cdots l} = \frac{1}{2}\left\{\mathrm{y}^{[1^l]}\theta_{1\ldots l} \pm (-i)^l \mathrm{y}^{[1^l]}\theta_{(2l)\cdots(l+1)}\right\}. \tag{2.13}
$$

Therefore, when $l = N/2$ the representation space $\mathrm{T}_{\mu}^{[\lambda]}$ can be divided to the self-dual and the anti-self-dual tensor subspaces with the same dimension. Notice that the combinations by the Young operators and the dual transformations (2.7) and (2.13) are all real except that the dual transformation (2.13) with $N = 4l + 2$ is complex.

In conclusion, the traceless tensor subspace $\mathrm{T}_{\mu}^{[\lambda]}$ corresponds to a representation $[\lambda]$ of the SO(N), where the row number l of Young pattern $[\lambda]$ is less than $N/2$. When $l = N/2$ the traceless tensor subspace $\mathrm{T}_{\mu}^{[\lambda]}$ can be decomposed into the self-dual tensor subspace $\mathrm{T}_{\mu}^{[+\lambda]}$ and anti-self-dual tensor subspace $\mathrm{T}_{\mu}^{[-\lambda]}$ corresponding to the representation $[\pm\lambda]$, respectively. All irreducible representations $[\lambda]$ and $[\pm\lambda]$ are real except for $[\pm\lambda]$ with $N = 4l + 2$.

As far as the orthonormal irreducible basis tensors of the SO(N), we are going to address two problems. The first is how to decompose the standard tensor Young tableaux into a sum of the traceless basis tensors. The second is how to combine the basis tensors such that they are the common eigenfunctions of H_j and orthonormal to each other. The advantage of the method based on the standard tensor Young tableaux is that the basis tensors are known explicitly and the multiplicity of any weight is equivalent to the number of the standard tensor Young tableaux with the weight.

For group SO(N), the key issue for finding the orthonormal irreducible basis is to find the common eigenstates of H_i and the highest weight state in an irreducible representation. For odd and even N, i.e., the groups SO($2l + 1$) and SO($2l$), the generators T_{ab} of the self-representation satisfy

$$\begin{aligned}
&[T_{ab}]_{cd} = -i(\delta_{ac}\delta_{bd} - \delta_{ad}\delta_{bc}),\\
&[T_{ab}, T_{cd}] = -i(\delta_{bc}T_{ad} + \delta_{ad}T_{bc} - \delta_{bd}T_{ac} - \delta_{ac}T_{bd}).
\end{aligned} \tag{2.14}$$

The bases H_i in the Cartan subalgebra can be written as

$$H_i = T_{(2i-1)(2i)}, \quad i \in [1, N/2]. \tag{2.15}$$

As what follows, we are going to study the irreducible basis tensors of the SO($2l + 1$) and SO($2l$), respectively.

4.2 *Irreducible Basis Tensors of the* SO($2l + 1$)

It is known that the Lie algebra of the SO($2l + 1$) is B_l. The simple roots of the SO($2l + 1$) are given by [139, 140]

$$r_\nu = e_\nu - e_{\nu+1}, \quad \nu \in [1, l-1], \qquad r_l = e_l, \tag{2.16}$$

where r_ν are the longer roots with $d_\nu = 1$ and r_l is the shorter root with $d_l = 1/2$. Based on the definition of the Chevalley bases, which include $3l$ bases E_ν, F_ν, and H_ν for the generators,

$$\frac{E_{\mathbf{r}_\nu}}{\sqrt{d_\nu}} \to E_\nu, \qquad \frac{E_{-\mathbf{r}_\nu}}{\sqrt{d_\nu}} \to F_\nu, \qquad \frac{1}{d_\nu}\sum_{i=1}^{l}(\mathbf{r}_\nu)_i H_i \equiv \frac{1}{d_\nu}\mathbf{r}_\nu \cdot \mathbf{H} \to H_\nu, \tag{2.17}$$

one is able to calculate the Chevalley bases of the SO(2l + 1) in the self-representation as follows:

$$\begin{aligned}
H_\nu &= T_{(2\nu-1)(2\nu)} - T_{(2\nu+1)(2\nu+2)},\\
E_\nu &= \frac{1}{2}[T_{(2\nu)(2\nu+1)} - iT_{(2\nu-1)(2\nu+1)} - iT_{(2\nu)(2\nu+2)} - T_{(2\nu-1)(2\nu+2)}],\\
F_\nu &= \frac{1}{2}[T_{(2\nu)(2\nu+1)} + iT_{(2\nu-1)(2\nu+1)} + iT_{(2\nu)(2\nu+2)} - T_{(2\nu-1)(2\nu+2)}],\\
H_l &= 2T_{(2l-1)(2l)},\\
E_l &= T_{(2l)(2l+1)} - iT_{(2l-1)(2l+1)},\\
F_l &= T_{(2l)(2l+1)} + iT_{(2l-1)(2l+1)}.
\end{aligned} \tag{2.18}$$

Note that θ_a are not the common eigenvectors of H_ν. By generalizing the spherical harmonic basis vectors for the SO(3) group, we may define the spherical harmonic basis vectors for the self-representation of the SO(2l + 1) as follows:

$$\phi_\beta = \begin{cases} (-1)^{l-\beta+1}\sqrt{\frac{1}{2}}(\theta_{2\beta-1} + i\theta_{2\beta}), & \beta \in [1, l],\\ \theta_{2l+1}, & \beta = l+1,\\ \sqrt{\frac{1}{2}}(\theta_{4l-2\beta+3} - i\theta_{4l-2\beta+4}), & \beta \in [l+2, 2l+1], \end{cases} \tag{2.19}$$

which are orthonormal and complete. In the spherical harmonic basis vectors ϕ_β, the nonvanishing matrix entries in the Chevalley bases are given by

$$\begin{aligned}
&H_\nu\phi_\nu = \phi_\nu, && H_\nu\phi_{\nu+1} = -\phi_{\nu+1},\\
&H_\nu\phi_{2l-\nu+1} = \phi_{2l-\nu+1}, && H_\nu\phi_{2l-\nu+2} = -\phi_{2l-\nu+2},\\
&H_l\phi_l = 2\phi_l, && H_l\phi_{l+2} = -2\phi_{l+2},\\
&E_\nu\phi_{\nu+1} = \phi_\nu, && E_\nu\phi_{2l-\nu+2} = \phi_{2l-\nu+1},\\
&E_l\phi_{l+1} = \sqrt{2}\phi_l, && E_l\phi_{l+2} = \sqrt{2}\phi_{l+1},\\
&F_\nu\phi_\nu = \phi_{\nu+1}, && F_\nu\phi_{2l-\nu+1} = \phi_{2l-\nu+2},\\
&F_l\phi_l = \sqrt{2}\phi_{l+1}, && F_l\phi_{l+1} = \sqrt{2}\phi_{l+2},
\end{aligned} \tag{2.20}$$

where $\nu \in [1, l-1]$. That is to say, the diagonal matrices of H_ν and H_l in the spherical harmonic basis vectors ϕ_β are expressed as follows:

$$\begin{aligned}
H_\nu &= \text{diag}\{\underbrace{0,\ldots,0}_{\nu-1}, 1, -1, \underbrace{0,\ldots,0}_{2l-2\nu-1}, 1, -1, \underbrace{0,\ldots,0}_{\nu-1}\},\\
H_l &= \text{diag}\{\underbrace{0,\ldots,0}_{l-1}, 2, 0, -2, \underbrace{0,\ldots,0}_{l-1}\}.
\end{aligned} \tag{2.21}$$

The spherical harmonic basis tensor $\phi_{\beta_1\cdots\beta_n}$ of rank n for the SO(2l + 1) becomes the direct product of n spherical harmonic basis vectors $\phi_{\beta_1}\cdots\phi_{\beta_n}$. The standard tensor Young tableaux $\mathcal{Y}_\nu^{[\lambda]}\phi_{\beta_1\cdots\beta_n}$ are the common eigenstates of the H_ν, but generally neither orthonormal nor traceless. The eigenvalue of H_ν in the standard tensor

Young tableaux $\mathcal{Y}_\nu^{[\lambda]}\phi_{\beta_1\cdots\beta_n}$ is equal to the number of the digits ν and $(2l-\nu+1)$ in the tableau, minus the number of $(\nu+1)$ and $(2l-\nu+2)$. The eigenvalue of H_l in the standard tensor Young tableau is equal to the number of l in the tableau, minus the number of $(l+2)$, and then multiplied with a factor 2. The action F_ν on the standard tensor Young tableau is equal to the sum of all possible tensor Young tableaux, each of which can be obtained from the original one through replacing one filled digit ν by the digit $(\nu+1)$, or through replacing one filled digit $(2l-\nu+1)$ by the digit $(2l-\nu+2)$. But the action of the F_l on the standard tensor Young tableau is equal to the sum, multiplied with a factor $\sqrt{2}$, of all possible tensor Young tableaux, each of which can be obtained from the original one through replacing one filled digit l by $(l+1)$ or through replacing one filled $(l+1)$ by $(l+2)$. However, the actions of E_ν and E_l on the standard tensor Young tableau are opposite to those of F_ν and F_l. Even though the obtained tensor Young tableaux may be not standard, they can be transformed into the sum of the standard tensor Young tableaux by symmetry.

Two standard tensor Young tableaux with different sets of filled digits are orthogonal to each other. For a given irreducible representation $[\lambda]$ of the SO$(2l+1)$, where the row number of Young pattern $[\lambda]$ is not larger than l, the highest weight state corresponds to the standard tensor Young tableau, in which each box in the βth row is filled with the digit β because each raising operator E_ν annihilates it. The highest weight $\mathbf{M}=\sum_\nu \omega_\nu M_\nu$ can be calculated from (2.20) as follows:

$$M_\nu=\lambda_\nu-\lambda_{\nu+1}, \quad \nu\in[1,l), \qquad M_l=2\lambda_l. \tag{2.22}$$

The tensor representation $[\lambda]$ of the SO$(2l+1)$ with even M_l is a single-valued representation, while the representation with odd M_l becomes a double-valued (spinor) representation.

The standard tensor Young tableaux $\mathcal{Y}_\nu^{[\lambda]}\phi_{\beta_1\cdots\beta_n}$ are generally not traceless, but the standard tensor Young tableau with the highest weight is traceless because it only contains ϕ_β with $\beta<l+1$ as shown in Eq. (2.19). For example, the tensor basis $\theta_1\theta_1$ is not traceless, but $\phi_1\phi_1$ is traceless. Since the highest weight is simple, the highest weight state is orthogonal to any other standard tensor Young tableau in the irreducible representation. Therefore, one is able to obtain the remaining orthonormal and traceless basis tensors in the representation $[\lambda]$ of the SO$(2l+1)$ from the highest weight state by the lowering operators F_ν based on the method of the block weight diagram.

4.3 Irreducible Basis Tensors of the SO($2l$)

The Lie algebra of the SO$(2l)$ is D_l and its simple roots are given by

$$r_\nu=e_\nu-e_{\nu+1}, \quad \nu\in[1,l-1], \qquad r_l=e_{l-1}+e_l. \tag{2.23}$$

The lengths of all simple roots are same, $d_\nu = 1$. Similarly, based on the definition of the Chevalley bases (2.17), we find that its Chevalley bases in the self-representation are same as those of the SO$(2l+1)$ except for $\nu = l$,

$$\begin{aligned} H_l &= T_{(2l-3)(2l-2)} + T_{(2l-1)(2l)}, \\ E_l &= \frac{1}{2}[T_{(2l-2)(2l-1)} - iT_{(2l-3)(2l-1)} + iT_{(2l-2)(2l)} + T_{(2l-3)(2l)}], \\ F_l &= \frac{1}{2}[T_{(2l-2)(2l-1)} + iT_{(2l-3)(2l-1)} - iT_{(2l-2)(2l)} + T_{(2l-3)(2l)}]. \end{aligned} \tag{2.24}$$

Likewise, θ_a are not the common eigenvectors of the H_ν. By generalizing the spherical harmonic basis vectors for the SO(4) group, we define the spherical harmonic basis vectors for the self-representation of the SO$(2l)$ as follows:

$$\phi_\beta = \begin{cases} (-1)^{l-\beta}\sqrt{\frac{1}{2}}(\theta_{2\beta-1} + i\theta_{2\beta}), & \beta \in [1, l], \\ \sqrt{\frac{1}{2}}(\theta_{4l-2\beta+1} - i\theta_{4l-2\beta+2}), & \beta \in [l+1, 2l], \end{cases} \tag{2.25}$$

which are orthonormal and complete. In these basis vectors, the nonvanishing matrix entries of the Chevalley bases are given by

$$\begin{aligned} &H_\nu\phi_\nu = \phi_\nu, && H_\nu\phi_{\nu+1} = -\phi_{\nu+1}, \\ &H_\nu\phi_{2l-\nu} = \phi_{2l-\nu}, && H_\nu\phi_{2l-\nu+1} = -\phi_{2l-\nu+1}, \\ &H_l\phi_{l-1} = \phi_{l-1}, && H_l\phi_l = \phi_l, \\ &H_l\phi_{l+1} = -\phi_{l+1}, && H_l\phi_{l+2} = -\phi_{l+2}, \\ &E_\nu\phi_{\nu+1} = \phi_\nu, && E_\nu\phi_{2l-\nu+1} = \phi_{2l-\nu}, \\ &E_l\phi_{l+1} = \phi_{l-1}, && E_l\phi_{l+2} = \phi_l, \\ &F_\nu\phi_\nu = \phi_{\nu+1}, && F_\nu\phi_{2l-\nu} = \phi_{2l-\nu+1}, \\ &F_l\phi_{l-1} = \phi_{l+1}, && F_l\phi_l = \phi_{l+2}, \end{aligned} \tag{2.26}$$

where $\nu \in [1, l-1]$. As a result, the diagonal matrices of the H_ν and H_l in the spherical harmonic basis vectors ϕ_β are calculated as:

$$\begin{aligned} H_\nu &= \text{diag}\{\underbrace{0,\ldots,0}_{\nu-1}, 1, -1, \underbrace{0,\ldots,0}_{2l-2\nu-2}, 1, -1, \underbrace{0,\ldots,0}_{\nu-1}\}, \\ H_l &= \text{diag}\{\underbrace{0,\ldots,0}_{l-2}, 1, 1, -1, -1, \underbrace{0,\ldots,0}_{l-2}\}. \end{aligned} \tag{2.27}$$

The spherical harmonic basis tensor $\phi_{\beta_1\cdots\beta_n}$ of rank n for the SO$(2l)$ is the direct product of n spherical harmonic basis vectors $\phi_{\beta_1}\cdots\phi_{\beta_n}$. The standard tensor Young tableaux $\mathcal{Y}_\nu^{[\lambda]}\phi_{\beta_1\cdots\beta_n}$ are the common eigenstates of the H_ν, but in general neither orthonormal nor traceless. The eigenvalue of H_ν in the standard tensor Young tableaux $\mathcal{Y}_\nu^{[\lambda]}\phi_{\beta_1\cdots\beta_n}$ is equal to the number of the digits ν and $(2l-\nu)$ in the tableau, minus the number of $(\nu+1)$ and $(2l-\nu+1)$. The eigenvalue of H_l in the standard tensor Young tableau is equal to the number of the digits $(l-1)$ and l in the tableau, minus the number of $(l+1)$ and $(l+2)$. The eigenvalues form the weight $\mathbf{m}$ of standard tensor Young tableau. The action of F_ν on the standard tensor Young tableau is equal to the sum of all possible tensor Young tableaux, each of which can be obtained from the original one through replacing one filled digit $(2l-\nu)$ by the digit $(2l-\nu+1)$. The action of F_l on the standard tensor Young tableau is equal to the sum of all possible tensor Young tableaux, each of which is obtained from the original one through replacing one filled digit $(l-1)$ by the digit $(l+1)$ or through replacing one filled l by the digit $(l+2)$. However, the actions of E_ν and E_l are opposite to those of F_ν and F_l. The obtained tensor Young tableaux may be not standard, but they can be transformed into the sum of the standard tensor Young tableaux by symmetry.

Two standard tensor Young tableaux with different weights are orthogonal to each other. For a given irreducible representation $[\lambda]$ or $[+\lambda]$ of the SO($2l$), where the row number of Young pattern $[\lambda]$ is not larger than l, the highest weight state corresponds to the standard tensor Young tableau where each box in the βth row is filled with the digit β because every raising operator E_ν annihilates it. In the standard tensor Young tableau with the highest weight of the representation $[-\lambda]$, the box in the βth row is filled with the digit β, but the box in the lth row with the digit $(l+1)$. The highest weight $\mathbf{M}=\sum_\nu \omega_\nu M_\nu$ is calculated from (2.20) as

$$\begin{aligned}
&M_\nu=\lambda_\nu-\lambda_{\nu+1}, && && \nu\in[1,l-1),\\
&M_{l-1}=M_l=\lambda_{l-1}, && && \lambda_l=0,\\
&M_{l-1}=\lambda_{l-1}-\lambda_l, && M_l=\lambda_{l-1}+\lambda_l, && \text{for } [+\lambda],\\
&M_{l-1}=\lambda_{l-1}+\lambda_l, && M_l=\lambda_{l-1}-\lambda_l, && \text{for } [-\lambda].
\end{aligned} \tag{2.28}$$

The tensor representation $[\lambda]$ of the SO($2l$) with even $(M_{l-1}+M_l)$ is a single-valued representation. However, the representation with odd $(M_{l-1}+M_l)$ is a double-valued (spinor) representation.

The standard tensor Young tableaux are generally not traceless, but the standard tensor Young tableau with the highest weight is traceless because it only contains ϕ_β with $\beta<l+2$. Furthermore, l and $l+1$ do not appear in the tableau simultaneously as illustrated in Eq. (2.25). Since the highest weight is simple, the highest weight state is orthogonal to any other standard tensor Young tableau in the irreducible representation. Hence, we can obtain the remaining orthonormal and traceless basis tensors in the irreducible representation of the SO($2l$) from the highest weight state by the lowering operators F_ν in light of the method of the block weight diagram. The multiplicity of a weight in the representation can be easily obtained by counting the number of the traceless tensor Young tableaux with this weight.

4.4 Dimensions of Irreducible Tensor Representations

The dimension $d_{[\lambda]}$ of the representation $[\lambda]$ of the SO(N) can be calculated by hook rule [139, 140]. The dimension is expressed as a quotient, where the numerator and the denominator are denoted by the symbols $Y_T^{[\lambda]}$ and $Y_h^{[\lambda]}$, respectively:

$$\begin{aligned} d_{[\pm\lambda]}[\mathrm{SO}(2l)] &= \frac{Y_T^{[\lambda]}}{2Y_h^{[\lambda]}}, \qquad \text{when } \lambda_l \neq 0, \\ d_{[\lambda]}[\mathrm{SO}(N)] &= \frac{Y_T^{[\lambda]}}{2Y_h^{[\lambda]}}, \qquad \text{others.} \end{aligned} \tag{2.29}$$

The first formula in Eq. (2.29) corresponds to the case where the row number of the Young pattern $[\lambda]$ is equal to $N/2$. The hook path (i, j) in the Young pattern $[\lambda]$ is defined as a path which enters the Young pattern at the rightmost of the ith row, goes leftward in the i row, turns downward at the j column, goes downward in the j column, and leaves from the Young pattern at the bottom of the j column. The inverse hook path denoted by $\overline{(i, j)}$ is the same path as the hook path (i, j), but with opposite direction. The number of boxes contained in the path (i, j), as well as in its inverse, is the hook number h_{ij}. The $Y_h^{[\lambda]}$ represents a tableau of the Young pattern $[\lambda]$ where the box in the jth column of the ith row is filled with the hook number H_{ij}. However, the $Y_T^{[\lambda]}$ is a tableau of the Young pattern $[\lambda]$ where each box is filled with the sum of the digits which are respectively filled in the same box of each tableau $Y_{T_b}^{[\lambda]}$ in the series. The notation $Y_T^{[\lambda]}$ means the product of the filled digits in it, so does the notation $Y_h^{[\lambda]}$. Here, the tableaux $Y_{T_b}^{[\lambda]}$ can be obtained by the following rules:

- $Y_{T_0}^{[\lambda]}$ is a tableau of the Young pattern $[\lambda]$, where the box in the jth column of the ith row is filled with the digit $(N + j - i)$.
- Let $[\lambda^{(1)}] = [\lambda]$. Starting with $[\lambda^{(1)}]$, define recursively the Young pattern $[\lambda^{(b)}]$ by removing the first row and the first column of the Young pattern $[\lambda^{(b-1)}]$ until $[\lambda^{(b)}]$ contains less two columns.
- If $[\lambda^{(b)}]$ contains more than one column, define $Y_{T_b}^{[\lambda]}$ as a tableau of the Young pattern $[\lambda]$ where the boxes in the first $(b-1)$ row and in the first $(b-1)$ column are filled with 0, and the remaining part of the Young pattern is $[\lambda^{(b)}]$. Let $[\lambda^{(b)}]$ have r rows. Fill the first r boxes along the hook path $(1, 1)$ of the Young pattern $[\lambda^{(b)}]$, starting with the box on the rightmost, with the digits $(\lambda_1^{(b)} - 1), (\lambda_2^{(b)} - 1), \ldots, (\lambda_r^{(b)} - 1)$, box by box, and fill the first $(\lambda_i^{(b)} - 1)$ boxes in each inverse hook path $\overline{(i, 1)}$ of the Young pattern $[\lambda^{(b)}]$, $i \in [1, r]$ with "-1". The remaining boxes are filled with 0. If several "-1" are filled in the same box, the digits are summed. The sum of all filled digits in the pattern $Y_{T_b}^{[\lambda]}$ with $b > 0$ is equal to 0.

4.5 Adjoint Representation of the SO(N)

We are going to study the adjoint representation of the SO(N) by replacing the tensors. The $N(N-1)/2$ generators T_{ab} in the self-representation of the SO(N) construct the complete bases of N-dimensional antisymmetric matrices. Denote T_{cd} by T_A for convenience, $A \in [1, N(N-1)/2]$. Then we have

$$\mathrm{Tr}(T_A T_B) = 2\delta_{AB}. \tag{2.30}$$

Based on the adjoint representation $D^{\mathrm{ad}}(G)$ satisfying

$$D(R) I_B D(R)^{-1} = \sum_D I_D D^{\mathrm{ad}}_{DB}(R), \quad R \in \mathrm{SO}(N), \tag{2.31}$$

where R is an infinitesimal element, we have

$$R T_A R^{-1} = \sum_{B=1}^{N(N-1)/2} T_B D^{\mathrm{ad}}_{BA}(R). \tag{2.32}$$

The antisymmetric tensor T_{ab} of rank 2 of the SO(N) satisfies a similar relation in the SO(N) transformation R

$$(O_R T)_{cd} = \sum_{ij} R_{ci} T_{ij} (R^{-1})_{jd} = (R T R^{-1})_{cd}, \tag{2.33}$$

where T_{ab} like an antisymmetric matrix can be expanded by $(T_A)_{ab}$ as follows:

$$T_{cd} = \sum_{A=1}^{N(N-1)/2} (T_A)_{cd} F_A, \quad F_A = \frac{1}{2} \sum_{cd} (T_A)_{dc} T_{cd}, \tag{2.34}$$

where the coefficient F_A is a tensor that transforms in the SO(N) transformation R as follows:

$$\begin{aligned}
(O_R T)_{cd} &= (R T R^{-1})_{cd} \\
&= \sum_A (R T_A R^{-1})_{cd} F_A \\
&= \sum_B (T_B)_{cd} \left\{ \sum_A D^{\mathrm{ad}}_{BA}(R) F_A \right\}, \\
(O_R T)_{cd} &= \sum_B (T_B)_{cd} O_R F_B.
\end{aligned} \tag{2.35}$$

Thus, in terms of the adjoint representation of the SO(N) we can transform F_A in such a way

$$(O_R F)_B = \sum_A D^{\mathrm{ad}}_{BA}(R) F_A. \tag{2.36}$$

The adjoint representation of the SO(N) is equal to the antisymmetric tensor representation $[1, 1]$ of rank 2. The adjoint representation of the SO(N) for $N = 3$ or $N > 4$ is irreducible. Except for $N = 2, 4$, the SO(N) is a simple Lie group.

4.6 Tensor Representations of the Groups O(N)

It is known that the group O(N) is a mixed Lie group with two disjoint regions corresponding to $\det R = \pm 1$. Its invariant subgroup SO(N) has a connected group space corresponding to $\det R = 1$. The set of elements related to the $\det R = -1$ is the coset of SO(N). The property of the O(N) can be characterized completely by the SO(N) and a representative element in the coset [139, 140].

For odd $N = 2l + 1$, we may choose $\varepsilon = -1$ as the representative element in the coset since ε is self-inverse and commutes with every element in O($2l + 1$). Thus, the representation matrix $D(\varepsilon)$ in the irreducible representation of O($2l + 1$) is a constant matrix

$$D(\varepsilon) = c\mathbf{1}, \qquad D(\varepsilon)^2 = \mathbf{1}, \quad c = \pm 1. \tag{2.37}$$

Denote by R the element in SO($2l+1$) and by $R' = \varepsilon R$ the element in the coset. From each irreducible representation $D^{[\lambda]}(\mathrm{SO}(2l + 1))$ one obtains two induced irreducible representations $D^{[\lambda]\pm}(\mathrm{O}(2l + 1))$,

$$D^{[\lambda]\pm}(R) = D^{[\lambda]}(R), \qquad D^{[\lambda]\pm}(\varepsilon R) = \pm D^{[\lambda]}(R). \tag{2.38}$$

Two representations $D^{[\lambda]\pm}(\mathrm{O}(2l + 1))$ are inequivalent because of different characters of the ε in two representations.

For even $N = 2l$, $\varepsilon = -1$ belongs to SO($2l$). We may choose the representative element in the coset to be a diagonal matrix σ, in which the diagonal entries are 1 except for $\sigma_{NN} = -1$. Even though $\sigma^2 = \mathbf{1}$, σ does not commute with some elements in O($2l$). Any tensor Young tableau $\mathcal{Y}_\nu^{[\lambda]}\theta_{\beta_1\cdots\beta_n}$ is an eigentensor of the σ with the eigenvalue 1 or -1 depending on whether the number of filled digits N in the tableau is even or odd. In the spherical harmonic basis tensors, σ interchanges the filled digits l and $l + 1$ in the tensor Young tableau $\mathcal{Y}_\nu^{[\lambda]}\phi_{\beta_1\cdots\beta_n}$. Therefore, the representation matrix $D^{[\lambda]}(\sigma)$ is known.

Denote by R the element in the SO($2l$) and by $R' = \sigma R$ the element in the coset. From each irreducible representation $D^{[\lambda]}(\mathrm{SO}(2l))$, where the row number of $[\lambda]$ is less than l, we obtain two induced irreducible representations $D^{[\lambda]\pm}(\mathrm{O}(2l))$,

$$D^{[\lambda]\pm}(R) = D^{[\lambda]}(R), \qquad D^{[\lambda]\pm}(\sigma R) = \pm D^{[\lambda]}(\sigma) D^{[\lambda]}(R). \tag{2.39}$$

Likewise, two representations $D^{[\lambda]\pm}(\mathrm{O}(2l))$ are inequivalent due to the different characters of the σ in two representations.

When $l = N/2$ there are two inequivalent irreducible representations $D^{[(\pm)\lambda]}$ of the SO($2l$). Their basis tensors are given in Eq. (2.12). Since two terms in Eq. (2.12) contain different numbers of the subscripts N, then σ changes the tensor Young tableau in $[\pm\lambda]$ to that in $[\mp\lambda]$, i.e., the representation spaces of both $D^{[\pm\lambda]}(\mathrm{SO}(2l))$ correspond to an irreducible representation $D^{[\lambda]}$ of the O($2l$),

$$D^{[\lambda]}(R) = D^{[+\lambda]}(R) \oplus D^{[-\lambda]}(R), \qquad D^{[\lambda]}(\sigma R) = D^{[\lambda]}(\sigma) D^{[\lambda]}(R), \tag{2.40}$$

where the representation matrix $D^{[\lambda]}(\sigma)$ is calculated by interchanging the filled digits l and $(l + 1)$ in the tensor Young tableau $\mathcal{Y}_\nu^{[\lambda]}\phi_{\beta_1\cdots\beta_n}$. Two representations with different signs of $D^{[\lambda]}(\sigma)$ are equivalent since they might be related by a similarity transformation

$$X = \begin{pmatrix} 1 & 0 \\ 0 & -1 \end{pmatrix}. \tag{2.41}$$

5 Γ Matrix Groups

Dirac generalized the Pauli matrices to four γ matrices, which satisfy the anticommutation relations. In terms of the γ matrices, Dirac established the Dirac equation to describe the relativistic particle with spin $1/2$. In the language of group theory, Dirac found the spinor representation of the Lorentz group. In this section we first generalize the γ matrices and find that the set of products of the γ matrices forms the matrix group Γ.

5.1 Fundamental Property of Γ *Matrix Groups*

First, let us review the property of the Γ matrix groups [88–90]. We define N matrices γ_a, which satisfy the following anticommutation relations

$$\{\gamma_a, \gamma_b\} = \gamma_a\gamma_b + \gamma_b\gamma_a = 2\delta_{ab}\mathbf{1}, \qquad a, b \in [1, N]. \tag{2.42}$$

That is, $\gamma_a^2 = \mathbf{1}$ and $\gamma_a\gamma_b = -\gamma_b\gamma_a$ for $a \neq b$. The set of all products of the γ_a matrices, in the multiplication rule of matrices, forms a group, denoted by Γ_N. In a product of γ_a matrices, two γ_b with the same subscript can be moved together and eliminated by Eq. (2.42) so that Γ_N is a finite matrix group.

We choose a faithful irreducible unitary representation of the Γ_N as its self-representation. It is known from Eq. (2.42) that γ_a is unitary and hermitian,

$$\gamma_a^\dagger = \gamma_a^{-1} = \gamma_a, \tag{2.43}$$

whose eigenvalue is 1 or -1.

Let

$$\gamma_\xi^{(N)} = \gamma_1\gamma_2\cdots\gamma_N, \qquad \left(\gamma_\xi^{(N)}\right)^2 = (-1)^{N(N-1)/2}\mathbf{1}. \tag{2.44}$$

For odd N, since $\gamma_\xi^{(N)}$ commutes with every γ_a matrix, then it is a constant matrix according to the Schur theorem (see Appendix B):

$$\gamma_\xi^{(N)} = \begin{cases} \pm\mathbf{1}, & \text{for } N = 4l + 1, \\ \pm i\mathbf{1}, & \text{for } N = 4l - 1. \end{cases} \tag{2.45}$$

Two groups with different $\gamma_\xi^{(4l+1)}$ are isomorphic through a one-to-one correspondence, say

$$\gamma_a \leftrightarrow \gamma_a', \quad a \in [1, 4l], \qquad \gamma_{4l+1} \leftrightarrow -\gamma_{4l+1}'. \tag{2.46}$$

On the other hand, for a given $\gamma_\xi^{(4l+1)}$, the $\gamma_{4l+1}^{(4l+1)}$ can be expressed as a product of other γ_a matrices. As a result, all elements both in Γ_{4l} and in Γ_{4l+1} can be expressed as the products of matrices $\gamma_a, a \in [1, 4l]$ so that they are isomorphic. In addition, since $\gamma_\xi^{(4l-1)}$ is equal to either $i\mathbf{1}$ or $-i\mathbf{1}$, Γ_{4l-1} is isomorphic onto a group composed of the Γ_{4l-2} and $i\Gamma_{4l-2}$,

$$\Gamma_{4l+1} \approx \Gamma_{4l}, \qquad \Gamma_{4l-1} \approx \{\Gamma_{4l-2}, i\Gamma_{4l-2}\}. \tag{2.47}$$

5.2 Case $N = 2l$

- Let us calculate the order $g^{(2l)}$ of the Γ_{2l}. Obviously, if $R \in \Gamma_{2l}$, then $-R \in \Gamma_{2l}$, too. If we choose one element in each pair of elements $\pm R$, then we obtain a set Γ'_{2l} containing $g^{(2l)}/2$ elements. Denote by $\mathcal{S}_n$ a product of n different γ_a. Since the number of different $\mathcal{S}_n$ contained in the set Γ'_{2l} is equal to the combinatorics of n among $2l$, then we have

$$g^{(2l)} = 2\sum_{n=0}^{2l} \binom{2l}{n} = 2(1+1)^{2l} = 2^{2l+1}. \tag{2.48}$$

- For any element $\mathcal{S}_n \in \Gamma_{2l}$ except for ± 1, we may find a matrix γ_a which is anticommutable with $\mathcal{S}_n$. In fact, when n is even and γ appears in the product $\mathcal{S}_n$, one has $\gamma_a \mathcal{S}_n = -\mathcal{S}_n \gamma_a$. However, when n is odd there exists at least one γ_a which does not appear in the product $\mathcal{S}_n$ so that $\gamma_a \mathcal{S}_n = -\mathcal{S}_n \gamma_a$. Therefore, we find that

$$\operatorname{Tr} \mathcal{S}_n = \operatorname{Tr}(\gamma_a^2 \mathcal{S}_n) = -\operatorname{Tr}(\gamma_a \mathcal{S}_n \gamma_a) = -\operatorname{Tr} \mathcal{S}_n = 0. \tag{2.49}$$

  That is to say, the character of the element $\mathcal{S}$ in the self-representation of the Γ_{2l} is

$$\xi(\mathcal{S}) = \begin{cases} \pm d^{(2l)}, & \text{when } \mathcal{S} = \pm\mathbf{1}, \\ 0, & \text{when } \mathcal{S} \neq \pm\mathbf{1}, \end{cases} \tag{2.50}$$

  where $d^{(2l)}$ is the dimension of the γ_a. Since the self-representation of the Γ_{2l} is irreducible, we have

$$2\left(d^{(2l)}\right)^2 = \sum_{\mathcal{S} \in \Gamma_{2l}} |\xi(\mathcal{S})|^2 = g^{(2l)} = 2^{2l+1}, \qquad d^{(2l)} = 2^l. \tag{2.51}$$

  Based on Eqs. (2.43) and (2.50), we have $\det \gamma_a = 1$ for $l > 1$.
- Since $\gamma_\xi^{(2l)}$ is anticommutable with every γ_a, one may define $\gamma_f^{(2l)}$ by multiplying $\gamma_\xi^{(2l)}$ with a factor such that $\gamma_f^{(2l)}$ satisfies Eq. (2.42), i.e.,

$$\gamma_f^{(2l)} = (-i)^l \gamma_\xi^{(2l)} = (-i)^l \gamma_1 \gamma_2 \cdots \gamma_{2l}, \qquad \left(\gamma_f^{(2l)}\right)^2 = \mathbf{1}. \tag{2.52}$$

  Actually, $\gamma_f^{(2l)}$ can also be defined as the matrix γ_{2l+1} in Γ_{2l+1}.

- The matrices in the set Γ'_{2l} are linearly independent. Otherwise, there exists a linear relation $\sum_{\mathcal{S}} C(\mathcal{S})\mathcal{S} = 0$, $\mathcal{S} \in \Gamma'_{2l}$. By multiplying it with $R^{-1}/d^{(2l)}$ and taking the trace, one obtains any coefficient $C(R) = 0$. Thus, the set Γ'_{2l} contains 2^{2l} linear independent matrices of dimension $d^{(2l)} = 2^l$ so that they form a complete set of basis matrices. Any matrix M of dimension $d^{(2l)}$ can be expanded by $\mathcal{S} \in \Gamma'_{2l}$ as follows:

$$M = \sum_{\mathcal{S}\in\Gamma'_{2l}} C(\mathcal{S})\mathcal{S}, \qquad C(\mathcal{S}) = \frac{1}{d^{(2l)}} \mathrm{Tr}(\mathcal{S}^{-1}M). \tag{2.53}$$

- According to Eq. (2.42), the $\pm\mathcal{S}$ form a class, while $\mathbf{1}$ and $-\mathbf{1}$ form two classes, respectively. The Γ_{2l} group contains $(2^{2l}+1)$ classes. Their representation is one-dimensional. Arbitrary chosen n matrices γ_a correspond to 1 and the remaining matrices γ_b correspond to -1. The number of the one-dimensional non-equivalent representations is calculated as

$$\sum_{n=0}^{2l} \binom{2l}{n} = 2^{2l}. \tag{2.54}$$

The remaining irreducible representation of the Γ_{2l} must be $d^{(2l)}$-dimensional, which is faithful. The γ_a matrices in the representation are called the irreducible γ_a matrices, which may be written as:

$$\begin{aligned} \gamma_{2n-1} &= \underbrace{\mathbf{1}\times\cdots\times\mathbf{1}}_{n-1}\times\sigma_1\times\underbrace{\sigma_3\times\cdots\times\sigma_3}_{l-n}, \\ \gamma_{2n} &= \underbrace{\mathbf{1}\times\cdots\times\mathbf{1}}_{n-1}\times\sigma_2\times\underbrace{\sigma_3\times\cdots\times\sigma_3}_{l-n}, \\ \gamma_f^{(2l)} &= \underbrace{\sigma_3\times\cdots\times\sigma_3}_{l}. \end{aligned} \tag{2.55}$$

Since $\gamma_f^{(2l)}$ is diagonal, the forms of Eq. (2.55) are called the reduced spinor representations. Remember that the eigenvalues ± 1 are arranged in the diagonal line of the $\gamma_f^{(2l)}$ in mixed way.
- Let us mention an equivalent theorem for the γ_a matrices.

Theorem 2.1 *Two sets of $d^{(2l)}$-dimensional matrices γ_a and $\bar{\gamma_a}$ satisfying the anti-commutation relation* (2.42), *where $N = 2l$, are equivalent*

$$\bar{\gamma_a} = X^{-1}\gamma_a X, \quad a \in [1, 2l]. \tag{2.56}$$

The similarity transformation matrix X is determined up to a constant factor. If the determinant of the matrix X is constrained to be 1, there are $d^{(2l)}$ choices for the factor:

$$\exp[-i2n\pi/d^{(2l)}], \quad n \in [0, d^{(2l)}). \tag{2.57}$$

5.3 Case $N = 2l + 1$

Since $\gamma_f^{(2l)}$ and $(2l)$ matrices γ_a in Γ_{2l}, $a \in [1, 2l]$, satisfy the antisymmetric relation (2.42), then they can be defined to be the $(2l+1)$ matrices γ_a in Γ_{2l+1}. In this definition, $\gamma_\xi^{(2l+1)}$ in Γ_{2l+1} is chosen as

$$\gamma_{2l+1} = \gamma_f^{(2l)}, \qquad \gamma_\xi^{(2l+1)} = \gamma_1 \cdots \gamma_{2l+1} = i^l \mathbf{1}. \tag{2.58}$$

Obviously, the dimension $d^{(2l+1)}$ of the matrices in Γ_{2l+1} is the same as $d^{(2l)}$ in Γ_{2l},

$$d^{(2l+1)} = d^{(2l)} = 2^l. \tag{2.59}$$

For odd N, the equivalent theorem must be modified because the multiplication rule of elements in Γ_{2l+1} includes Eq. (2.45). A similarity transformation cannot change the sign of $\gamma_\xi^{(2l+1)}$, i.e., the equivalent condition for two sets of γ_a and $\bar{\gamma}_a$ has to include a new condition $\gamma_\xi = \bar{\gamma}_\xi$, in addition to those given in Theorem 2.1.

If we take $\bar{\gamma}_a = -(\gamma_a)^T$, then we have

$$\begin{aligned}
\bar{\gamma}_\xi^{(2l+1)} &= \bar{\gamma}_1 \cdots \bar{\gamma}_{2l+1} = -\{\gamma_{2l+1} \cdots \gamma_1\}^T \\
&= (-1)^{l+1} \{\gamma_\xi^{(2l+1)}\}^T \\
&= (-1)^{l+1} \gamma_\xi^{(2l+1)}.
\end{aligned} \tag{2.60}$$

6 Spinor Representations of the SO(N)

6.1 Covering Groups of the SO(N)

Based on a set of N irreducible unitary matrices γ_a satisfying the anticommutation relation (2.42), we define

$$\bar{\gamma}_a = \sum_{i=1}^{N} R_{ai} \gamma_i, \qquad R \in \mathrm{SO}(N). \tag{2.61}$$

Since R is a real orthogonal matrix, then $\bar{\gamma}_a$ satisfy

$$\begin{aligned}
\bar{\gamma}_a \bar{\gamma}_b + \bar{\gamma}_b \bar{\gamma}_a &= \sum_{ij} R_{ai} R_{bj} \{\gamma_i \gamma_j + \gamma_j \gamma_i\} \\
&= 2 \sum_i R_{ai} R_{bi} \mathbf{1} \\
&= 2\delta_{ab} \mathbf{1}.
\end{aligned} \tag{2.62}$$

Due to Eq. (2.42) and $\sum_a R_{1a} R_{2a} = 0$, we have

$$\sum_{c_1c_2} R_{1c_1}R_{2c_2}\gamma_{c_1}\gamma_{c_2} = \frac{1}{2}\sum_{c_1\neq c_2} R_{1c_1}R_{2c_2}(\gamma_{c_1}\gamma_{c_2} - \gamma_{c_2}\gamma_{c_1}), \tag{2.63}$$

$$\begin{aligned} \bar{\gamma}_1\bar{\gamma}_2\cdots\bar{\gamma}_N &= \sum_{c_1\cdots c_N} R_{1\,c_1}\cdots R_{N\,c_N}\gamma_{c_1}\cdots\gamma_{c_N} \\ &= \sum_{c_1\cdots c_N} R_{1\,c_1}\cdots R_{N\,c_N}\epsilon_{c_1\cdots c_N}\gamma_1\gamma_2\cdots\gamma_N \\ &= (\det R)\gamma_1\gamma_2\cdots\gamma_N = \gamma_1\gamma_2\cdots\gamma_N. \end{aligned} \tag{2.64}$$

From Theorem 2.1, we know that γ_a and $\bar{\gamma}_a$ are related by a unitary similarity transformation $D(R)$ with determinant 1,

$$D(R)^{-1}\gamma_a D(R) = \sum_{i=1}^{N} R_{ai}\gamma_i, \qquad \det D(R) = 1, \tag{2.65}$$

where $D(R)$ is determined up to a constant $\exp[-i2n\pi/d^{(N)}], n \in [0, d^{(N)})$. In terms of the definition of the group, the set of $D(R)$ defined in Eq. (2.65) and operated in the multiplication rule of matrices, forms a Lie group G'_N. There exists a $d^{(N)}$-to-one correspondence between the elements in G'_N and those in SO(N), and the correspondence keeps invariant in the multiplication of elements. Therefore, the G'_N is homomorphic to SO(N). Because the group space of the SO(N) is doubly-connected, its covering group is homomorphic to it by a two-to-one correspondence. As a result, the group space of the G'_N must fall into several disjoint pieces, where the piece containing the identity element E forms an invariant subgroup G_N of the G'_N. The G_N is a connected Lie group and becomes the covering group of the SO(N) . Since the group space of G_N is connected, based on the property of the infinitesimal elements, a discontinuous condition can be found to pick up G_N from the G'_N.

Let R be an infinitesimal element. We may expand R and $D(R)$ with respect to the infinitesimal parameters $\omega_{\alpha\beta}$ as follows

$$\begin{aligned} R_{ab} &= \delta_{ab} - i\sum_{\alpha<\beta}\omega_{\alpha\beta}(T_{\alpha\beta})_{ab} = \delta_{ab} - \omega_{ab}, \\ D(R) &= \mathbf{1} - i\sum_{\alpha<\beta}\omega_{\alpha\beta}S_{\alpha\beta}, \end{aligned} \tag{2.66}$$

where $T_{\alpha\beta}$ are the generators in the self-representation of the SO(N) as given in Eq. (2.14). The $S_{\alpha\beta}$ are the generators in G_N. From Eq. (2.65) one has

$$[\gamma_c, S_{\alpha\beta}] = \sum_d (T_{\alpha\beta})_{cd}\gamma_d = -i\{\delta_{\alpha c}\gamma_\beta - \delta_{\beta c}\gamma_\alpha\}, \tag{2.67}$$

from which we obtain

$$S_{\alpha\beta} = \frac{1}{4i}(\gamma_\alpha\gamma_\beta - \gamma_\beta\gamma_\alpha). \tag{2.68}$$

It is easy to prove that $S_{\alpha\beta}$ is hermitian since $D(R)$ is unitary.

Define[3]

$$C = \begin{cases} B^{(N)}, & \text{when } N = 4l + 1, \\ C^{(N)}, & \text{when } N \neq 4l + 1. \end{cases} \tag{2.69}$$

Based on this, we have

$$\begin{aligned} C^{-1} S_{\alpha\beta} C &= -(S_{\alpha\beta})^T = -S_{\alpha\beta}^*, \\ C^{-1} D(R) C &= \{D(R^{-1})\}^T = D(R)^*. \end{aligned} \tag{2.70}$$

This discontinuous condition restricts the factor in $D(R)$ such that there is a two-to-one correspondence between $\pm D(R)$ in G_N and R in SO(N) through relations (2.65) and (2.70). That is to say, the G_N is the covering group of SO(N),

$$\mathrm{SO}(N) \sim G_N, \tag{2.71}$$

where the G_N is the fundamental spinor representation denoted by $D^{[s]}(\mathrm{SO}(N))$. Therefore, the $S_{\alpha\beta}$ represent the spinor angular momentum operators [88–90]. The irreducible tensor representation $[\lambda]$ is a single-valued representation of the SO(N), but a non-faithful representation of G_N because its faithful representation is a double-valued representation of the SO(N).

Since the products S_n span a complete set of the $d^{(N)}$-dimensional matrices, this can be decided by checking the commutation relations of the S_n with the generators $S_{\alpha\beta}$ whether there exists a non-constant matrix commutable with all $S_{\alpha\beta}$. It is found that only $\gamma_\xi^{(N)}$ is commutable with all $S_{\alpha\beta}$. The $\gamma_\xi^{(2l+1)}$ is a constant matrix so that the fundamental spinor representation $D^{[s]}(\mathrm{SO}(2l+1))$ is irreducible and self-conjugate.

On the contrary, since $\gamma_\xi^{(2l)}$ is not a constant matrix so that the fundamental spinor representation $D^{[s]}(\mathrm{SO}(2l))$ is reducible. By a similarity transformation X, the $\gamma_f^{(2l)}$ can be transferred to $\sigma_3 \times \mathbf{1}$ and $D^{[s]}(\mathrm{SO}(2l))$ is reduced to the direct sum of two irreducible representations

$$X^{-1} D^{[s]}(R) X = \begin{pmatrix} D^{[+s]}(R) & 0 \\ 0 & D^{[-s]}(R) \end{pmatrix}. \tag{2.72}$$

Two representations $D^{[\pm s]}(\mathrm{SO}(2l))$ are proved inequivalent by leading to an absurdity. In fact, if $Z^{-1} D^{[-s]}(R) Z = D^{[+s]}(R)$ and $Y = \mathbf{1} \oplus Z$, then all generators $(XY)^{-1} S_{\alpha\beta} XY$ are commutable with $\sigma_1 \times \mathbf{1}$, but their product is not commutable with it,

$$2^l (XY)^{-1} (S_{12} S_{34} \cdots S_{(2l-1)(2l)}) XY = Y^{-1} [X^{-1} \gamma_f^{(2l)} X] Y = \sigma_3 \times \mathbf{1}, \tag{2.73}$$

which results in a contradiction.

---

[3] B^N is the strong space-time reflection matrix and C^N is the charge conjugation matrix, which are usually used in particle physics.

Introduce two project operators $P_\pm$ [139, 140],

$$\begin{aligned}
&P_\pm = \frac{1}{2}\left(\mathbf{1} \pm \gamma_f^{(2l)}\right), && P_\pm D^{[s]}(R) = D^{[s]} P_\pm, \\
&X^{-1} P_+ X = \begin{pmatrix} \mathbf{1} & 0 \\ 0 & 0 \end{pmatrix}, && X^{-1} P_+ D^{[s]}(R) X = \begin{pmatrix} D^{[+s]}(R) & 0 \\ 0 & 0 \end{pmatrix}, \\
&X^{-1} P_- X = \begin{pmatrix} 0 & 0 \\ 0 & \mathbf{1} \end{pmatrix}, && X^{-1} P_- D^{[s]}(R) X = \begin{pmatrix} 0 & 0 \\ 0 & D^{[-s]}(R) \end{pmatrix}.
\end{aligned} \tag{2.74}$$

From the following relation

$$(C^{(2l)})^{-1} \gamma_f^{(2l)} C^{(2l)} = (-i)^l (\gamma_1)^T (\gamma_2)^T \cdots (\gamma_{2l})^T = (-1)^l (\gamma_f^{(2l)})^T, \tag{2.75}$$

where $C^{(2l)}$ and T denote the charge conjugation matrix and the transpose of the matrix, respectively, one has

$$C^{-1} D^{[s]}(R) P_\pm C = \begin{cases} D^{[s]}(R)^* P_\pm, & \text{when } N = 4l, \\ D^{[s]}(R)^* P_\mp, & \text{when } N = 4l + 2. \end{cases} \tag{2.76}$$

Two non-equivalent representations $D^{[\pm s]}(R)$ are conjugate to each other when $N = 4l+2$, while they are self-conjugate when $N = 4l$. The dimension of the irreducible spinor representations of the SO(N) is calculated as

$$d_{[s]}[\mathrm{SO}(2l+1)] = 2^l, \qquad d_{[\pm s]}[\mathrm{SO}(2l)] = 2^{(l-1)}. \tag{2.77}$$

6.2 *Fundamental Spinors of the* SO(N)

For an SO(N) transformation R, Ψ is called the fundamental spinor of the SO(N) if it transforms through the fundamental spinor representation $D^{[s]}(R)$:

$$(O_R \Psi)_\nu = \sum_\mu D^{[s]}_{\nu\mu}(R) \Psi_\mu, \qquad O_R \Psi = D^{[s]}(R) \Psi, \tag{2.78}$$

where Ψ is a column matrix with $d_{[s]}$ components.

The Chevalley bases $H_\nu(S)$, $E_\nu(S)$ and $F_\nu(S)$ with respect to the spinor angular momentum can be obtained from Eqs. (2.18) and (2.24) through replacing T_{ab} by S_{ab}. In the chosen forms of γ_a given in Eq. (2.55), the Chevalley bases for the SO($2l+1$) group are given by

$$\begin{aligned}
H_\nu(S) &= \underbrace{\mathbf{1} \times \cdots \times \mathbf{1}}_{\nu-1} \times \frac{1}{2}\{\sigma_3 \times \mathbf{1} - \mathbf{1} \times \sigma_3\} \times \underbrace{\mathbf{1} \times \cdots \times \mathbf{1}}_{l-\nu-1}, \\
H_l(S) &= \underbrace{\mathbf{1} \times \cdots \times \mathbf{1}}_{l-1} \times \sigma_3, \\
E_\nu(S) &= \underbrace{\mathbf{1} \times \cdots \times \mathbf{1}}_{\nu-1} \times \{\sigma_+ \times \sigma_-\} \times \underbrace{\mathbf{1} \times \cdots \times \mathbf{1}}_{l-\nu-1} = F_\nu(S)^T, \\
E_l(S) &= \underbrace{\sigma_3 \times \cdots \times \sigma_3}_{l-1} \times \sigma_+ = F_l(S)^T,
\end{aligned} \tag{2.79}$$

where $\nu \in [1, l)$. The Chevalley bases for the SO(2l) group are the same as those for the SO(2$l+1$) except for $\nu = l$,

$$\begin{aligned} H_l(S) &= \underbrace{\mathbf{1} \times \cdots \times \mathbf{1}}_{l-2} \times \frac{1}{2}\{\sigma_3 \times \mathbf{1} + \mathbf{1} \times \sigma_3\}, \\ E_l(S) &= -\underbrace{\mathbf{1} \times \cdots \times \mathbf{1}}_{l-2} \times \{\sigma_+ \times \sigma_+\} = F_l(S)^T. \end{aligned} \tag{2.80}$$

The basis spinor $\xi[\mathbf{m}]$ of the SO(N) can also be expressed as a direct product of l two-dimensional basis spinors $\xi(\beta)$,

$$\xi[\mathbf{m}] = \xi(\beta_1, \beta_2, \ldots, \beta_l) = \xi(\beta_1)\xi(\beta_2)\ldots\xi(\beta_l), \tag{2.81}$$

$$\xi(+) = \begin{pmatrix} 1 \\ 0 \end{pmatrix}, \qquad \xi(-) = \begin{pmatrix} 0 \\ 1 \end{pmatrix}. \tag{2.82}$$

For even N, the fundamental spinor space can be decomposed into two subspaces by the project operators $P_\pm$, $\Psi_\pm = P_\pm \Psi$, corresponding to irreducible spinor representations $D^{[\pm s]}$. The basis spinor in the representation space of $D^{[+s]}$ contains even number of factors $\xi(-)$, and that of $D^{[-s]}$ contains odd number of factors $\xi(-)$. The highest weight states $\xi[\mathbf{M}]$ and their highest weights $\mathbf{M}$ are given by

$$\begin{aligned} &\underbrace{\xi(+)\ldots\xi(+)}_{l-1}\xi(+), && \mathbf{M} = [\underbrace{0, \ldots, 0}_{l-1}, 1], && [s] \text{ of the SO}(2l+1), \\ &\underbrace{\xi(+)\ldots\xi(+)}_{l-1}\xi(+), && \mathbf{M} = [\underbrace{0, \ldots, 0}_{l-2}, 0, 1], && [+s] \text{ of the SO}(2l), \\ &\underbrace{\xi(+)\ldots\xi(+)}_{l-1}\xi(-), && \mathbf{M} = [\underbrace{0, \ldots, 0}_{l-2}, 1, 0], && [-s] \text{ of the SO}(2l). \end{aligned} \tag{2.83}$$

The remaining basis states are calculated by the applications of lowering operators $F_\nu(S)$.

6.3 Direct Products of Spinor Representations

Since the spinor representation is unitary so that we have

$$O_R \Psi^\dagger = \Psi^\dagger D^{[s]}(R)^{-1}, \tag{2.84}$$

$$\begin{aligned} \Psi^\dagger \Psi &= \textstyle\sum_\mu \Psi_\mu^* \Psi_\mu = \sum_{\mu\nu} \Psi_\mu^* \delta_{\mu\nu} \Psi_\nu, \\ O_R(\Psi^\dagger \Psi) &= \Psi^\dagger D^{[s]}(R)^{-1} D^{[s]}(R) \Psi = \Psi^\dagger \Psi, \end{aligned} \tag{2.85}$$

which means that $\Psi^\dagger \Psi$ keeps invariant in the SO(N) transformations and is a scalar of the SO(N). In other words, the products of $\Psi_\mu^\dagger$ and Ψ_ν span an invariant linear space, corresponding to the direct product representation $D^{[s]*} \times D^{[s]}$ of the SO(N). In the reduction of $D^{[s]*} \times D^{[s]}$ there is an identical representation where the CGCs are $\delta_{\mu\nu}$. In general, one has

$$\begin{aligned} O_R(\Psi^\dagger \gamma_{a_1} \cdots \gamma_{a_n} \Psi) &= \Psi^\dagger D^{[s]}(R)^{-1} \gamma_{a_1} \cdots \gamma_{a_n} D^{[s]}(R) \Psi \\ &= \sum_{c_1 \ldots c_n} R_{a_1 c_1} \cdots R_{a_n c_n} \Psi^\dagger \gamma_{c_1} \cdots \gamma_{c_n} \Psi, \end{aligned} \tag{2.86}$$

where $\Psi^\dagger \gamma_{a_1} \cdots \gamma_{a_n} \Psi$ is an antisymmetric tensor of rank n of the SO(N) corresponding to the Young pattern $[1^n]$ with $n \leq N$. Otherwise, the respective γ_a can be moved together and eliminated.

When $N = 2l+1$, the $\gamma_f^{(2l+1)}$ is a constant matrix so that the product of $(N-n)$ matrices γ_a can be changed to a product of n matrices γ_a. Thus, the rank n of the tensor (2.86) is less than $N/2$, and the Clebsch-Gordan series is given by

$$[s]^* \times [s] \simeq [s] \times [s] \simeq [0] \oplus [1] \oplus [1^2] \oplus \cdots \oplus [1^l], \quad \text{for SO}(2l+1). \tag{2.87}$$

The matrix entries of product of γ_a are the CGCs. The highest weight in product space is given by $\mathbf{M} = [0, \ldots, 0, 2]$ corresponding to representation $[1^l]$.

When $N = 2l$, according to the property of the project operators $P_\pm$,

$$\begin{aligned} &P_+ P_- = P_- P_+ = 0, \qquad P_\pm P_\pm = P_\pm, \qquad \gamma_f^{(2l)} P_\pm = \pm P_\pm, \\ &P_\mp \gamma_{b_1} \cdots \gamma_{b_{2n}} P_\pm = 0, \qquad P_\pm \gamma_{b_1} \cdots \gamma_{b_{2n+1}} P_\pm = 0, \end{aligned} \tag{2.88}$$

the product of the $(N-n)$ matrices γ_b can still be changed to a product of n matrices γ_b. If $n = l$, we have

$$\begin{aligned} &\gamma_1 \gamma_2 \cdots \gamma_l = (-i)^l \gamma_{2l} \gamma_{2l-1} \cdots \gamma_{l+1} \gamma_f^{(2l)}, \\ &\gamma_1 \gamma_2 \cdots \gamma_l P_\pm = \frac{1}{2} \{ \gamma_1 \gamma_2 \cdots \gamma_l \pm (-i)^l \gamma_{2l} \gamma_{2l-1} \cdots \gamma_{l+1} \} P_\pm. \end{aligned} \tag{2.89}$$

If $N = 4l$, we have

$$\begin{aligned} &[\pm s]^* \times [\pm s] \simeq [\pm s] \times [\pm s] \simeq [0] \oplus [1^2] \oplus [1^4] \oplus \cdots \oplus [(\pm 1) 1^{2l}], \\ &[\mp s]^* \times [\pm s] \simeq [\mp s] \times [\pm s] \simeq [1] \oplus [1^3] \oplus [1^5] \oplus \cdots \oplus [1^{2l-1}]. \end{aligned} \tag{2.90}$$

If $N = 4l+2$, one has

$$\begin{aligned} &[\pm s]^* \times [\pm s] \simeq [\mp s] \times [\pm s] \simeq [0] \oplus [1^2] \oplus [1^4] \oplus \cdots \oplus [1^{2l}], \\ &[\mp]^* \times [\pm s] \simeq [\pm s] \times [\pm s] \simeq [1] \oplus [1^3] \oplus [1^5] \oplus \cdots \oplus [(\pm) 1^{2l+1}]. \end{aligned} \tag{2.91}$$

The self-dual and anti-self-dual representations occur in the reduction of the direct product $[\pm s] \times [\pm s]$, but not in the reduction of $[+s] \times [-s]$. The highest weights are $\mathbf{M} = [0, \ldots, 0, 0, 2]$ in the product space $[+s] \times [+s]$, $\mathbf{M} = [0, \ldots, 0, 2, 0]$ in $[-s] \times [-s]$, and $\mathbf{M} = [0, \ldots, 0, 1, 1]$ in $[+s] \times [-s]$.

6.4 Spinor Representations of Higher Ranks

In the SO(3) group, $D^{1/2}$ is a fundamental spinor representation. The spinor representations D^j of higher ranks can be obtained by reducing the direct product of the fundamental spinor representation and a tensor representation,

$$D^{1/2} \times D^l \simeq D^{l+1/2} \oplus D^{l-1/2}. \tag{2.92}$$

The spinor representations of higher ranks of the SO(N) can also be obtained in a similar way.

A spinor $\Psi_{a_1\dots a_n}$ with the tensor indices is called a spin-tensor if it transforms in $R \in \mathrm{SO}(N)$ as follows:

$$(O_R\Psi)_{a_1\cdots a_n} = \sum_{c_1\cdots c_n} R_{a_1c_1}\cdots R_{a_nc_n} D^{[s]}(R)\Psi_{c_1\cdots c_n}. \tag{2.93}$$

The tensor part of the spin-tensor can be decomposed into a direct sum of the traceless tensors with different ranks. Each traceless tensor subspace can be reduced by the projection of the Young operators. Thus, the reduced subspace of the traceless tensor part of the spin-tensor is denoted by a Young pattern $[\lambda]$ or $[\pm\lambda]$ where the row number of $[\lambda]$ is not larger than $N/2$. However, this subspace of the spin-tensor corresponds to the direct product of the fundamental spinor representation $[s]$ and the irreducible tensor representation $[\lambda]$ or $[\pm\lambda]$, and it is still reducible. It is required to find a new restriction to pick up the irreducible subspace like the subspace of $D^{l+1/2}$ in Eq. (2.92) for the SO(3) group. The restriction is from the so-called trace of the second kind of the spin-tensor which keeps invariant in the SO(N) transformations:

$$\Phi_{a_1\cdots a_{i-1}a_{i+1}\cdots a_n} = \sum_{c=1}^{N} \gamma_c \Psi_{a_1\cdots a_{i-1}\,c\,a_{i+1}\cdots a_n}, \tag{2.94}$$

and

$$\begin{aligned}
&(O_R\Phi)_{a_1\cdots a_{i-1}a_{i+1}\cdots a_n}\\
&\quad = \sum_{c_1\cdots c_n c'} R_{a_1c_1}\cdots R_{a_nc_n}\left[\sum_c \gamma_c R_{cc'}\right] D^{[s]}(R)\Psi_{c_1\cdots c_{i-1}c'c_{i+1}\cdots c_n}\\
&\quad = \sum_{c_1\cdots c_n} R_{a_1c_1}\cdots R_{a_nc_n} D^{[s]}(R)\left[\sum_{c'} \gamma_{c'}\Psi_{c_1\cdots c_{i-1}c'c_{i+1}\cdots c_n}\right]\\
&\quad = \sum_{c_1\cdots c_n} R_{a_1c_1}\cdots R_{a_nc_n} D^{[s]}(R)\Psi_{c_1\cdots c_{i-1}c_{i+1}\cdots c_n}.
\end{aligned} \tag{2.95}$$

The irreducible subspace of the SO(N) contained in the spin-tensor space, in addition to the projection of a Young operator, satisfies the usual traceless conditions of tensors and the traceless conditions of the second kind

$$\sum_d \psi_{a\cdots d\cdots d\cdots c} = 0, \qquad \sum_d \gamma_d \psi_{a\cdots d\cdots c} = 0. \tag{2.96}$$

The highest weight $\mathbf{M}$ of the irreducible representation is the highest weight in the direct product space. The irreducible representation is denoted by $[s,\lambda]$ for the SO($2l+1$)

$$\begin{cases} [s]\times[\lambda] \simeq [s,\lambda]\oplus\cdots, \\ \mathbf{M} = [(\lambda_1-\lambda_2),\dots,(\lambda_{l-1}-\lambda_l),(2\lambda_l+1)], \end{cases} \tag{2.97}$$

and $[\pm s, \lambda]$ for the SO($2l$)

$$\begin{cases} [+s]\times[\lambda] \text{ or } [+s]\times[+\lambda] \simeq [+s,\lambda]\oplus\cdots, \\ \mathbf{M}=[(\lambda_1-\lambda_2),\ldots,(\lambda_{l-1}-\lambda_l),(\lambda_{l-1}+\lambda_\lambda+1)], \end{cases}$$
$$\begin{cases} [-s]\times[\lambda] \text{ or } [-s]\times[-\lambda] \simeq [-s,\lambda]\oplus\cdots, \\ \mathbf{M}=[(\lambda_1-\lambda_2),\ldots,(\lambda_{l-1}+\lambda_l+1),(\lambda_{l-1}-\lambda_l)], \end{cases}$$
$$\begin{cases} [+s]\times[-\lambda] \simeq [-s,\lambda_1,\lambda_2,\ldots,\lambda_{l-1},(\lambda_l-1)]\oplus\cdots, \\ \mathbf{M}=[(\lambda_1-\lambda_2),\ldots,(\lambda_{l-1}+\lambda_l),(\lambda_{l-1}-\lambda_l+1)], \end{cases}$$
$$\begin{cases} [-s]\times[+\lambda] \simeq [+s,\lambda_1,\lambda_2,\ldots,\lambda_{l-1},(\lambda_l-1)]\oplus\cdots, \\ \mathbf{M}=[(\lambda_1-\lambda_2),\ldots,(\lambda_{l-1}-\lambda_l+1),(\lambda_{l-1}+\lambda_l)]. \end{cases} \tag{2.98}$$

These irreducible representations $[s, \lambda]$ of the SO($2l+1$) and $[\pm s, \lambda]$ of the SO($2l$) are called the spinor representations of higher ranks. It should be noted that the row number of the Young pattern $[\lambda]$ in the spinor representation of higher rank is not larger than l. Otherwise, the space is null.

The remaining representations in the Clebsch-Gordan series (2.97) and (2.98) are calculated by the method of dominant weight diagram. For example, when $[\lambda]$ is a one-row Young diagram, one has

$$\begin{aligned} &\mathrm{SO}(2l+1): \quad [s]\times[\lambda,0,\ldots,0]\simeq[s,\lambda,0,\ldots,0]\oplus[s,\lambda-1,0,\ldots,0], \\ &\mathrm{SO}(2l): \quad [\pm s]\times[\lambda,0,\ldots,0]\simeq[\pm s,\lambda,0,\ldots,0]\oplus[\mp s,\lambda-1,0,\ldots,0], \end{aligned} \tag{2.99}$$

where $[\mp s, \lambda-1, 0, \ldots, 0]$ appears because the factor γ_b in Eq. (2.96) is anticommutable with γ_f in $P_\pm$.

6.5 Dimensions of the Spinor Representations

In a similar way, the dimension of a spinor representation $[s, \lambda]$ of the SO($2l+1$) or $[\pm s, \lambda]$ of the SO($2l$) can be calculated by hook rule. The dimension is expressed as a quotient multiplied with the dimension of the fundamental spinor representation, where the numerator and the denominator are denoted by the symbols $Y_S^{[\lambda]}$ and $Y_h^{[\lambda]}$, respectively:

$$\begin{aligned} d_{[s,\lambda]}[\mathrm{SO}(2l+1)] &= 2^l\,\frac{Y_S^{[\lambda]}}{Y_h^{[\lambda]}}, \\ d_{[\pm s,\lambda]}[\mathrm{SO}(2l)] &= 2^{l-1}\,\frac{Y_S^{[\lambda]}}{Y_h^{[\lambda]}}. \end{aligned} \tag{2.100}$$

The concepts of a hook path (i, j) and an inverse hook path $\overline{i, j}$ have been discussed above. The number of boxes contained in the hook path (i, j) is the hook number h_{ij} of the box in the jth column of the ith row. The $Y_h^{[\lambda]}$ is a tableau of the Young pattern $[\lambda]$ where the box in the jth column of the ith row is filled with the hook number h_{ij}. The $Y_S^{[\lambda]}$ is a tableau of the Young pattern $[\lambda]$ where each box is filled with the sum of the digits which are respectively filled in the same box of each tableau $Y^{[\lambda]}_{S_b}$ in the series. The notation $Y^{[\lambda]}_{S}$ means the product of the filled digits in it, so does the notation $Y^{[\lambda]}_{h}$. The tableaux $Y^{[\lambda]}_{S_b}$ are defined by the following rules:

- $Y^{[\lambda]}_{S_0}$ is a tableau of the Young pattern $[\lambda]$ where the box in the jth column of the ith row is filled with the digit $(N-1+j-i)$.
- Let $[\lambda^{(1)}]=[\lambda]$. Staring with $[\lambda^{(1)}]$, we define recursively the Young pattern $[\lambda^{(b)}]$ by removing the first row and the first column of the Young pattern $[\lambda^{(b-1)}]$ until $[\lambda^{(b)}]$ contains less two columns.
- If $[\lambda^{(b)}]$ contains more than one column, we define $Y^{[\lambda]}_{S_b}$ as the tableau of the Young pattern $[\lambda]$ where the boxes in the first $(b-1)$ row and column are filled with 0, and the remaining part of the Young pattern is $[\lambda^{(b)}]$. Let $[\lambda^{(b)}]$ have r rows. Fill the first r boxes along the hook path $(1,1)$ of the Young pattern $[\lambda^{(b)}]$, starting with the box on the rightmost, with the digits $(\lambda_1^{(b)}-1)$, $(\lambda_2^{(b)}-1),\ldots,(\lambda_r^{(b)}-1)$, box by box, and fill the first $(\lambda_i^{(b)}-1)$ boxes in each inverse hook path $\overline{(i,1)}$ of the Young pattern $[\lambda^{(b)}]$, $i\in[2,r]$ with "-1". The remaining boxes are filled with 0. If several "-1" are filled in the same box, the digits are summed. The sum of all filled digits in the pattern $Y^{[\lambda]}_{S_b}$ with $b>0$ is equal to 0.

7 Concluding Remarks

In this Chapter we have sketched some basic properties for the Lie group SO(N) since it shall be very helpful in successive several Chapters. The tensor and spinor representations of the SO(N) group, the calculation of the dimensions of irreducible tensor and spinor representations have been addressed. The more information about the properties of the Lie groups and Lie algebras, in particular the SO(N) group as well as the corresponding Lie algebra may refer to textbooks [136, 138–140].

Chapter 3
Rotational Symmetry and Schrödinger Equation in D-Dimensional Space

1 Introduction

It is well known that in classical mechanics an image has rotational symmetry if there is a center point around which the object is turned a certain number of degrees and the object still looks the same, i.e., it matches itself a number of times while it is being rotated. In the language of quantum mechanics, isotropy of space means that the system Hamiltonian keeps invariant by a rotation. In our case the Schrödinger equation with the spherically symmetric fields possesses this kind of property. If the Hamiltonian has rotational symmetry, we can show that the angular momentum operators $\mathbf{L}$ commute with the Hamiltonian, which means that the angular momentum is a conserved quantity, i.e., $d\mathbf{L}/dt = 0$. Thus, this constant of the motion enables us to reduce the D-dimensional Schrödinger equation to a radial differential equation. This may be explained well from the rotation group theory as below.

This Chapter is organized as follows. In Sect. 2 we give a brief review of the rotation operator. In Sect. 3 we are going to study the generalized orbital angular momentum operators in higher dimensions. The linear momentum operators and radial momentum operator are to be studied in Sects. 4 and 5, respectively. The generalized spherical harmonic polynomials shall be discussed in Sect. 6. We are going to study the Schrödinger equation for a two-body system in Sect. 7. We shall give some concluding remarks in Sect. 8.

2 Rotation Operator

We define the rotation operator $O_R(\vec{\varphi})$ in such a way that the rotated scalar state $\psi_{\alpha'}(\mathbf{r},t)$ follows from the initial state $\psi_\alpha(\mathbf{r},t)$ by

$$\psi_{\alpha'}(\mathbf{r},t) = O_R(\vec{\varphi})\psi_\alpha(\mathbf{r},t). \tag{3.1}$$

S.-H. Dong, *Wave Equations in Higher Dimensions*,
DOI 10.1007/978-94-007-1917-0_3, © Springer Science+Business Media B.V. 2011

Consequently, the following expression has to be valid

$$\begin{aligned} i\hbar\frac{\partial\psi_{\alpha'}(\mathbf{r},t)}{\partial t} &= i\hbar\frac{\partial O_R(\vec{\varphi})\psi_\alpha(\mathbf{r},t)}{\partial t} \\ &= O_R(\vec{\varphi})\left(i\hbar\frac{\partial\psi_\alpha(\mathbf{r},t)}{\partial t}\right) = O_R(\vec{\varphi})H\psi_\alpha(\mathbf{r},t) \\ &= O_R(\vec{\varphi})HO_R^{-1}(\vec{\varphi})\psi_{\alpha'}(\mathbf{r},t) = H\psi_{\alpha'}(\mathbf{r},t), \end{aligned} \tag{3.2}$$

where the last step expresses the requirement of invariance of the Schrödinger equation concerning rotations or, equivalently, isotropy of configuration space. Certainly, we have used the fact that the initial state $\psi_\alpha(\mathbf{r},t)$ satisfies the Schrödinger equation $i\hbar\partial\psi_\alpha(\mathbf{r},t)/\partial t = H\psi_\alpha(\mathbf{r},t)$.

Therefore, for arbitrary rotation vector $\vec{\varphi}$, we have

$$O_R(\vec{\varphi})HO_R^{-1}(\vec{\varphi}) = H \tag{3.3}$$

or

$$[O_R, H] = 0. \tag{3.4}$$

Since $\vec{\varphi}$ is arbitrary, we have

$$[\mathbf{L}, H] = 0, \tag{3.5}$$

which means the conservation of angular momentum.

In order to construct the explicit rotation operator, we consider a rotation by an infinitesimal rotation angle $\delta\varphi$ about z axis. The rotation operator can be written as

$$O_{R_{e_z}}(\delta\varphi) = \mathbf{1} - \frac{i\delta\varphi}{\hbar}L_z, \tag{3.6}$$

where L_z is the generator of infinitesimal rotations to be determined. Applying it to a position eigenstate leads to

$$O_{R_{e_z}}(\delta\varphi)|x, y\rangle = |x - y\delta\varphi, y + x\delta\varphi\rangle. \tag{3.7}$$

In a similar way, we may obtain the following relation

$$\langle x, y|\mathbf{1} - \frac{i\delta\varphi}{\hbar}L_z|\psi\rangle = \psi(x + y\delta\varphi, y - x\delta\varphi). \tag{3.8}$$

Expanding this equation in a Taylor series yields

$$\psi(x, y) - \frac{i\delta\varphi}{\hbar}\langle x, y|L_z|\psi\rangle = \psi(x, y) + y\delta\varphi\frac{\partial\psi}{\partial x} - x\delta\varphi\frac{\partial\psi}{\partial y}, \tag{3.9}$$

from which we have

$$\langle x, y|L_z|\psi\rangle = -i\hbar\left(x\frac{\partial}{\partial y} - y\frac{\partial}{\partial x}\right)\psi. \tag{3.10}$$

As a result, we find

$$\begin{aligned} L_z &= -i\hbar\left(x\frac{\partial}{\partial y} - y\frac{\partial}{\partial x}\right) \\ &= xp_y - yp_x. \end{aligned} \tag{3.11}$$

The finite rotation operator is obtained by repeating many infinitesimally small rotations, i.e.,

$$O_{R_{e_z}}(\varphi) = \lim_{n\to\infty}\left(\mathbf{1} - \frac{i}{\hbar}\frac{\varphi}{n}L_z\right)^n = e^{-\frac{i\varphi}{\hbar}L_z}. \tag{3.12}$$

This means that the z component of angular momentum L_z is a conserved quantity. Similarly, we could do infinitesimal rotations about the x or y axis and shall show that all the components of the angular momentum operator commute with the Hamiltonian, i.e., satisfying (3.5). Thus, the rotation operator can be written out

$$O_{R_{\mathbf{n}}}(\varphi) = e^{-\frac{i}{\hbar}\varphi\mathbf{L}\cdot\mathbf{n}}. \tag{3.13}$$

3 Orbital Angular Momentum Operators

In this section, we study the orbital angular momentum operators [13, 14, 16, 19, 20]. The relations between the Cartesian coordinates x_i and the hyperspherical coordinates r and θ_b in D-dimensional space are defined by

$$\begin{aligned}
x_1 &= r\cos\theta_1\sin\theta_2\cdots\sin\theta_{D-1},\\
x_2 &= r\sin\theta_1\sin\theta_2\cdots\sin\theta_{D-1},\\
x_b &= r\cos\theta_{b-1}\sin\theta_b\cdots\sin\theta_{D-1},\\
x_D &= r\cos\theta_{D-1},
\end{aligned} \tag{3.14}$$

where $b \in [3, D-1]$. The unit vector $\hat{\mathbf{x}}$ along $\mathbf{x}$ is usually denoted by $\hat{\mathbf{x}} = \mathbf{x}/r$. The sum of the squares of Eqs. (3.14) gives

$$r^2 = \sum_{i=1}^{D} x_i^2. \tag{3.15}$$

Thus r is the radius of a D-dimensional sphere. The Laplacian is given in terms of polar coordinates by

$$\nabla_D^2 = \frac{1}{h}\sum_{j=0}^{D-1}\frac{\partial}{\partial\theta_j}\left(\frac{h}{h_j^2}\frac{\partial}{\partial\theta_j}\right), \tag{3.16}$$

where

$$\theta_0 = r, \qquad h = \prod_{j=0}^{D-1} h_j, \qquad h_j^2 = \sum_{i=1}^{D}\left(\frac{\partial x_i}{\partial\theta_j}\right)^2. \tag{3.17}$$

By direct calculation we have

$$\begin{aligned}
h_0 &= 1, \\
h_1 &= r \sin\theta_2 \sin\theta_3 \cdots \sin\theta_{D-1}, \\
h_2 &= r \sin\theta_3 \sin\theta_4 \cdots \sin\theta_{D-1}, \\
h_3 &= r \sin\theta_4 \sin\theta_5 \cdots \sin\theta_{D-1}, \\
&\vdots \\
h_j &= r \sin\theta_{j+1} \sin\theta_{j+2} \cdots \sin\theta_{D-1}, \\
&\vdots \\
h_{D-2} &= r \sin\theta_{D-1}, \\
h_{D-1} &= r, \\
h &= r^{D-1} \sin\theta_2 \sin^2\theta_3 \sin^3\theta_4 \cdots \sin^{j-1}\theta_j \cdots \sin^{D-2}\theta_{D-1},
\end{aligned} \tag{3.18}$$

for $D \geq 3$; $h_0 = 1$, $h_1 = r$ when $D = 2$. By substituting these results into Eq. (3.16), we are able to obtain the following polar coordinate form for the D-dimensional Laplacian

$$\begin{aligned}
\nabla_D^2 &= \frac{1}{r^{D-1}} \frac{\partial}{\partial r}\left(r^{D-1} \frac{\partial}{\partial r}\right) \\
&\quad + \frac{1}{r^2} \sum_{j=1}^{D-2} \frac{1}{\sin^2\theta_{j+1} \sin^2\theta_{j+2} \cdots \sin^2\theta_{D-1}} \\
&\quad \times \left\{\frac{1}{\sin^{j-1}\theta_j} \left(\frac{\partial}{\partial\theta_j} \sin^{j-1}\theta_j \frac{\partial}{\partial\theta_j}\right)\right\} \\
&\quad + \frac{1}{r^2} \left\{\frac{1}{\sin^{D-1}\theta_{D-1}} \left(\frac{\partial}{\partial\theta_{D-1}} \sin^{D-2}\theta_{D-1} \frac{\partial}{\partial\theta_{D-1}}\right)\right\}.
\end{aligned} \tag{3.19}$$

The volume element of the configuration space is calculated as

$$\prod_{a=1}^{D} dx_a = r^{D-1} dr\, d\Omega, \quad d\Omega = \prod_{a=1}^{D-1} (\sin\theta_a)^{a-1} d\theta_a, \tag{3.20}$$

where $r \in [0, \infty)$, $\theta_1 \in [-\pi, \pi]$, and $\theta_b \in [0, \pi]$, $b \in [2, D-1]$. The orbital angular momentum operators L_{ab} are the generators of the transformation operators O_R for the scalar function, $R \in \mathrm{SO}(D)$, defined as

$$\begin{aligned}
L_{ab} &= -i x_a \frac{\partial}{\partial x_b} + i x_b \frac{\partial}{\partial x_a} \\
&= i\left(x_b \frac{\partial}{\partial x_a} - x_a \frac{\partial}{\partial x_b}\right) \\
&= i[x_b \partial_a - x_a \partial_b],
\end{aligned} \tag{3.21}$$

where $\partial_a = \partial / \partial_{x_a}$.

It should be noted that there exists a set of commutable angular momentum operators

$$
\begin{aligned}
L_1^2 &= L_{12}^2 = -\frac{\partial^2}{\partial\theta_1^2},\\
L_2^2 &= -\left\{\frac{1}{\sin\theta_2}\frac{\partial}{\partial\theta_2}\left(\sin\theta_2\frac{\partial}{\partial\theta_2}\right) - \frac{L_1^2}{\sin^2\theta_2}\right\},\\
&\vdots\\
L_j^2 &= \sum_{a<b=2}^{j+1} L_{ab}^2 = -\left\{\frac{1}{\sin^{j-1}\theta_j}\frac{\partial}{\partial\theta_j}\left(\sin^{j-1}\theta_j\frac{\partial}{\partial\theta_j}\right) - \frac{L_{j-1}^2}{\sin^2\theta_j}\right\},\\
&\vdots\\
L_{D-1}^2 &= -\left\{\frac{1}{\sin^{D-2}\theta_{D-1}}\frac{\partial}{\partial\theta_{D-1}}\left(\sin^{D-2}\theta_{D-1}\frac{\partial}{\partial\theta_{D-1}}\right) - \frac{L_{D-2}^2}{\sin^2\theta_{D-1}}\right\},\\
j &\in [2, D-1], \qquad L^2 \equiv L_{D-1}^2,
\end{aligned} \tag{3.22}
$$

where the atomic unit $\hbar = 1$ is used. The first two operators are recognized as the ordinary orbital angular momentum operators L_z^2 and L^2 for the three-dimensional case where $\theta_1 = \varphi$ and $\theta_2 = \theta$. It is interesting to note that the generalized orbital angular momentum operators are independent of the dimensionality of space in the sense that we obtain the same expression for L_j^2 independently of whether it is calculated in a $(j+1)-$, $(j+2)-$, $(j+3)-, \ldots,$ dimensional space.

Now, the Laplacian may be written as

$$
\nabla_D^2 = \frac{1}{r^{D-1}}\frac{\partial}{\partial r}\left(r^{D-1}\frac{\partial}{\partial r}\right) - \frac{L_{D-1}^2}{r^2}. \tag{3.23}
$$

As what follows, we are going to show more properties about the generalized angular momentum operators L_{ab}. Let us first calculate the following commutation relations:

$$
\begin{aligned}
[L_{ab}, L_{cd}] &= i^2[(x_b\partial_a - x_a\partial_b), (x_d\partial_c - x_c\partial_d)]\\
&= i^2[x_b\partial_a x_d\partial_c - x_b\partial_a x_c\partial_d + x_a\partial_b x_c\partial_d - x_a\partial_b x_d\partial_c\\
&\quad - x_d\partial_c x_b\partial_a + x_d\partial_c x_a\partial_b + x_c\partial_d x_b\partial_a - x_c\partial_d x_a\partial_b]\\
&= i^2[x_b\delta_{ad}\partial_c + x_b x_d\partial_a\partial_c - x_b\delta_{ac}\partial_d - x_b x_c\partial_a\partial_d\\
&\quad - x_a\delta_{bd}\partial_c - x_a x_d\partial_b\partial_c + x_a\delta_{bc}\partial_d + x_a x_c\partial_b\partial_d\\
&\quad - x_d\delta_{cb}\partial_a - x_d x_b\partial_c\partial_a + x_d\delta_{ca}\partial_b + x_d x_a\partial_c\partial_b\\
&\quad + x_c\delta_{db}\partial_a + x_c x_b\partial_d\partial_a - x_c\delta_{da}\partial_b - x_c x_a\partial_d\partial_b].
\end{aligned} \tag{3.24}
$$

Since $[x_a, x_b] = [\partial_a, \partial_b] = 0$, the terms of the form $x_a x_b\partial_a\partial_b$ cancel pairwise. Moreover, from the property of the Kronecker delta $\delta_{ab} = \delta_{ba}$ we have

$$
\begin{aligned}
[L_{ab}, L_{cd}] &= i^2[\delta_{ad}(x_b\partial_c - x_c\partial_b) - \delta_{ac}(x_b\partial_d - x_d\partial_b)\\
&\quad - \delta_{bd}(x_a\partial_c - x_c\partial_a) + \delta_{bc}(x_a\partial_d - x_d\partial_a)].
\end{aligned} \tag{3.25}
$$

With the help of the angular momentum relation $L_i \times L_j = \varepsilon_{ijk} L_k$, where the Levi-Civita symbol ε_{ijk} is defined as

$$\varepsilon_{ijk} = \begin{cases} 1, & \text{for } (ijk) \text{ an even permutation of } (123), \\ -1, & \text{for } (ijk) \text{ an odd permutation}, \\ 0, & \text{otherwise}, \end{cases} \tag{3.26}$$

we get

$$[L_{ab}, L_{cd}] = i[\delta_{ad} L_{cb} - \delta_{ac} L_{db} - \delta_{bd} L_{ca} + \delta_{bc} L_{da}]. \tag{3.27}$$

By using the antisymmetry of the operators $L_{ab} = -L_{ba}$, we obtain the standard form

$$[L_{ab}, L_{cd}] = i[\delta_{ac} L_{bd} + \delta_{bd} L_{ac} - \delta_{ad} L_{bc} - \delta_{bc} L_{ad}]. \tag{3.28}$$

This means that the orbital angular momentum operators L_{ab} satisfy the same commutation relations as Eq. (2.14) for the generators T_{ab} in the self-representation of the group SO(D). What will we do next step is the calculation of the Casimir operator, which commutes with all group elements L_{ab}.

Define the Casimir operator as the sum over all squares of the group elements

$$C_2 = \frac{1}{2} \sum_{a,b}^{D} (L_{ab})^2 = \frac{1}{2} \delta^{ac} \delta^{bd} L_{ab} L_{cd}. \tag{3.29}$$

Thus, we may calculate the following commutation relation

$$\begin{aligned} [C_2, L_{fg}] &= \frac{1}{2} \delta^{ac} \delta^{bd} [L_{ab} L_{cd}, L_{fg}] \\ &= \frac{1}{2} \delta^{ac} \delta^{bd} (L_{ab} L_{cd} L_{fg} - L_{fg} L_{ab} L_{cd}) \\ &= \frac{1}{2} \delta^{ac} \delta^{bd} \{L_{ab}[L_{cd}, L_{fg}] + [L_{ab}, L_{fg}] L_{cd}\}. \end{aligned} \tag{3.30}$$

By inserting the commutation relation (3.28), one gets

$$\begin{aligned} [C_2, L_{fg}] &= \frac{i}{2} \delta^{ac} \delta^{bd} \big\{L_{ab}(\delta_{cf} L_{dg} + \delta_{dg} L_{cf} - \delta_{cg} L_{df} - \delta_{df} L_{cg}) \\ &\quad + (\delta_{af} L_{bg} + \delta_{bg} L_{af} - \delta_{ag} L_{bf} - \delta_{bf} L_{ag}) L_{cd}\big\} \\ &= \frac{i}{2} \big\{L_{ab}(\delta^{a}{}_{f} \delta^{bd} L_{dg} + \delta^{ac} \delta^{b}{}_{g} L_{cf} - \delta^{a}{}_{g} \delta^{bd} L_{df} - \delta^{ac} \delta^{b}{}_{f} L_{cg}) \\ &\quad + (\delta^{c}{}_{f} \delta^{bd} L_{bg} + \delta^{ac} \delta^{d}{}_{g} L_{af} - \delta^{c}{}_{g} \delta^{bd} L_{bf} - \delta^{ac} \delta^{d}{}_{f}) L_{cd}\big\} \\ &= 0. \end{aligned} \tag{3.31}$$

This implies that the Casimir operator commutes with all group elements L_{ab}.

Let us calculate the eigenvalue of the Casimir operator C_2 in algebraic way. As we know, the Casimir operator C_2 can also be written explicitly as

$$
\begin{aligned}
C_2 &= \frac{1}{2}L^{ab}L_{ab} \\
&= -\frac{1}{2}(x^b\partial^a - x^a\partial^b)(x_b\partial_a - x_a\partial_b) \\
&= -\frac{1}{2}(x^b\partial^a x_b\partial_a - x^b\partial^a x_a\partial_b - x^a\partial^b x_b\partial_a + x^a\partial^b x_a\partial_b) \\
&= -\frac{1}{2}(x^b\delta^a{}_b\partial_a + x^b x_b\partial^a\partial_a - x^b\delta^a{}_a\partial_b - x^b x_a\partial^a\partial_b \\
&\quad - x^a\delta^b{}_b\partial_a - x^a x_b\partial^b\partial_a + x^a\delta^b{}_a\partial_b + x^a x_a\partial^b\partial_b) \\
&= -(x^a\partial_a + x^b x_b\partial^a\partial_a - Dx^a\partial_a - x^b x^a\partial_a\partial_b).
\end{aligned}
\tag{3.32}
$$

If we define the Euler operator as

$$J = \sum_{i=1}^{D} x^i \frac{\partial}{\partial x_i} = x^a\partial_a, \tag{3.33}$$

then we have

$$
\begin{aligned}
J^2 &= x^a\partial_a x^b\partial_b \\
&= x^a\delta_a{}^b\partial_b + x^a x^b\partial_a\partial_b \\
&= J + x^a x^b\partial_a\partial_b.
\end{aligned}
\tag{3.34}
$$

As a result, we may rewrite Eq. (3.32) as

$$
\begin{aligned}
C_2 &= -[J + x^b x_b\partial_a\partial_a - DJ - J(J-1)] \\
&= -[x^b x_b\partial_a\partial_a - J(J+D-2)].
\end{aligned}
\tag{3.35}
$$

If we define a Hilbert space $f(\mathbf{r})$ as the space of all homogeneous polynomials of degree l satisfying the Laplacian, the eigenvalue of the Casimir operator C_2 is calculated as

$$C_2 f(\mathbf{r}) = l(l+D-2)f(\mathbf{r}). \tag{3.36}$$

This result can also be obtained by the one-row Young pattern as shown in Sect. 6.

4 The Linear Momentum Operators

Let us first examine the linear momentum operator ($\hbar = 1$)

$$p_k = -i\frac{\partial}{\partial x_k}. \tag{3.37}$$

According to the polar coordinates, we have

$$\frac{\partial}{\partial x_k} = \sum_{j=0}^{D-1} \frac{\partial\theta_j}{\partial x_k}\frac{\partial}{\partial\theta_j}. \tag{3.38}$$

It is known that the calculations of the first derivatives $\partial\theta_j/\partial x_k$ are very difficult since we would need the inverse transformation of Eqs. (3.14). However, it follows from the transformation equations and the orthogonality of our coordinate system that

$$\sum_{i=0}^{D-1}\frac{\partial\theta_i}{\partial x_k}\frac{\partial x_l}{\partial\theta_i}=\delta_{lk},\qquad \sum_{l=0}^{D-1}\frac{\partial x_l}{\partial\theta_i}\frac{\partial x_l}{\partial\theta_j}=\delta_{ij}h_i^2, \tag{3.39}$$

respectively. On multiplying the second equation by $\partial\theta_i/\partial x_k$ and summing over i, we obtain

$$\frac{\partial\theta_j}{\partial x_k}=\frac{1}{h_j^2}\frac{\partial x_k}{\partial\theta_j}. \tag{3.40}$$

Thus, we find

$$p_k=-i\sum_{j=0}^{D-1}\left(\frac{1}{h_j^2}\frac{\partial x_k}{\partial\theta_j}\right)\frac{\partial}{\partial\theta_j}. \tag{3.41}$$

Explicitly, we may obtain

$$p_D=-i\left\{\cos\theta_{D-1}\frac{\partial}{\partial r}-\frac{\sin\theta_{D-1}}{r}\frac{\partial}{\partial\theta_{D-1}}\right\},\qquad D=3,4,5,\ldots. \tag{3.42}$$

5 Radial Momentum Operator

Let us study the radial momentum operator [170]

$$p_r=-\frac{i}{2}[\nabla\cdot(\hat{\mathbf{r}}\cdots)+\hat{\mathbf{r}}\cdot\nabla], \tag{3.43}$$

where the notation $\nabla\cdot(\hat{\mathbf{r}}\cdots)$ indicates that ∇ differentiates $\hat{\mathbf{r}}$ and everything on its right.[1] On basis of the following identity

---

[1]As an illustration, we present it in three-dimensional case. The unit vectors in spherical polar coordinates are given by

$$\hat{\mathbf{r}}=\vec{i}\sin\theta\cos\varphi+\vec{j}\sin\theta\sin\varphi+\vec{k}\cos\theta, \tag{3.44}$$

and

$$\begin{aligned}\frac{\partial}{\partial x}&=\sin\theta\cos\varphi\frac{\partial}{\partial r}+\cos\theta\cos\varphi\frac{\partial}{\partial\theta}-\frac{\sin\varphi}{r\sin\theta}\frac{\partial}{\partial\varphi},\\ \frac{\partial}{\partial y}&=\sin\theta\sin\varphi\frac{\partial}{\partial r}+\cos\theta\sin\varphi\frac{\partial}{\partial\theta}+\frac{\cos\varphi}{r\sin\theta}\frac{\partial}{\partial\varphi},\\ \frac{\partial}{\partial z}&=\cos\theta\frac{\partial}{\partial r}-\sin\theta\frac{1}{r}\frac{\partial}{\partial\varphi}.\end{aligned} \tag{3.45}$$

$$\nabla(\hat{\mathbf{r}}\Psi) = \hat{\mathbf{r}} \cdot \nabla\Psi + \Psi\nabla \cdot \hat{\mathbf{r}}, \tag{3.47}$$

we have

$$p_r = -\frac{i}{2}[\nabla \cdot \hat{\mathbf{r}} + 2\hat{\mathbf{r}} \cdot \nabla], \tag{3.48}$$

where

$$\hat{\mathbf{r}} = \frac{\mathbf{r}}{r} = \frac{\sum_{a=1}^{D} x_a \hat{\mathbf{x}}_a}{\sqrt{\sum_{a=1}^{D}(x_a)^2}}. \tag{3.49}$$

First, we calculate

$$\begin{aligned} \nabla \cdot \hat{\mathbf{r}} &= \sum_{a=1}^{D} \frac{\partial}{\partial x_a}\left(\frac{x_a}{r}\right) \\ &= \sum_{a=1}^{D}\left[\frac{1}{r} - \left(\frac{1}{r^2}\right) x_a \frac{\partial r}{\partial x_a}\right] \\ &= \frac{D}{r} - \frac{1}{r^2}\sum_{a=1}^{D} x_a \frac{\partial r}{\partial x_a}. \end{aligned} \tag{3.50}$$

From Eq. (3.15), we have

$$\frac{\partial r}{\partial x_a} = \frac{x_a}{r}. \tag{3.51}$$

Thus, Eq. (3.50) is simplified to

$$\nabla \cdot \hat{\mathbf{r}} = \frac{D-1}{r}. \tag{3.52}$$

By definition $\partial/\partial r = \hat{\mathbf{r}} \cdot \nabla$, we might express the radial momentum operator in D dimensions as

$$p_r = -i\left(\frac{\partial}{\partial r} + \frac{D-1}{2r}\right). \tag{3.53}$$

If we take into the following identity account

$$\left[\frac{\partial}{\partial r}, \frac{D-1}{2r}\right] = -\frac{D-1}{2r^2}, \tag{3.54}$$

then we have

$$p_r^2 = -\left\{\frac{\partial^2}{\partial r^2} + \frac{D-1}{r}\frac{\partial}{\partial r} + \frac{(D-1)(D-3)}{4r^2}\right\}. \tag{3.55}$$

If we take $\psi(r,\theta,\varphi) = R(r)\Theta(\theta)\Phi(\varphi)$, then we have

$$\begin{aligned} \nabla \cdot \hat{\mathbf{r}}\psi(r,\theta,\varphi) &= \left(\frac{\partial}{\partial r} + \frac{2}{r}\right)\psi(r,\theta,\varphi), \\ \hat{\mathbf{r}} \cdot \nabla\psi(r,\theta,\varphi) &= \frac{\partial}{\partial r}\psi(r,\theta,\varphi). \end{aligned} \tag{3.46}$$

6 Spherical Harmonic Polynomials

The Chevalley bases $H_\nu(L)$, $E_\nu(L)$, and $F_\nu(L)$ can be obtained from (2.18) and (2.24) through replacing T_{ab} by L_{ab}. Since

$$L_{ab}x_d = \sum_{i=1}^{D} x_i (T_{ab})_{id}, \qquad O_R\theta_d = \sum_{i=1}^{D} \theta_i R_{id}, \tag{3.56}$$

then we may obtain the common eigenfunctions X_β of $H_\nu(L)$ in terms of ϕ_β given in Eq. (2.19) for the SO($2l+1$) and in Eq. (2.25) for the SO($2l$) through replacing the basis vector θ_a by the rectangular coordinate x_a.

The spherical harmonics $Y^{[\lambda]}_{\mathbf{m}}(\hat{\mathbf{x}})$ are the eigenfunctions of the orbital angular momentum operators $H_\nu(L)$ for a single particle,

$$H_\nu(L)Y^{[\lambda]}_{\mathbf{m}}(\hat{\mathbf{x}}) = m_\nu Y^{[\lambda]}_{\mathbf{m}}(\hat{\mathbf{x}}), \qquad L^2 Y^{[\lambda]}_{\mathbf{m}}(\hat{\mathbf{x}}) = C_2([\lambda]) Y^{[\lambda]}_{\mathbf{m}}(\hat{\mathbf{x}}), \tag{3.57}$$

where $C_2([\lambda])$ is the second-order invariant Casimir operator. For an irreducible tensor representation of the SO(D), the $C_2([\lambda])$ or $C_2([\pm\lambda])$ can be worked out

$$C_2([\lambda]) = \sum_{i=1}^{l} \lambda_i(\lambda_i + D - 2i), \qquad D = 2l+1 \text{ or } 2l. \tag{3.58}$$

Since there is only one coordinate vector $\mathbf{x}$, the representation $[\lambda]$ for the spherical harmonics $Y^{[\lambda]}_{\mathbf{m}}(\hat{\mathbf{x}})$ must be the totally symmetric representation denoted by the one-row Young pattern $[\lambda] = [\lambda, 0, \ldots, 0]$. $C_2([\lambda] = [\lambda, 0, \ldots, 0]) = \lambda(\lambda + D - 2)$ for the SO(D).

Generally, the highest weight state $Y^{[\lambda]}_{\mathbf{M}}(\hat{\mathbf{x}})$ with $[\lambda] = [\lambda, 0, \ldots, 0]$ and $\mathbf{M} = [\lambda, 0, \ldots, 0]$ is expressed by

$$\begin{aligned} Y^{[\lambda]}_{\mathbf{M}}(\hat{\mathbf{x}}) &= C_{D,\lambda} r^{-\lambda}\left\{\frac{(-1)^s(x_1 + ix_2)}{\sqrt{2}}\right\}^{\lambda}, \\ C_{D,\lambda} &= \begin{cases} [\frac{(2l+2\lambda-1)!}{2^{(2l+\lambda)}\pi^l \lambda!(\lambda+l-1)!}]^{1/2}, & \text{when } D = 2l+1, \\ [\frac{(\lambda+l-1)!}{2^{(1-\lambda)}\pi^l \lambda!}]^{1/2}, & \text{when } D = 2l, \end{cases} \end{aligned} \tag{3.59}$$

where $C_{D,\lambda}$ is the normalization factor and the s is defined as

$$s = \begin{cases} l, & D = 2l+1, \\ l-1, & D = 2l. \end{cases} \tag{3.60}$$

The remaining spherical harmonics $Y^{[\lambda]}_{\mathbf{m}}(\hat{\mathbf{x}})$ with the weight $\mathbf{m}$ can be calculated by the lowering operators $F_\nu(L)$.

In fact, $r^l Y^{[\lambda]}_{\mathbf{m}}(\hat{\mathbf{x}})$ is a homogeneous polynomial of order λ with respect to the Cartesian coordinates x_a and satisfies the Laplace equation

$$\nabla^2[r^\lambda Y^{[\lambda]}_{\mathbf{m}}(\hat{\mathbf{x}})] = 0, \qquad \nabla^2 = \sum_{a=1}^{D} \frac{\partial^2}{\partial x_a^2}. \tag{3.61}$$

For a given angular momentum l, the degeneracy of the eigenfunctions of L^2 is calculated as

$$d_l(D) = \frac{(D+2l-2)(D+l-3)!}{l!(D-2)!}. \tag{3.62}$$

7 Schrödinger Equation for a Two-Body System

An isolated two-body quantum system keeps invariant in the translation of space-time and the spatial rotation. After separating the motion of the center-of-mass, there is only one Jacobi coordinate vector for a two-body system in D dimensions, denoted by $\mathbf{R}_1 \equiv \mathbf{x}$ for simplicity. Note that a factor of the square root of mass has been included in $\mathbf{R}_1$. The eigenfunction of angular momentum has to be proportional to the spherical harmonics $Y^{[\lambda]}_{\mathbf{m}}(\hat{\mathbf{x}})$, where $[\lambda] = [\lambda, 0, \ldots, 0]$ is a one-row Young pattern

$$\psi^{[\lambda]}_{\mathbf{m}}(\mathbf{x}) = \phi^{[\lambda]}(r) Y^{[\lambda]}_{\mathbf{m}}(\hat{\mathbf{x}}) = \frac{\phi^{[\lambda]}(r)}{r^\lambda} Y^{[\lambda]}_{\mathbf{m}}(\mathbf{x}), \quad \hat{\mathbf{x}} = \frac{\mathbf{x}}{x}, \tag{3.63}$$

where $\phi^{[\lambda]}(r)$ is the radial function. Since the quantum system is spherically symmetric, we can study the wavefunction with the highest weight $\mathbf{M}$. The calculation can be simplified by using the harmonic polynomials

$$\begin{aligned}
\nabla^2[\phi^{[\lambda]}(r) Y^{[\lambda]}_{\mathbf{M}}(\hat{\mathbf{x}})] &= \nabla^2 \left[\frac{\phi^{[\lambda]}(r)}{r^\lambda} Y^{[\lambda]}_{\mathbf{M}}(\mathbf{x})\right] \\
&= Y^{[\lambda]}_{\mathbf{M}}(\mathbf{x}) \nabla^2 \left[\frac{\phi^{[\lambda]}(r)}{r^\lambda}\right] + 2\nabla_x \left[\frac{\phi^{[\lambda]}(r)}{r^\lambda}\right] \cdot \nabla Y^{[\lambda]}_{\mathbf{M}}(\mathbf{x}) \\
&= Y^{[\lambda]}_{\mathbf{M}}(\mathbf{x}) \left\{ r^{1-D} \frac{d}{dr} r^{D-1} \frac{d}{dr} \left[\frac{\phi^{[\lambda]}(r)}{r^\lambda}\right]\right\} \\
&\quad + 2\frac{d}{dr}\left[\frac{\phi^{[\lambda]}(r)}{r^\lambda}\right]\left[\frac{\mathbf{x}}{r} \cdot \nabla Y^{[\lambda]}_{\mathbf{M}}(\mathbf{x})\right] \\
&= Y^{[\lambda]}_{\mathbf{M}}(\hat{\mathbf{x}}) \left\{ \frac{d^2}{dr^2}\phi^{[\lambda]}(r) + \frac{D-2\lambda-1}{r}\frac{d}{dr}\phi^{[\lambda]}(r) \right. \\
&\quad \left. - \frac{\lambda(D-\lambda-2)}{r^2}\phi^{[\lambda]}(r)\right\} \\
&\quad + 2\left\{-\frac{\lambda}{r}\phi^{[\lambda]}(r) + \frac{d}{dr}\phi^{[\lambda]}(r)\right\}\frac{\lambda}{r} Y^{[\lambda]}_{\mathbf{M}}(\hat{\mathbf{x}}) \\
&= Y^{[\lambda]}_{\mathbf{M}}(\hat{\mathbf{x}}) \left\{ \frac{d^2}{dr^2}\phi^{[\lambda]}(r) + \frac{D-1}{r}\frac{d}{dr}\phi^{[\lambda]}(r) \right. \\
&\quad \left. - \frac{\lambda(D+\lambda-2)}{r^2}\phi^{[\lambda]}(r)\right\}.
\end{aligned} \tag{3.64}$$

Substituting $\psi^{[\lambda]}_{\mathbf{M}}(\mathbf{x})$ into the Schrödinger equation in the coordinate system of the center-of-mass ($\hbar = m = 1$)

$$-\frac{1}{2}\nabla^2\psi^{[\lambda]}_{\mathbf{M}}(\mathbf{x}) + V(r)\psi^{[\lambda]}_{\mathbf{M}}(\mathbf{x}) = E\psi^{[\lambda]}_{\mathbf{M}}(\mathbf{x}), \tag{3.65}$$

we obtain the following radial equation

$$-\frac{1}{2}\left\{\frac{d^2}{dr^2} + \frac{D-1}{r}\frac{d}{dr} - \frac{\lambda(D+\lambda-2)}{r^2}\right\}\phi^{[\lambda]}(r) = [E - V(r)]\phi^{[\lambda]}(r). \tag{3.66}$$

On the other hand, if we take the wavefunction of the form

$$\psi^{[\lambda]}_{\mathbf{M}}(\mathbf{x}) = r^{-\frac{D-1}{2}}\phi^{[\lambda]}(r)Y^{[\lambda]}_{\mathbf{M}}(\hat{\mathbf{x}}), \tag{3.67}$$

then substituting this into (3.65) allows us to obtain

$$\begin{aligned}&-\frac{1}{2}\left\{\frac{d^2}{dr^2} - \frac{\lambda(D+\lambda-2) + (D-1)(D-3)/4}{r^2}\right\}\phi^{[\lambda]}(r)\\ &= [E - V(r)]\phi^{[\lambda]}(r).\end{aligned} \tag{3.68}$$

Let us turn to physical quantum systems. For a given angular momentum l, the wavefunction $\psi^{l}_{l_{D-2}\dots l_1}(\mathbf{x})$ can be now decomposed as a product of the radial function $R_l(r)$ and the generalized spherical harmonics $Y^{l}_{l_{D-2}\dots l_1}(\hat{\mathbf{x}})$

$$\psi^{l}_{l_{D-2}\dots l_1}(\mathbf{x}) = r^{-\frac{D-1}{2}}R_l(r)Y^{l}_{l_{D-2}\dots l_1}(\hat{\mathbf{x}}). \tag{3.69}$$

Substitution of Eq. (3.69) into Eq. (3.66) leads to D-dimensional radial Schrödinger equation:

$$\begin{aligned}&\frac{d^2}{dr^2}R_l(r) - \frac{l(l+D-2) + (D-1)(D-3)/4}{r^2}R_l(r)\\ &+ 2[E - V(r)]R_l(r) = 0.\end{aligned} \tag{3.70}$$

8 Concluding Remarks

It is known that an isolated two-body quantum system keeps invariant in the translation of space-time and the spatial rotation in D-dimensional space. Such a quantum system possesses an SO(D) symmetry. The ration operator, orbital angular momentum operators, linear momentum operator, radial momentum operator and generalized spherical harmonic polynomials are constructed. The Schrödinger equation with central potentials has been studied in D-dimensional space.

Chapter 4
Dirac Equation in Higher Dimensions

1 Introduction

It is well known that the exact solutions of quantum systems in real three-dimensional space play an important role in physics. As mentioned above, a number of works have been contributed to the Schrödinger equation case. On the contrary, the studies of the Dirac equation in higher dimensions are less than those of the Schrödinger equation case except for the works in usual three- [171–176], two- [177] and one-dimensional [178] space.

This Chapter is organized as follows. In Sect. 2 we show how to generalize the Dirac equation to $(D+1)$ space-time and discuss the conserved angular momentum operators and their quantum numbers [91]. In Sect. 3 we calculate the eigenfunctions of the total angular momentums for both odd $(2N+1)$ and even $2N$ cases in terms of the technique of group theory and present the radial equations. In Sect. 4 we shall deal with the hydrogen-like atoms by the series method. The exact solutions are expressed by the confluent hypergeometric functions. The eigenvalues as well as their fine structure energy are also studied. Finally we summarize this Chapter in Sect. 5.

2 Dirac Equation in $(D+1)$ Dimensions

The Dirac equation ($\hbar = c = 1$) in $(D+1)$ dimensions can be expressed as [179]

$$i\sum_{\mu=0}^{D} \gamma_\mu(\partial_\mu + ieA_\mu)\Psi(\mathbf{x},t) = M\Psi(\mathbf{x},t), \tag{4.1}$$

where M is the mass of the particle, and $(D+1)$ matrices γ_μ satisfy the anticommutation relation (2.42).

Discuss the special case where only the zero component of A_μ is non-vanishing and spherically symmetric:

$$eA_0 = V(r), \quad A_a = 0, \ \text{when}\ a \neq 0. \tag{4.2}$$

S.-H. Dong, *Wave Equations in Higher Dimensions*,
DOI 10.1007/978-94-007-1917-0_4, © Springer Science+Business Media B.V. 2011

The Hamiltonian $H(\mathbf{x})$ of the system is expressed as

$$\begin{aligned} &i\partial_0\Psi(\mathbf{x},t) = H(\mathbf{x})\Psi(\mathbf{x},t), \\ &H(\mathbf{x}) = \sum_{j=1}^{D} \gamma_0\gamma_j p_j + V(r) + \gamma_0 M, \\ &p_j = -i\partial_j = -i\frac{\partial}{\partial x_j}, \quad j \in [1, D]. \end{aligned} \tag{4.3}$$

The orbital angular momentum operators L_{ab}, the spinor operators S_{ab}, and the total angular momentum operators J_{ab} are defined as follows:

$$\begin{aligned} &L_{ab} = -L_{ba} = ix_a\partial_b - ix_b\partial_a, \qquad S_{ab} = -S_{ba} = \frac{i}{2}\gamma_a\gamma_b, \\ &J_{ab} = L_{ab} + S_{ab}, \qquad 1 \le a < b \le D, \\ &J^2 = \sum_{a<b=2}^{D} J_{ab}^2, \quad L^2 = \sum_{a<b=2}^{D} L_{ab}^2, \quad S^2 = \sum_{a<b=2}^{D} S_{ab}^2. \end{aligned} \tag{4.4}$$

The eigenvalue of J^2 (L^2 or S^2) is denoted by the Casimir operator $C_2(\mathbf{M})$ to be determined, where $\mathbf{M}$ is the highest weight of the representation to which the total (orbital or spinor) wavefunction belongs. We will discuss the Casimir operator in next section. It is easy to show by the standard method [179] that J_{ab} and κ commute with the Hamiltonian $H(\mathbf{x})$,

$$\begin{aligned} \kappa &= \gamma_0\left\{\sum_{a<b} i\gamma_a\gamma_b L_{ab} + (D-1)/2\right\} \\ &= \gamma_0\{J^2 - L^2 - S^2 + (D-1)/2\}. \end{aligned} \tag{4.5}$$

3 The Radial Equations

As we know, for spherically symmetric potential $V(r)$ the symmetry group of the system is SO(D) group. To derive the Dirac equation in higher dimensions, we also need the hyperspherical coordinates in the real D-dimensional space (3.14). It is known in Chap. 2 that the Lie algebras of the SO($2N+1$) group and the SO($2N$) group are B_N and D_N, respectively, and the Chevalley bases of the SO($2N$) in the self-representation are the same as those of the SO($2N+1$) except for $\nu = N = l$ as displayed in Eqs. (2.19) and (2.25) [139, 140, 180, 181]. We do not repeat them for simplicity. The operator J_{ab} can be replaced by L_{ab} or S_{ab} depending on the studied wavefunction. $H_\nu(J)$ span the Cartan subalgebra, and their eigenvalues for an eigenstate $|\mathbf{m}\rangle$ in a given irreducible representation are the components of a weight vector $\mathbf{m} = (m_1, \ldots, m_n)$:

$$H_\nu(J)|\mathbf{m}\rangle = m_\nu|\mathbf{m}\rangle, \quad \nu \in [1, N]. \tag{4.6}$$

If the eigenstates $|\mathbf{m}\rangle$ for a given weight $\mathbf{m}$ are degeneracy, this weight is called a multiple weight, otherwise a simple one. E_ν are called the raising operators and F_ν

the lowering ones. For an irreducible representation there is a highest weight **M**, which is a simple weight and can be used to describe the irreducible representation. Usually, the irreducible representation is also called the highest weight representation and directly denoted by **M**. The Casimir $C_2(\mathbf{M})$ can be calculated by the formula as follows [182]:

$$C_2(\mathbf{M}) = \mathbf{M} \cdot (\mathbf{M} + 2\boldsymbol{\rho}) = \sum_{\mu,\nu=1}^{N} M_\mu d_\mu (A^{-1})_{\mu\nu} (M_\nu + 2), \tag{4.7}$$

where $\boldsymbol{\rho}$ is the half sum of the positive roots in the Lie algebra, A^{-1} is the inverse of the Cartan matrix, and d_μ are the half square lengths of the simple roots.

As shown above, we have known that the orbital wavefunction in D-dimensional space is usually expressed by the spherical harmonic $Y^{[\lambda]}_{\mathbf{m}}(\hat{\mathbf{x}})$, which belongs to the weight **m** of the highest weight representation $[\lambda] \equiv [\lambda, 0, \ldots, 0]$. For the highest weight state $Y^{[\lambda]}_{\mathbf{M}}(\hat{\mathbf{x}})$ where $\mathbf{M} = [\lambda]$, we have obtained it as Eq. (3.59). Its partners $Y^{[\lambda]}_{\mathbf{m}}(\hat{\mathbf{x}})$ can be calculated from $Y^{[\lambda]}_{\mathbf{M}}(\hat{\mathbf{x}})$ by lowering operators $F_\nu(L)$. The Casimir for the spherical harmonic $Y^{[\lambda]}_{\mathbf{m}}(\hat{\mathbf{x}})$ is calculated by Eq. (4.7) as

$$L^2 Y^{[\lambda]}_{\mathbf{m}}(\hat{\mathbf{x}}) = C_2([\lambda]) Y^{[\lambda]}_{\mathbf{m}}(\hat{\mathbf{x}}), \quad C_2([\lambda]) = \lambda(\lambda + D - 2). \tag{4.8}$$

Since the spinor wavefunctions as well as those for the total angular momentum are different for $D = 2N + 1$ and $D = 2N$, we are going to study them separately.

3.1 The SO(2N + 1) Case

For $D = 2N + 1$ we define

$$\gamma_0 = \sigma_3 \times \mathbf{1}, \qquad \gamma_a = (i\sigma_2) \times \beta_a, \quad a \in [1, 2N + 1], \tag{4.9}$$

where σ_a is the Pauli matrix, **1** denotes the 2^N-dimensional unit matrix, and $(2N + 1)$ matrices β_a satisfy the anticommutation relations [183]

$$\beta_a \beta_b + \beta_b \beta_a = 2\delta_{ab} \mathbf{1}, \quad a, b = 1, 2, \ldots, (2N + 1), \tag{4.10}$$

which are the same as those γ_a given in Eq. (2.42). The dimension of β_a matrices is 2^l. Thus, the spinor operator S_{ab} becomes a block matrix

$$S_{ab} = \mathbf{1} \times \overline{S}_{ab}, \quad \overline{S}_{ab} = -\frac{i}{2} \beta_a \beta_b. \tag{4.11}$$

The relation between S_{ab} and $\overline{S}_{ab}$ is similar to that between the spinor operators for the Dirac spinors and for the Pauli spinors. The operator κ becomes

$$\kappa = \sigma_3 \times \overline{\kappa}, \quad \overline{\kappa} = -i \sum_{a<b} \beta_a \beta_b L_{ab} + \frac{D-1}{2}. \tag{4.12}$$

The fundamental spinor $\xi(\mathbf{m})$ belongs to the fundamental spinor representation $[s] \equiv [0, \ldots, 0, 1]$. From Eq. (4.7) the Casimir for the representation $[s]$ is calculated as $C_2([s]) = (2N^2 + N)/4$.

The product of $Y^{[\lambda]}_{\mathbf{m}}(\hat{\mathbf{x}})$ and $\xi(\mathbf{m}')$ belongs to the direct product of two representations $[\lambda]$ and $[s]$, which is a reducible representation:

$$[\lambda]\times[s]\simeq[\lambda,0,\ldots,0,1]\oplus[\lambda-1,0,\ldots,0,1]. \tag{4.13}$$

Besides, in order to construct a wavefunction belonging to the representation $[j]\equiv[\lambda,0,\ldots,0,1]$ there are two different ways: the combination of $Y^{[\lambda]}_{\mathbf{m}}(\hat{\mathbf{x}})\xi(\mathbf{m}')$ and that of $Y^{[\lambda+1]}_{\mathbf{m}}(\hat{\mathbf{x}})\xi(\mathbf{m}')$. They are different in eigenvalues of $\overline{\kappa}$. Since the system is spherically symmetric, we only need to calculate the highest weight state for the representation $[j]$ in terms of the CGCs

$$\begin{aligned}
&\phi_{|K|,[j]}(\hat{\mathbf{x}})=Y^{[\lambda]}_{[\lambda]}(\hat{\mathbf{x}})\xi([s])=N_{D,\lambda}r^{-\lambda}(x^1+ix^2)^{\lambda}\xi([s]),\\
&|K|=C_2([j])-C_2([\lambda])-C_2([s])+N=\lambda+N.
\end{aligned} \tag{4.14}$$

$$\begin{aligned}
\phi_{-|K|,[j]}(\hat{\mathbf{x}})&=\sum_{\mathbf{m}}Y^{[\lambda+1]}_{\mathbf{m}}(\hat{\mathbf{x}})\xi([j]-\mathbf{m})\\
&\quad\times\langle[\lambda+1],\mathbf{m},[s],[j]-\mathbf{m}|[j],[j]\rangle\\
&=N_{2N+1,\lambda}r^{-\lambda-1}(x^1+ix^2)^{\lambda}\{x^{2N+1}\xi([s])\\
&\quad+(x^{2N-1}+ix^{2N})\xi([0,\ldots,0,1,\overline{1}])\\
&\quad+(x^{2N-3}+ix^{2N-2})\xi([0,\ldots,0,1,\overline{1},1])+\cdots\\
&\quad+(x^3+ix^4)\xi([1,\overline{1},0,\ldots,0,1])\\
&\quad+(x^1+ix^2)\xi([\overline{1},0,\ldots,0,1])\},\\
-|K|&=C_2([j])-C_2([\lambda+1])-C_2([s])+N=-\lambda-N.
\end{aligned} \tag{4.15}$$

The wavefunction $\Psi_{K,[j]}(\mathbf{x})$ of the total angular momentum belonging to the irreducible representation $[j]$ can be expressed as the wave function $\Psi_{K,[j]}(\mathbf{x})$ of the total angular momentum belonging to the irreducible representation $[j]$ can be expressed as

$$\begin{aligned}
&\Psi_{K,[j]}(\mathbf{x},t)=r^{-N}e^{-iEt}\begin{pmatrix}F(r)\phi_{K,[j]}(\hat{\mathbf{x}})\\ iG(r)\phi_{-K,[j]}(\hat{\mathbf{x}})\end{pmatrix},\\
&H_1(J)\Psi_{K,[j]}(\mathbf{x})=l\Psi_{K,[j]}(\mathbf{x}),\\
&H_N(J)\Psi_{K,[j]}(\mathbf{x})=\Psi_{K,[j]}(\mathbf{x}),\\
&H_\nu(J)\Psi_{K,[j]}(\mathbf{x})=0,\quad \nu\in[2,N-1],\\
&\kappa\Psi_{K,[j]}(\mathbf{x})=K\Psi_{K,[j]}(\mathbf{x}),\quad K=\pm(l+N).
\end{aligned} \tag{4.16}$$

Their partners can be calculated by the lowering operators F_ν.

The radial equation depends on the explicit forms of β_a (γ_a) matrices. Therefore, by considering two introduced operators

$$\boldsymbol{\beta}\cdot\hat{\mathbf{x}}=\frac{1}{r}\sum_{i=1}^{2N+1}\beta_i x_i,\qquad \boldsymbol{\beta}\cdot\nabla=\sum_{i=1}^{2N+1}\beta_i\frac{\partial}{\partial x_i}, \tag{4.17}$$

we obtain

$$\begin{aligned}(\vec{\beta}\cdot\hat{\mathbf{x}})\phi_{K,[j]}(\hat{\mathbf{x}}) &= \frac{1}{r}\sum_{a=1}^{2N+1}\beta_a x^a \phi_{K,[j]}(\hat{\mathbf{x}}) = \phi_{-K,[j]}(\hat{\mathbf{x}}),\\ (\vec{\beta}\cdot\vec{\mathbf{p}})\frac{1}{r^N}\phi_{K,[j]}(\hat{\mathbf{x}}) &= \sum_{a=1}^{2N+1}\beta_a p_a \frac{1}{r^N}\phi_{K,[j]}(\hat{\mathbf{x}}) = iK\frac{1}{r^{N+1}}\phi_{-K,[j]}(\hat{\mathbf{x}}).\end{aligned} \tag{4.18}$$

Substituting $\Psi_{K,[j]}(\mathbf{x})$ into the Dirac equation (4.3) allows us to obtain the radial equations

$$\begin{aligned}\frac{dG(r)}{dr} + \frac{K}{r}G(r) &= [E - V(r) - M]F(r),\\ -\frac{dF(r)}{dr} + \frac{K}{r}F(r) &= [E - V(r) + M]G(r).\end{aligned} \tag{4.19}$$

3.2 *The* SO(2*N*) *Case*

As is well known, the spinor representation of the SO(2N) group is reducible and can be reduced to two non-equivalent fundamental spinor representations $[+s] \equiv [1, 0, \ldots, 0, 1]$ and $[-s] \equiv [1, 0, \ldots, 0, 1, 0]$. From Eq. (4.7) the Casimir for both spinor representations are calculated as $C_2([\pm s]) = (2N^2 - N)/4$. In terms of the β_a matrices, we define the γ_μ matrices for $N = 2l$:

$$\gamma_0 = \beta_{2N+1}, \qquad \gamma_a = \beta_{2N+1}\beta_a, \quad a \in [1, 2N]. \tag{4.20}$$

γ_0 is a diagonal matrix where half of the diagonal elements are equal to $+1$ and the remaining to -1. Because the spinor operator S_{ab} and the operator κ commutes with γ_0, each of them becomes a direct sum of two matrices, referring to the rows with the eigenvalues $+1$ and -1 of the γ_0, respectively. The fundamental spinors $\xi_\pm(\mathbf{m})$ belong to the fundamental spinor representations $[+s]$ and $[-s]$, respectively, and satisfy

$$\gamma_0\xi_\pm(\mathbf{m}) = \pm\xi_\pm(\mathbf{m}). \tag{4.21}$$

The product of $Y^{[\lambda]}_{\mathbf{m}}(\hat{\mathbf{x}})$ and $\xi_\pm(\mathbf{m}')$ belongs to the direct product of two representation $[\lambda]$ and $[\pm s]$, which is a reducible representation:

$$\begin{aligned}[][\lambda] \times [+s] &\simeq [\lambda, 0, \ldots, 0, 1] \oplus [\lambda - 1, 0, \ldots, 0, 1, 0],\\ [\lambda] \times [-s] &\simeq [\lambda, 0, \ldots, 0, 1, 0] \oplus [\lambda - 1, 0, \ldots, 0, 1].\end{aligned} \tag{4.22}$$

There are two kinds of representations for the total angular momentum: the representation $[j_1] \equiv [\lambda, 0, \ldots, 0, 1]$ and the representation $[j_2] \equiv [\lambda, 0, \ldots, 0, 1, 0]$. Their Casimirs are the same:

$$C_2([j_1]) = C_2([j_2]) = \lambda(\lambda + 2N - 1) + \frac{2N^2 - N}{4}. \tag{4.23}$$

There are two different ways to construct wavefunction belonging to the representation $[j_1]$: the combination of $Y^{[\lambda]}_{\mathbf{m}}(\hat{\mathbf{x}})\xi_+(\mathbf{m}')$ and that of $Y^{[\lambda+1]}_{\mathbf{m}}(\hat{\mathbf{x}})\xi_-(\mathbf{m}')$. Due to the spherical symmetry, we only calculate the highest weight state for the representation $[j_1]$ by the CGCs:

$$\begin{aligned}
\phi_{K,[j_1]}(\hat{\mathbf{x}}) &= Y^{[\lambda]}_{[\lambda]}(\hat{\mathbf{x}})\xi_+([+s]) = N_{2N,\lambda} r^{-\lambda}(x^1+ix^2)^\lambda \xi_+([+s]),\\
\phi_{-K,[j_1]}(\hat{\mathbf{x}}) &= \sum_{\mathbf{m}} Y^{[\lambda+1]}_{\mathbf{m}}(\hat{\mathbf{x}})\xi_-([j_1]-\mathbf{m})\\
&\quad \times \langle [\lambda+1],\mathbf{m},[+s],[j_1]-\mathbf{m}|[j_1],[j_1]\rangle\\
&= N_{2N,\lambda} r^{-\lambda-1}(x^1+ix^2)^\lambda \{x^{2N-1}+ix^{2N}\xi_-([-s])\\
&\quad + (x^{2N-3}+ix^{2N-2})\xi_-([0,\ldots,0,1,\overline{1},0])\\
&\quad + (x^{2N-5}+ix^{2N-4})\xi_-([0,\ldots,0,1,\overline{1},0,1]) + \cdots\\
&\quad + (x^3+ix^4)\xi_-([1,\overline{1},0,\ldots,0,1])\\
&\quad + (x^1+ix^2)\xi_-([\overline{1},0,\ldots,0,1])\},\\
K &= C_2([j_1]) - C_2([\lambda+1]) - C_2([+s]) + N - \frac{1}{2}\\
&= \lambda + N - \frac{1}{2}.
\end{aligned} \tag{4.24}$$

For the representation $[j_2] \equiv [\lambda, 0, \ldots, 0, 1, 0]$, we have

$$\begin{aligned}
\phi_{K,[j_2]}(\hat{\mathbf{x}}) &= \sum_{\mathbf{m}} Y^{[\lambda+1]}_{\mathbf{m}}(\hat{\mathbf{x}})\xi_+([j_2]-\mathbf{m})\\
&\quad \times \langle [\lambda+1],\mathbf{m},[-s],[j_2]-\mathbf{m}|[j_2],[j_2]\rangle\\
&= N_{2N,\lambda} r^{-\lambda-1}(x^1+ix^2)^l \{x^{2N-1}-ix^{2N}\xi([+s])\\
&\quad + (x^{2N-3}+ix^{2N-2})\xi_+([0,\ldots,0,1,0,\overline{1}])\\
&\quad + (x^{2N-5}+ix^{2N-4})\xi_+([0,\ldots,0,1,\overline{1},1,0]) + \cdots\\
&\quad + (x^3+ix^4)\xi_+([1,\overline{1},0,\ldots,0,1,0])\\
&\quad + (x^1+ix^2)\xi_+([\overline{1},0,\ldots,0,1,0])\},\\
\phi_{-K,[j_2]}(\hat{\mathbf{x}}) &= Y^{[\lambda]}_{[\lambda]}(\hat{\mathbf{x}})\xi_-([-s]) = N_{D,\lambda} r^{-\lambda}(x^1+ix^2)^\lambda \xi_-([-s]),\\
K &= C_2([j_2]) - C_2([\lambda+1]) - C_2([+s]) + N - \frac{1}{2}\\
&= -\lambda - N + \frac{1}{2}.
\end{aligned} \tag{4.25}$$

In terms of the explicit forms of β_a we obtain

$$\begin{aligned}
(\vec{\beta}\cdot\hat{\mathbf{x}})\phi_{K,[j_\omega]}(\hat{\mathbf{x}}) &= \frac{1}{r}\sum_{a=1}^{2N}\beta_a x^a \phi_{K,[j_\omega]}(\hat{\mathbf{x}}) = \phi_{-K,[j_\omega]}(\hat{\mathbf{x}}),\\
(\vec{\beta}\cdot\vec{\mathbf{p}}) r^{-N+1/2}\phi_{K,[j_\omega]}(\hat{\mathbf{x}}) &= \sum_{a=1}^{2N}\beta_a p_a r^{-N+1/2}\phi_{K,[j_\omega]}(\hat{\mathbf{x}})\\
&= iKr^{-N-1/2}\phi_{-K,[j_\omega]}(\hat{\mathbf{x}}),\\
\omega &= 1 \text{ or } 2.
\end{aligned} \tag{4.26}$$

The wavefunction $\Psi_{K,[j_\omega]}(\mathbf{x})$ of the total angular momentum belonging to the irreducible representation $[j_\omega]$ can be expressed as

$$\begin{aligned}
&\Psi_{|K|,[j_1]}(\mathbf{x},t) = r^{-N+1/2}e^{-iEt}\{F(r)\phi_{|K|,[j_1]}(\hat{\mathbf{x}}) + iG(r)\phi_{-|K|,[j_1]}(\hat{\mathbf{x}})\},\\
&\Psi_{-|K|,[j_2]}(\mathbf{x},t) = r^{-N+1/2}e^{-iEt}\{F(r)\phi_{-|K|,[j_2]}(\hat{\mathbf{x}}) + iG(r)\phi_{|K|,[j_2]}(\hat{\mathbf{x}})\},\\
&\kappa\Psi_{K,[j_\omega]}(\mathbf{x}) = K\Psi_{K,[j_\omega]}(\mathbf{x}), \quad |K| = \lambda + N - 1/2, \ \omega = 1 \text{ or } 2,\\
&H_1(J)\Psi_{K,[j_\omega]}(\mathbf{x}) = \lambda\Psi_{K,[j_1]}(\mathbf{x}), \qquad (4.27)\\
&H_{N-1}(J)\Psi_{K,[j_1]}(\mathbf{x}) = 0, \qquad H_N(J)\Psi_{K,[j_1]}(\mathbf{x}) = \Psi_{K,[j_1]}(\mathbf{x}),\\
&H_{N-1}(J)\Psi_{K,[j_2]}(\mathbf{x}) = \Psi_{K,[j_2]}(\mathbf{x}), \qquad H_N(J)\Psi_{K,[j_2]}(\mathbf{x}) = 0,\\
&H_\nu(J)\Psi_{K,[j_\omega]}(\mathbf{x}) = 0, \quad \nu \in [2, N-2].
\end{aligned}$$

Their partners can be calculated by the lowering operators F_ν.

By substituting $\Psi_{K,[j_\omega]}(\mathbf{x})$ into the Dirac equation (4.3), we obtain the radial equations

$$\begin{aligned}
&\frac{d}{dr}G(r) + \frac{K}{r}G(r) = (E - V(r) - M)F(r),\\
&-\frac{d}{dr}F(r) + \frac{K}{r}F(r) = (E - V(r) + M)G(r),
\end{aligned} \qquad (4.28)$$

which are the same as those in $D = 2N + 1$ case.

4 Application to Hydrogen Atom

Although the wavefunctions and the eigenvalues K are different for the cases $D = 2N + 1$ and $D = 2N$, the radial equations are unified as

$$\begin{aligned}
&\frac{d}{dr}G_{KE}(r) + \frac{K}{r}G_{KE}(r) = [E - V(r) - M]F_{KE}(r),\\
&-\frac{d}{dr}F_{KE}(r) + \frac{K}{r}F_{KE}(r) = [E - V(r) + M]G_{KE}(r),\\
&K = \pm\frac{1}{2}(2l + D - 1).
\end{aligned} \qquad (4.29)$$

For definiteness we discuss the attractive Coulomb potential

$$V(r) = -\frac{\xi}{r}, \qquad \xi = Z\alpha > 0, \qquad (4.30)$$

where and later $\alpha = 1/137$ is the fine structure constant. It is easy to see that the solution for the repulsive potential can be obtained from that for the attractive potential by interchanging

$$F_{KE} \longleftrightarrow G_{-K-E}, \qquad V(r) \longleftrightarrow -V(r). \qquad (4.31)$$

From the Sturm-Liouville theorem [184], there are bound states with the energy less than and near M for the attractive Coulomb potential and with the energy larger than and near $-M$ for the repulsive potential, if the interaction is not too strong.

For convenience, we introduce a new variable ρ in Eq. (4.29) for bound states:

$$\rho = 2r\sqrt{M^2 - E^2}, \qquad E/M \in (0, 1]. \tag{4.32}$$

Solving $F(\rho)$ from Eq. (4.29),

$$F_{KE}(\rho) = \left(-\frac{1}{2}\sqrt{\frac{M-E}{M+E}} + \frac{\xi}{\rho}\right)^{-1} \left[\frac{d}{d\rho}G_{KE}(\rho) + \frac{K}{\rho}G_{KE}(\rho)\right], \tag{4.33}$$

we obtain a second-order differential equation of $G_{KE}(\rho)$:

$$\begin{aligned} &\frac{d^2}{d\rho^2}G_{KE}(\rho) + \left\{-\frac{1}{4} - \frac{K^2 - \xi^2 + K}{\rho^2} + \frac{E\xi}{\rho\sqrt{M^2 - E^2}}\right\}G_{KE}(\rho) \\ &\quad + \left\{\rho - \frac{\rho^2}{2\xi}\sqrt{\frac{M-E}{M+E}}\right\}^{-1}\left[\frac{d}{d\rho}G_{KE}(\rho) + \frac{K}{\rho}G_{KE}(\rho)\right] = 0. \end{aligned} \tag{4.34}$$

From the behavior of $G_{KE}(\rho)$ at the origin and at infinity, define

$$\begin{aligned} &G_{KE}(\rho) = \rho^{\lambda} e^{-\rho/2} R(\rho), \quad \lambda = \sqrt{K^2 - \xi^2} > 0, \\ &\omega = \frac{1}{2\xi}\sqrt{\frac{M-E}{M+E}}, \qquad \tau = \frac{E\xi}{\sqrt{M^2 - E^2}}. \end{aligned} \tag{4.35}$$

Substitution of them into Eq. (4.34) leads to

$$\begin{aligned} &(\rho - \omega\rho^2)\frac{d^2}{d\rho^2}R(\rho) + [\omega\rho^2 - (2\lambda\omega + 1)\rho + 2\lambda + 1]\frac{d}{d\rho}R(\rho) \\ &\quad + [\omega(\lambda - \tau)\rho + \omega(K + \lambda) + \tau - \lambda - 1/2]R(\rho) = 0. \end{aligned} \tag{4.36}$$

Equation (4.36) can be solved by the power series expansion method, which was used in [175, 176] for $(3+1)$ dimensions. The results are calculated as

$$\begin{aligned} \left.\begin{matrix} F_{KE}(\rho) \\ G_{KE}(\rho) \end{matrix}\right\} &= \frac{(M^2 - E^2)^{1/4}}{\Gamma(2\lambda + 1)}\sqrt{\frac{(M \pm E)E\Gamma(n' + 2\lambda + 1)}{2M^2\tau(j + \tau M/E)n'!}} \\ &\quad \times \rho^{\lambda} e^{-\rho/2}\{(K + \tau M/E)_1F_1(-n'; 2\lambda + 1; \rho) \\ &\quad \mp n'{}_1F_1(1 - n'; 2\lambda + 1; \rho)\} \\ &\quad \times \int_0^{\infty} (|F_{KE}(\rho)|^2 + |G_{KE}(\rho)|^2)dr = 1, \end{aligned} \tag{4.37}$$

$$n' = \tau - \lambda = 0, 1, 2, \ldots. \tag{4.38}$$

When $n' = 0$, K has to be positive. Introduce a principal quantum number

$$n = |K| - (D-3)/2 + n' = |K| - (D-3)/2 + \tau - \lambda = 1, 2, \ldots. \tag{4.39}$$

The n can be equal to 1 only for $K = (D-1)/2$ and equal to other positive integers for both signs of K. The energy E can be calculated as

$$E = M\left\{1 + \frac{\xi^2}{(\sqrt{K^2 - \xi^2} + n - |K| + (D-3)/2)^2}\right\}^{-1/2}. \tag{4.40}$$

Expanding Eq. (4.40) in powers of ξ^2, we have

$$E \simeq M\left\{1 - \frac{\xi^2}{2[n+(D-3)/2]^2} - \frac{\xi^4}{2[n+(D-3)/2]^4} \right.$$
$$\left. \times \left(\frac{n+(D-3)/2}{|K|} - \frac{3}{4}\right)\right\}, \qquad (4.41)$$

where the first term on the right hand side is the rest energy M ($c^2 = 1$), the second one coincides with the energy from the solutions of the Schrödinger equation, and the third one is the fine structure energy, which removes the degeneracy between the states with the same n.

5 Concluding Remarks

In this Chapter we have generalized the Dirac equation to $(D+1)$-dimensional space-time. The conserved angular momentum operators and their quantum numbers are discussed. The eigenfunctions of the total angular momentums are calculated for both odd $D = 2N + 1$ and even $D = 2N$ cases, respectively. The unified radial equations for a spherically symmetric system are obtained. As an illustration, we have dealt with Coulomb potential problem by the series approach. The exact solutions are expressed by the confluent hypergeometric functions. The eigenvalues as well as their fine structure energy have also been studied.

Chapter 5
Klein-Gordon Equation in Higher Dimensions

1 Introduction

It is known that the exact solutions of non-relativistic and relativistic equations in the spherically central fields have become an important subject in quantum mechanics. As illustrated above, the main contributions have been made to the Schrödinger and Dirac equations. During the past several decades, however, the Klein-Gordon equation with the Coulomb potential has been studied in three dimensions such as the operator analysis [185], in an intense laser field [186], in two dimensions [187] and in one dimension [188–191]. On the other hand, the Klein-Gordon equation with a Coulomb potential in $(D+1)$ dimensions has been discussed by the different approaches like the large-N expansion approximate method [93]. The purpose of this Chapter is to present the Klein-Gordon equation in arbitrary dimensions and solve the hydrogen-like atom problem.

2 The Radial Equations

For a particle moving in a spherically symmetric potential $V(r)$, the symmetric group of the quantum system is the SO(D) group. The time-independent Klein-Gordon equation ($\hbar = c = 1$) is written as

$$(-\nabla^2 + M^2)\Psi(\mathbf{r}) = [E - V(r)]^2\Psi(\mathbf{r}), \tag{5.1}$$

where M and E denote the mass and the energy of the particle, respectively. As shown in Eq. (3.69), we take wavefunction as

$$\Psi(\mathbf{r}) = r^{-\frac{D-1}{2}} R_l(r) Y^l_{l_{D-2}\cdots l_1}(\hat{\mathbf{x}}). \tag{5.2}$$

Substitution of this into Eq. (5.1) allows us to obtain the D-dimensional radial Klein-Gordon equation

$$\left(\frac{d^2}{dr^2} - \frac{\kappa^2 - 1/4}{r^2}\right) R_l(r) = -\{[E - V(r)]^2 - M^2\} R_l(r), \tag{5.3}$$

S.-H. Dong, *Wave Equations in Higher Dimensions*,
DOI 10.1007/978-94-007-1917-0_5,

where

$$\kappa = |l - 1 + D/2| \tag{5.4}$$

depends on the angular momentum l and the spatial dimension D. $Y^{l}_{l_{N-2}\dots l_1}(\hat{\mathbf{x}})$ in Eq. (5.2) is called the generalized spherical harmonics.

3 Application to Hydrogen-like Atom

As an illustration, we take the symmetric potential $V(r)$ as the Coulomb-type potential

$$V(r) = -\frac{\xi}{r}, \quad \xi = Z\alpha. \tag{5.5}$$

We will discuss the weak potential, say $|\xi| < 1/2$. The radial equation (5.3) thus becomes

$$\frac{d^2}{dr^2}R_l(r) + \frac{\xi^2 - \kappa^2 + 1/4}{r^2}R_l(r) + \frac{2E\xi}{r}R_l(r) + (E^2 - M^2)R_l(r) = 0. \tag{5.6}$$

Take a new variable ρ for the bound states:

$$\rho = 2r\sqrt{M^2 - E^2}, \quad |E| < M. \tag{5.7}$$

Equation (5.6) thus changes to

$$\frac{d^2}{d\rho^2}R_l(\rho) + \frac{1}{\rho^2}\left(\xi^2 - \kappa^2 + \frac{1}{4}\right)R_l(\rho) + \frac{\tau}{\rho}R_l(\rho) - \frac{1}{4}R_l(\rho) = 0, \tag{5.8}$$

with

$$\tau = \frac{E\xi}{\sqrt{M^2 - E^2}}. \tag{5.9}$$

From the behaviors of radial function at the origin and at infinity, we take the wavefunction of the form

$$R_l(\rho) = \rho^{\lambda + 1/2}e^{-\rho/2}\phi(\rho), \quad \lambda = \sqrt{\kappa^2 - \xi^2} > 0, \tag{5.10}$$

where we assume $\kappa^2 > \xi^2$. The $\phi(\rho)$ satisfies the following confluent hypergeometric equation

$$\rho\frac{d^2}{d\rho^2}\phi(\rho) + (2\lambda + 1 - \rho)\frac{d}{d\rho}\phi(\rho) + \left(\tau - \lambda - \frac{1}{2}\right)\phi(\rho) = 0. \tag{5.11}$$

Thus, the radial function can be written out

$$R_l(\rho) = N_l\rho^{\lambda + 1/2}e^{-\rho/2}{}_1F_1(\lambda - \tau + 1/2; 2\lambda + 1; \rho), \tag{5.12}$$

where N_l is the normalization factor to be determined.

We now discuss the eigenvalues. From the finiteness of the solutions at infinity, the general quantum condition is obtained from Eq. (5.12)

$$\tau - \lambda - \frac{1}{2} = n' = 0, 1, 2, \dots. \tag{5.13}$$

By introducing a principal quantum number

$$n = n' + \kappa - \frac{N}{2} + 2 = n' + l + 1, \tag{5.14}$$

we obtain

$$\frac{E\xi}{\sqrt{M^2 - E^2}} = \tau = n - l - \frac{1}{2} + \sqrt{\kappa^2 - \xi^2} > 0. \tag{5.15}$$

Therefore, we can obtain E with the same sign as ξ

$$E(n, l, D) = M\frac{\xi}{|\xi|}\left\{1 + \frac{\xi^2}{(n - l - \frac{1}{2} + \sqrt{\kappa^2 - \xi^2})^2}\right\}^{-1/2}, \tag{5.16}$$

which essentially coincides with that of [87] except that the factor $\xi/|\xi|$ was not considered there.

For a large D, we have

$$E(n, D) = M\frac{\xi}{|\xi|}[1 - 2\xi^2 D^{-2} + 4\xi^2(2n - 3)D^{-3} - \cdots]. \tag{5.17}$$

For a small ξ, we have

$$\begin{aligned} E(n, l, D) = M\frac{\xi}{|\xi|}\Bigg\{1 &- \frac{\xi^2}{2[n + (D - 3)/2]^2} \\ &+ \frac{\xi^4(D + 6l - 4n)}{4(2l - 2 + D)[n + (D - 3)/2]^4}\Bigg\}, \end{aligned} \tag{5.18}$$

where the first term on the right hand side is the rest energy M, the second one is from the solutions of the Schrödinger equation with this potential, and the third one is the fine structure energy, which removes the degeneracy between the states with same n.

We now calculate the normalization factor N_l from the normalization condition

$$\int_0^\infty R_l(\rho)^2 dr = 1. \tag{5.19}$$

Since $n' = \tau - \lambda - 1/2$ is a non-negative integer, we can express the confluent hypergeometric functions ${}_1F_1(-n'; \beta + 1; \rho)$ by the associated Laguerre polynomial $L^\beta_{n'}(\rho)$. In terms of the following formulas [192]:

$$\rho\, {}_1F_1(\alpha + 1; \beta + 1; \rho) = \beta\, {}_1F_1(\alpha + 1; \beta; \rho) - \beta\, {}_1F_1(\alpha; \beta; \rho), \tag{5.20}$$

$$\begin{aligned} L^\beta_n(\rho) &= \frac{\Gamma(\beta + n + 1)}{n!\Gamma(\beta + 1)}\, {}_1F_1(-n; \beta + 1; \rho), \\ \int_0^\infty \rho^\beta e^{-\rho} L^\beta_n(\rho) L^\beta_m(\rho) d\rho &= \frac{\Gamma(n + \beta + 1)}{n!}\delta_{nm}, \end{aligned} \tag{5.21}$$

we obtain through a direct calculation

$$\begin{aligned} N_l &= \frac{(M^2 - E^2)^{1/4}}{\Gamma(2\lambda + 1)}\left[\frac{2\Gamma(n' + 2\lambda + 1)}{n'!(2n' + 2\lambda + 1)}\right]^{1/2}, \\ \lambda &= [(l - 1 + D/2)^2 - \xi^2]^{1/2}. \end{aligned} \tag{5.22}$$

4 Concluding Remarks

In this Chapter we have presented the D-dimensional Klein-Gordon equation and applied it to the Coulomb-like potential as an illustration. The eigenfunctions are analytically obtained and expressed by the confluent hypergeometric functions. The eigenvalues as well as their fine structure are also studied.

Part III
Applications in Non-relativistic Quantum Mechanics

Chapter 6
Harmonic Oscillator

1 Introduction

It is well known that the quantum harmonic oscillator is analog of the classical harmonic oscillator. It is one of the most important model systems in quantum mechanics. There are several reasons for its pivotal role. First, it represents one of few quantum mechanical systems for which the simple exact solutions are known. Second, as in classical mechanics, a wide variety of physical situations can be reduced to it either exactly or approximately. In particular, more complicated quantum systems can always be analyzed in terms of normal modes—formally equivalent to harmonic oscillators—of motion whenever the interaction forces are linear functions of the relative displacements. Therefore, it is not surprising that the harmonic oscillator has become very important for the quantum mechanical treatment of such physical problems as the vibrations of individual atoms in molecules and in crystals, in which the linear harmonic oscillator describes vibrations in molecules and their counterparts in solids, the phonons. Third, the most eminent role of the harmonic oscillator is its linkage to the boson, one of the conceptual building blocks of microscopic physics. For example, bosons describe the modes of the electromagnetic field, providing the basis for its quantization. Even though the linear harmonic oscillator may represent rather non-elementary objects like a solid and a molecule, it provides a window into the most elementary structure of the physical world. The most likely reason for this connection with fundamental properties of matter is that the harmonic oscillator Hamiltonian is symmetric in momentum and position, both operators appearing as quadratic terms. On the other hand, the harmonic oscillator also provides the key to the quantum theory of the electromagnetic field, whose vibrations in a cavity can be analyzed into harmonic normal modes, each of which has energy levels of the harmonic oscillator type.

This Chapter is organized as follows. In Sect. 2 we first study exact solutions of harmonic oscillator in arbitrary dimensions. Section 3 is devoted to the recurrence relations for the radial wavefunction. We shall show the realization of dynamic algebra su$(1, 1)$ in Sect. 4. In Sect. 5 we carry out the generalized harmonic oscillator named the pseudoharmonic oscillator, whose exact solutions, ladder operators and

S.-H. Dong, *Wave Equations in Higher Dimensions*,
DOI 10.1007/978-94-007-1917-0_6, © Springer Science+Business Media B.V. 2011

recurrence relations are presented. The position and momentum information entropy is reviewed briefly in Sect. 6. Some concluding remarks are given in Sect. 7.

2 Exact Solutions of Harmonic Oscillator

Based on the results presented in Chap. 3, we are going to study the exact solutions of harmonic oscillator. Consider a D-dimensional Schrödinger equation with a harmonic potential

$$V(r) = \frac{1}{2}kr^2, \quad k = M\omega^2, \tag{6.1}$$

where M and ω denote the mass and vibration frequency of the particle, respectively.

Let

$$\phi^{[\lambda]}(r) = e^{-\alpha\xi}\xi^{\beta}R(\xi), \quad \xi = r^2, \tag{6.2}$$

where α and β are the parameters to be determined. Substituting Eqs. (6.1) and (6.2) into the radial equation (3.66) where $\lambda = l$ given in Chap. 3, we have ($\hbar = 1$)

$$\begin{aligned} &\xi\frac{d^2R(\xi)}{d\xi^2} + \left(2\beta + \frac{D}{2} - 2\alpha\xi\right)\frac{dR(\xi)}{d\xi} \\ &+ \left\{\xi^{-1}\left[\frac{\beta(2\beta + D - 2)}{2} - \frac{l(l + D - 2)}{4}\right]\right. \\ &\left. + \xi\left(\alpha^2 - \frac{Mk}{4}\right) + \left[\frac{ME}{2} - 2\alpha\left(\beta + \frac{D}{4}\right)\right]\right\}R(\xi) = 0. \end{aligned} \tag{6.3}$$

Take the parameters α and β to remove the terms of ξ^{-1} and ξ,

$$\alpha = \frac{\sqrt{Mk}}{2}, \quad \beta = \frac{l}{2}. \tag{6.4}$$

Further defining a variable by

$$\rho = 2\alpha\xi, \tag{6.5}$$

we obtain the confluent hypergeometric differential equation [192]:

$$\rho\frac{d^2R(\rho)}{d\rho^2} + \left(l + \frac{D}{2} - \rho\right)\frac{dR(\rho)}{d\rho} + \left\{E\sqrt{\frac{M}{4k}} - \frac{(2l + D)}{4}\right\}R(\rho) = 0. \tag{6.6}$$

Therefore, the solutions to the D-dimensional Schrödinger equation with a harmonic potential are given by

$$\begin{aligned} &\psi^l_{l_{D-1},\ldots,l_1}(\mathbf{x}) = \mathcal{N}_{E,l}e^{-\rho/2}\rho^{l/2}\,{}_1F_1(-n_r; l + D/2; \rho)Y^l_{l_{D-1},\ldots,l_1}(\hat{\mathbf{x}}), \\ &\rho = r^2\sqrt{Mk}, \quad n_r = E\sqrt{\frac{M}{4k}} - \frac{(2l + D)}{4} = 0, 1, \ldots, \end{aligned} \tag{6.7}$$

where $\mathcal{N}_{E,l}$ is the normalization factor. The consideration of the finiteness of the solutions at infinity leads to a quantum condition that n_r has to be a non-negative integer. Hence, the energy E can be quantized by a principal quantum number n:

$$E_{n,l} = \sqrt{k/M}(n + D/2), \quad n = 2n_r + l = 0, 1, 2, \ldots, \tag{6.8}$$

from which we note that

$$l = \begin{cases} 0, 2, 4, \ldots, n, & \text{when } n \text{ is even,} \\ 1, 3, 5, \ldots, n, & \text{when } n \text{ is odd.} \end{cases} \tag{6.9}$$

The degeneracy of the states with the energy $E_{n,l}$ is calculated as

$$m = \sum_l d_l(D) = \frac{(n + D - 1)!}{n!(D - 1)!}. \tag{6.10}$$

We now determine the normalization factor $\mathcal{N}_{E,l}$ from the normalization condition

$$\int |\psi^l_{l_{D-1},\ldots,l_1}(\mathbf{x})|^2 \prod_{a=1}^{D} dx_a = 1. \tag{6.11}$$

Notice that when n_r is a non-negative integer, the confluent hypergeometric functions ${}_1F_1(-n_r; \beta + 1; \rho)$ can be expressed by the associated Laguerre polynomial $L^{\beta}_{n_r}(\rho)$. Using Eq. (5.21), through a direct calculation we obtain the normalization factor

$$\mathcal{N}_{E,l} = \frac{(Mk)^{D/8}}{\Gamma(l + D/2)} \left[\frac{2\Gamma(n_r + l + D/2)}{n_r!}\right]^{1/2}. \tag{6.12}$$

3 Recurrence Relations for the Radial Function

It is shown from Eq. (3.66) that the Hamiltonian of the D-fold degenerate oscillator for a given angular momentum as follows ($\hbar = 1$):

$$H = -\frac{1}{2M}\left\{\frac{d^2}{dr^2} + \frac{D-1}{r}\frac{d}{dr} - \frac{l(D+l-2)}{r^2}\right\} + \frac{1}{2}M\omega^2 r^2, \tag{6.13}$$

with the property

$$HR(n,l) = E_n R(n,l), \quad E_n = \omega(n + D/2). \tag{6.14}$$

For convenience, define a new variable $\varrho = \sqrt{M\omega}r$. As a result, the Hamiltonian is modified as

$$H(l) = -\frac{1}{2}\left\{\frac{d^2}{d\varrho^2} + \frac{D-1}{\varrho}\frac{d}{d\varrho} - \frac{l(D+l-2)}{\varrho^2} - \varrho^2\right\}, \tag{6.15}$$

with the property

$$H(l)R(n,l) = \varepsilon R(n,l), \quad \varepsilon = (n + D/2) \tag{6.16}$$

and

$$H = \omega H(l), \qquad E_n = \omega \varepsilon. \tag{6.17}$$

To obtain the recurrence relations for the radial function, we define the following operators [14]

$$\begin{aligned} A^+ &= -\frac{d}{d\varrho} + \frac{l}{\varrho} + \varrho, \\ A^- &= \frac{d}{d\varrho} + \frac{(l+D-2)}{\varrho} + \varrho, \\ a^+ &= \frac{d}{d\varrho} - \frac{l}{\varrho} + \varrho, \\ a^- &= -\frac{d}{d\varrho} - \frac{(l+D-2)}{\varrho} + \varrho. \end{aligned} \tag{6.18}$$

Based on these operators, we may obtain the following identities on the Hamiltonian

$$2H(l) = \begin{cases} A^-(l+1)A^+(l) - (2l+D), \\ A^+(l-1)A^-(l) - (2l+D-4), \\ a^-(l+1)a^+(l) + (2l+D), \\ a^+(l-1)a^-(l) + (2l+D-4). \end{cases} \tag{6.19}$$

By multiplying $A^\pm$ and $a^\pm$ appropriately on both sides of Eqs. (6.19), we are able to obtain

$$\begin{aligned} H(l\pm 1)A^\pm(l) - A^\pm(l)H(l) &= \pm A^\pm(l), \\ H(l\pm 1)a^\pm(l) - a^\pm(l)H(l) &= \mp a^\pm(l). \end{aligned} \tag{6.20}$$

Operating on the radial function $R(n,l)$ allows us to obtain

$$\begin{aligned} H(l\pm 1)A^\pm(l)R(n,l) &= (\varepsilon \pm 1)A^\pm(l)R(n,l), \\ H(l\pm 1)a^\pm(l)R(n,l) &= (\varepsilon \pm 1)a^\pm(l)R(n,l), \end{aligned} \tag{6.21}$$

from which we have

$$\begin{aligned} A^\pm(l)R(n,l) &= B^\pm R(n\pm 1, l\pm 1), \\ a^\pm(l)R(n,l) &= C^\pm R(n\mp 1, l\pm 1), \end{aligned} \tag{6.22}$$

where

$$\begin{aligned} B^+ &= -\sqrt{2(n+l+D)}, \\ B^- &= -\sqrt{2(n+l+D-2)}, \\ C^+ &= \sqrt{2(n-l)}, \\ C^- &= \sqrt{2(n-l+2)}. \end{aligned} \tag{6.23}$$

In terms of these relations, one is able to derive a few useful recurrence relations as follows:

$$\varrho R(n,l) = \begin{cases} -\sqrt{\frac{n+l+D}{2}} R(n+1,l+1) \\ \quad +\sqrt{\frac{n-l}{2}} R(n-1,l+1), \\ -\sqrt{\frac{n+l+D-2}{2}} R(n-1,l-1) \\ \quad +\sqrt{\frac{n-l+2}{2}} R(n+1,l-1), \end{cases} \tag{6.24}$$

$$\frac{D+2l-2}{\varrho} R(n,l) = \begin{cases} -\sqrt{2(n+l+D)} R(n+1,l+1) \\ \quad -\sqrt{2(n-l+2)} R(n+1,l-1), \\ -\sqrt{2(n+l+D-2)} R(n-1,l-1) \\ \quad -\sqrt{2(n-l)} R(n-1,l+1), \end{cases} \tag{6.25}$$

$$(D+2l-2)\frac{dR(n,l)}{d\varrho} = \begin{cases} (D+l-2)\{\sqrt{\frac{n+l+D}{2}} R(n+1,l+1) \\ \quad +\sqrt{\frac{n-l}{2}} R(n-1,l+1)\}, \\ -l\{\sqrt{\frac{n+l+D-2}{2}} R(n-1,l-1) \\ \quad +\sqrt{\frac{n-l+2}{2}} R(n+1,l-1)\}. \end{cases} \tag{6.26}$$

Before ending this section, let us review the results given by Coulson and Joseph [89]. In a D-dimensional Euclidean space we define the Hamiltonian

$$H = \frac{1}{2}(p^2 + r^2), \tag{6.27}$$

where the atomic units $\hbar = M = \omega$ are used. Denote its eigenfunctions by ψ_l, where l is the total orbital angular momentum quantum number. If $\boldsymbol{\sigma}_{(D+1)}$ is a vector in D-space with components $\sigma_{i(D+1)}$ $(i = 1, \ldots, D)$, then we let

$$A_{(D+1)} = \mathbf{r} \cdot \boldsymbol{\sigma}_{(D+1)}, \qquad B_{(D+1)} = -\mathbf{p} \cdot \boldsymbol{\sigma}_{(D+1)}, \tag{6.28}$$

where σ are the generalized Pauli spin operators with the properties [88]

$$\begin{aligned} &\sigma_{ij} = -\sigma_{ji}, \qquad \sigma_{ij} = \sigma_{ij}^{\dagger}, \qquad \sigma_{ij}^2 = 1, \\ &\sigma_{ij}\sigma_{ik} = i\sigma_{jk}, \qquad [\sigma_{ij}, \sigma_{kl}] = 0. \end{aligned} \tag{6.29}$$

From Eqs. (6.27) and (6.28) as well as the properties of $\boldsymbol{\sigma}_{(D+1)}$, it may be verified that

$$[H, A_{(D+1)}] = iB_{(D+1)}, \qquad [H, iB_{(D+1)}] = A_{(D+1)}, \tag{6.30}$$

which are the Fourier transforms of each other apart from an irrelevant constant.

Combining them leads to an important result

$$[H, (A_{(D+1)} \pm iB_{(D+1)})] = \pm[A_{(D+1)} \pm iB_{(D+1)}], \tag{6.31}$$

which means that $(A_{(D+1)} \pm iB_{(D+1)})$ are the ladder operators for the Hamiltonian H. We should point out that the ladder operators given by Coulson and Joseph are self-adjoint ones unlike those (6.18), which cannot be self-adjoint. This is because the adjoint of either ladder operator must be that operator which steps the eigenfunctions in the opposite sense.

Up to this stage the discussion of the ladder operators has been purely algebraic without reference to any particular set of coordinates. Undoubtedly, it shall be very interesting to determine the analytical form assumed by $A_{(D+1)}$ and $B_{(D+1)}$ when expressed in terms of spherical polar coordinates. The results are given by [89, 193]

$$[A_{(D+1)} \pm i B_{(D+1)}] = \hat{A}_{(D+1)}[r \mp (\partial/\partial r - r^{-1} l_n)], \tag{6.32}$$

where

$$\begin{aligned} \hat{A}_{(D+1)} &= \frac{A_{(D+1)}}{\|A_{(D+1)}\|}, \qquad A_{(D+1)} = \sum_{i=1}^{D} x_i \sigma_{i(D+1)}, \\ l_n &= \sum_{m=2}^{D} L_m, \qquad L_{(D+1)} = \sum_{i=1}^{D} \sigma_{i(D+1)} L_{i(D+1)}. \end{aligned} \tag{6.33}$$

This means that the operator may be decomposed into two parts. One of these $\hat{A}_{(D+1)}$ steps the angular part of wavefunction. The other steps the radial part.

4 Realization of Dynamic Group SU(1, 1)

In this section, we are going to realize the dynamic group SU(1, 1) for the radial Schrödinger equation [194]. For simplicity, we may write down the D-dimensional radial function as follows:

$$R_{n_r l}^{D}(r) = \sqrt{\frac{2 n_r!}{\Gamma(n_r + l + D/2)}}\, r^l e^{-r^2/2} L_{n_r}^{l+(D-2)/2}(r^2). \tag{6.34}$$

Making use of the following relations for the Laguerre functions [192]

$$x \frac{d}{dx} L_n^{\beta}(x) = \begin{cases} n L_n^{\beta}(x) - (n+\beta) L_{n-1}^{\beta}(x), \\ (n+1) L_{n+1}^{\beta}(x) - (n+\beta+1-x) L_n^{\beta}(x), \end{cases} \tag{6.35}$$

as well as the relation (C.29) for arbitrary operators A and B (see Appendix C), we may obtain the raising and lowering operators for the quantum number n_r

$$L_{\pm} = \frac{1}{2}\left(\pm r \frac{d}{dr} - r^2 + H(l) \pm \frac{D}{2} \right), \tag{6.36}$$

where $H(l)$ is the radial Hamiltonian (6.15).

By defining $L_0 = H(l)/2$, it is found that $L_{\pm}$ and L_0 satisfies the commutation relations of an su(1, 1) algebra

$$[L_0, L_{\pm}] = \pm L_{\pm}, \qquad [L_+, L_-] = -2L_0, \tag{6.37}$$

with the following properties

$$L_+R^D_{n_rl}(r) = \sqrt{(n_r+1)\left(n_r+l+\frac{D}{2}\right)}R^D_{(n_r+1)l}(r),$$
$$L_-R^D_{n_rl}(r) = \sqrt{n_r\left(n_r+l+\frac{D}{2}-1\right)}R^D_{(n_r-1)l}(r), \tag{6.38}$$
$$L_0R^D_{n_rl}(r) = \left(n_r+\frac{l}{2}+\frac{D}{4}\right)R^D_{n_rl}(r).$$

The Casimir operator is calculated as

$$C = L_0^2 - L_1^2 - L_2^2 = L_0^2 - \frac{1}{2}(L_+L_- + L_-L_+) \tag{6.39}$$

with eigenvalue $k(k-1)$. After calculating we find the Casimir eigenvalue

$$CR^D_{n_rl}(r) = \frac{2l+D}{4}\left(\frac{2l+D}{4}-1\right)R^D_{n_rl}(r), \tag{6.40}$$

from which we have two possible solutions $k = (2l+D)/4$ or $k = -[(2l+D)/4-1]$. However, we are interested only in the positive discrete representations of the su(1, 1), $D^+(k)$. Thus, we take the former $k = (2l+D)/4$.

5 Generalization to Pseudoharmonic Oscillator

5.1 Introduction

It is well known that the real molecular vibrations are anharmonic even though the harmonic oscillator model is widely used as mentioned above. For instance, some anharmonic oscillator molecular potentials such as the Morse potential and Pöschl-Teller potential represent two typical model potentials to describe the molecular vibrations. In this section we want to study the pseudoharmonic oscillator proposed by Goldman and Krivchenkov in the early 1960s and expressed as $V(x) = V_0(x/a - a/x)^2$ [195], where V_0 and a represent two potential parameters. It is obvious to see that this potential is the sum of the harmonic oscillator and inversely quadratic potential. For the inverse squared interaction potential, Post proposed it in 1956 when he studied the one-dimensional many-identical-particle problem for pair-force interaction between the particles [196]. Since 1961 such a quantum system has been studied by many authors [2, 195–209]. For example, Landau and Lifshitz studied its exact solutions in three dimensions [2]. Hurley found that this kind of pseudoharmonic oscillator interaction between the particles can be exactly solved when he studied three-body problem in one dimension [198]. Several years later, Calogero studied the one-dimensional three- and many-body problems interacting pairwise via harmonic and inverse square (centrifugal) potential [199, 200]. Also, this potential was generalized by Camiz and Dodonov *et al.* to the non-stationary (varying frequency) pseudoharmonic oscillator potential [201–203]. In addition to these, such a physical problem was discussed in arbitrary dimensions D [201–205]. Recently, Sage has studied its vibrations and rotations in order to describe the diatomic molecule [208], in which he briefly reinvestigated some properties of this oscillator to study the pseudogaussian oscillator. On the other hand, we notice that the harmonic oscillator plus inverse squared potential is essentially equal to a pseudoharmonic oscillator except for some unimportant parameters as studied in [210]. Consequently, it is unnecessary to restudy this case for simplicity.

5.2 Exact Solutions

As shown in Ref. [3], the pseudoharmonic oscillator is taken as

$$V(r) = \frac{1}{8}\kappa r_0^2\left(\frac{r}{r_0} - \frac{r_0}{r}\right)^2, \tag{6.41}$$

where κ is a force constant and the r_0 represents the equilibrium bond length. It should be noted that the potential taken here is slightly different from that of Ref. [207], but they are the same essentially except for an unimportant constant.

Let

$$\phi(r) = e^{-\lambda\xi}\xi^{\tau}R(\xi), \quad \xi = r^2. \tag{6.42}$$

In a similar way, we may obtain the radial equation from Eqs. (3.66) as follows:

$$\begin{aligned}
&\xi\frac{d^2R(\xi)}{d\xi^2} + \left(2\tau + \frac{D}{2} - 2\lambda\xi\right)\frac{dR(\xi)}{d\xi}\\
&\quad + \left\{\xi^{-1}\left[\frac{\tau(2\tau + D - 2)}{2} - \frac{l(l + D - 2)}{4} - \frac{M\kappa r_0^4}{16}\right]\right.\\
&\quad \left. + \xi\left(\lambda^2 - \frac{M\kappa}{16}\right) + \left[\frac{ME}{2} + \frac{M\kappa r_0^2}{8} - 2\lambda\left(\tau + \frac{D}{4}\right)\right]\right\}R(\xi) = 0.
\end{aligned} \tag{6.43}$$

Take the parameters λ and τ to remove the terms of ξ^{-1} and ξ,

$$\lambda = \frac{\sqrt{M\kappa}}{4}, \qquad \tau = -\frac{D-2}{4} + \frac{1}{4}\left[(2l + D - 2)^2 + M\kappa r_0^4\right]^{1/2}. \tag{6.44}$$

Further define a new variable $\rho = 2\lambda\xi$. We may obtain the confluent hypergeometric equation [192]:

$$\begin{aligned}
&\rho\frac{d^2R(\rho)}{d\rho^2} + \left(2\tau + \frac{D}{2} - \rho\right)\frac{dR(\rho)}{d\rho}\\
&\quad + \left\{E\sqrt{\frac{M}{\kappa}} + \frac{r_0^2\sqrt{M\kappa}}{4} - \tau - \frac{D}{4}\right\}R(\rho) = 0.
\end{aligned} \tag{6.45}$$

Therefore, the solutions to the D-dimensional Schrödinger equation with a pseudoharmonic potential are written as

$$\psi^{l}_{l_{D-1},\ldots,l_1}(\mathbf{x}) = N_{E,l}e^{-\rho/2}\rho^{\tau}{}_1F_1(-n_r; 2\tau + D/2; \rho)Y^{l}_{l_{D-1},\ldots,l_1}(\hat{\mathbf{x}}), \tag{6.46}$$

where

$$\begin{aligned}
\rho &= \frac{\sqrt{M\kappa}}{2} r^2, \\
n_r &= E\sqrt{\frac{M}{\kappa}} + \frac{r_0^2\sqrt{M\kappa}}{4} - \tau - \frac{D}{4} = 0, 1, \ldots, \\
N_{E,l} &= \frac{(M\kappa)^{D/8}}{\Gamma(2\tau + D/2)} \left[\frac{\Gamma(n_r + 2\tau + D/2)}{2^{D/2-1} n_r!}\right]^{1/2}.
\end{aligned} \tag{6.47}$$

Likewise, considering the finiteness of the solutions at infinity leads to a quantum condition that n_r has to be a non-negative integer. The quantized energy E is calculated as

$$E_{n_r,l} = \sqrt{\frac{\kappa}{M}}\left(n_r - \frac{r_0^2\sqrt{M\kappa}}{4} + \tau + \frac{D}{4}\right). \tag{6.48}$$

The energy spectrum is equidistant for given constants τ and r_0.

When r_0 is very large, $r_0^2 \gg 2/\sqrt{M\kappa}$, we have

$$E_{n_r,l} = \sqrt{\frac{\kappa}{M}}\left\{n_r + \frac{1}{2} - \frac{(l + D/2 - 1)^2}{2r_0^2\sqrt{M\kappa}}\right\}, \tag{6.49}$$

and when $r_0^2 \ll 2/\sqrt{M\kappa}$, we have

$$E_{n_r,l} = \sqrt{\frac{\kappa}{M}}\left\{n_r + \frac{l}{2} + \frac{D}{4} - \frac{r_0^2\sqrt{M\kappa}}{4} + \frac{M\kappa r_0^4}{16(l + D/2 - 1)}\right\}. \tag{6.50}$$

5.3 Ladder Operators

During the past half century, the factorization method has played an important role in physics [3, 193, 211]. With this method, one can construct the ladder operators for certain potentials such as the Morse and modified Pöschl-Teller potentials and then constitute a suitable algebra su(2). This approach is different from the traditional one, where an auxiliary non-physical variable is introduced [212, 213]. In particular, the matrix elements of some related operators, which is of significance in physics, can be analytically evaluated from the ladder operators. Therefore, we are going to establish the ladder operators for pseudoharmonic oscillator.

By using the recurrence relation between the confluent hypergeometric functions and the associated Laguerre functions as well as the orthogonal relation of the associated Laguerre functions as Eq. (5.21), we may reexpress the radial wavefunction of pseudoharmonic oscillator as follows:

$$\begin{aligned}
R_n(\rho) &= N_n \rho^\tau e^{-\frac{\rho}{2}} L_n^{2\tau+D/2-1}(\rho), \\
N_n &= (M\kappa)^{D/8}\sqrt{\frac{n!}{2^{D/2-1}\Gamma(2\tau + n + D/2)}},
\end{aligned} \tag{6.51}$$

where $n = n_r$ is used.

Now, we study the ladder operators $\hat{M}_\pm$ with the following properties

$$\hat{M}_\pm = A_\pm(\rho)\frac{d}{d\rho} + B_\pm(\rho), \qquad \hat{M}_\pm R_n(\rho) = m_\pm R_{n\pm1}(\rho), \tag{6.52}$$

where we stress that these operators only depend on the physical variable ρ.

For this purpose, we begin by applying the operator $\frac{d}{d\rho}$ on the wavefunction (6.51)

$$\frac{d}{d\rho}R_n(\rho) = \left(\frac{\tau}{\rho} - \frac{1}{2}\right)R_n(\rho) + N_n\rho^\tau e^{-\frac{\rho}{2}}\frac{d}{d\rho}L_n^{2\tau+D/2-1}(\rho), \tag{6.53}$$

which is used to construct the ladder operators $\hat{M}_\pm$.

To this end, if we consider useful relations (6.35) and substitute them into (6.53), then we are able to obtain the following relations

$$\left(\frac{d}{d\rho} - \frac{n+\tau}{\rho} + \frac{1}{2}\right)R_n(\rho) = -\frac{n+2\tau+D/2-1}{\rho}\frac{N_n}{N_{n-1}}R_{n-1}(\rho), \tag{6.54}$$

$$\left(\frac{d}{d\rho} + \frac{n+\tau+D/2}{\rho} - \frac{1}{2}\right)R_n(\rho) = \frac{n+1}{\rho}\frac{N_n}{N_{n+1}}R_{n+1}(\rho), \tag{6.55}$$

from which, together with N_n given in Eq. (6.51), we can obtain the following ladder operators

$$\begin{aligned}\hat{M}_- &= -\rho\frac{d}{d\rho} + \tau + \hat{n} - \frac{\rho}{2},\\ \hat{M}_+ &= \rho\frac{d}{d\rho} + \tau + \hat{n} + \frac{D}{2} - \frac{\rho}{2},\end{aligned} \tag{6.56}$$

where we have introduced a number operator $\hat{n}$ with the property

$$\hat{n}R_n(\rho) = nR_n(\rho). \tag{6.57}$$

The ladder operators $\hat{M}_\pm$ have the following properties

$$\begin{aligned}\hat{M}_-R_n(\rho) &= m_-R_{n-1}(\rho),\\ \hat{M}_+R_n(\rho) &= m_+R_{n+1}(\rho),\end{aligned} \tag{6.58}$$

where

$$\begin{aligned}m_- &= \sqrt{n(n+2\tau+D/2-1)},\\ m_+ &= \sqrt{(n+1)(n+2\tau+D/2)}.\end{aligned} \tag{6.59}$$

On the other hand, we note that the radial wavefunction can be directly obtained by applying the creation operator $\hat{M}_+$ on the ground state $R_0(\rho)$, i.e.,

$$R_n(\rho) = \mathcal{N}_n\hat{M}_+^n R_0(\rho), \tag{6.60}$$

with

$$\mathcal{N}_n = \sqrt{\frac{\Gamma(2\tau+D/2)}{n!\Gamma(n+2\tau+D/2)}}, \qquad R_0(\rho) = \sqrt{\frac{1}{\Gamma(2\tau+D/2)}}\rho^\tau e^{-\frac{\rho}{2}}. \tag{6.61}$$

We now study a suitable Lie algebra associated with the operators $\hat{M}_+$ and $\hat{M}_-$. Based on the results (6.58) and (6.59), we can calculate the commutator $[\hat{M}_-, \hat{M}_+]$:

$$[\hat{M}_-, \hat{M}_+]R_n(\rho) = 2m_0 R_n(\rho), \tag{6.62}$$

where we have introduced the eigenvalue

$$m_0 = \left(n + \tau + \frac{D}{4}\right). \tag{6.63}$$

We can thus define the operator

$$\hat{M}_0 = \left(\hat{n} + \tau + \frac{D}{4}\right). \tag{6.64}$$

The operators $\hat{M}_\pm$ and $\hat{M}_0$ thus satisfy the commutation relations

$$[\hat{M}_-, \hat{M}_+] = 2\hat{M}_0, \qquad [\hat{M}_0, \hat{M}_\pm] = \pm\hat{M}_\pm, \tag{6.65}$$

which correspond to an su(1, 1) algebra.

As we know, there are four series of irreducible unitary representations for the SU(1, 1) group except for the identity representation [214]. Since this quantum system has ground state, the representation of the dynamic algebra su(1, 1) belongs to $D^+(j)$:

$$\begin{aligned} I_0|j, \nu\rangle &= \nu|j, \nu\rangle, \\ I_-|j, \nu\rangle &= \sqrt{(\nu + j)(\nu - j - 1)}|j, \nu - 1\rangle, \\ I_+|j, \nu - 1\rangle &= \sqrt{(\nu + j)(\nu - j - 1)}|j, \nu\rangle, \\ \nu &= -j + n, \quad n = 0, 1, 2, \ldots, j < 0. \end{aligned} \tag{6.66}$$

In comparison with Eqs. (6.58), (6.59), (6.63) and (6.64), we have $j = -(\tau + D/4)$, $\nu = n + \tau + D/4$, and $R_n(\rho) = |j, \nu\rangle$.

Finally, from the ladder operators $\hat{M}_\pm$ we may obtain two related functions

$$\rho = 2\hat{M}_0 - \hat{M}_- - \hat{M}_+, \qquad \rho\frac{d}{d\rho} = \frac{1}{2}\left(\hat{M}_+ - \hat{M}_- - \frac{D}{2}\right), \tag{6.67}$$

from which, together with Eqs. (6.58) and (6.59), we have

$$\begin{aligned} \int_0^\infty R_{n'}(r) r^2 R_n(r) dr = &\left(2n + 2\tau + \frac{D}{2}\right)\delta_{n',n} \\ &- \sqrt{n(n + 2\tau + D/2 - 1)}\delta_{n',(n-1)} \\ &- \sqrt{(n+1)(n + 2\tau + D/2)}\,\delta_{n',(n+1)}, \end{aligned} \tag{6.68}$$

and

$$\begin{aligned} \int_0^\infty R_{n'}(r)\frac{r}{2}\frac{d}{dr}R_n(r)dr = &\frac{1}{2}\sqrt{(n+1)(n + 2\tau + D/2)}\,\delta_{n',(n+1)} \\ &- \frac{1}{2}\sqrt{n(n + 2\tau + D/2 - 1)}\,\delta_{n',(n-1)} \\ &- \frac{D}{4}\delta_{n',n}. \end{aligned} \tag{6.69}$$

5.4 Recurrence Relation

To obtain the recurrence relations among the solutions $R_n(\rho)$ given in Eq. (6.51), it is convenient to employ the associated Laguerre functions satisfying the following recurrence relation

$$(n+1)L_{n+1}^{\beta}(x)+(n+\beta)L_{n-1}^{\beta}(x)+(x-2n-\beta-1)L_n^{\beta}(x)=0, \tag{6.70}$$

from which, together with the normalization constant N_n we obtain a useful three-term recurrence relation

$$\begin{aligned}&\sqrt{(n+1)(2\tau+n+D/2)}R_{n+1}(\rho)+\sqrt{n(2\tau+n+D/2-1)}R_{n-1}(\rho)\\&=(2n+2\tau+D/2-\rho)R_n(\rho).\end{aligned} \tag{6.71}$$

6 Position and Momentum Information Entropy

In this section we shall review the position and momentum information entropies of the D-dimensional harmonic oscillator [58]. An information measure closely related to the entropy is the Boltzmann-Shannon entropy defined by

$$S_\rho=-\int\rho(\mathbf{r})\ln\rho(\mathbf{r})d(\mathbf{r}) \tag{6.72}$$

in the position space and

$$S_\gamma=-\int\gamma(\mathbf{p})\ln\gamma(\mathbf{p})d(\mathbf{p}) \tag{6.73}$$

in the momentum space. The position and momentum single-particle densities are simply given by

$$\rho(\mathbf{r})=|\psi(\mathbf{r})|^2,\qquad \gamma(\mathbf{p})=|\psi(\mathbf{p})|^2, \tag{6.74}$$

where $\psi(\mathbf{p})$ is the Fourier transform of $\psi(\mathbf{r})$. The information entropies S_ρ and S_γ give a measure of the spread of the single-particle density in position and momentum space, respectively.

These two entropies have shown to play an important role in the quantum-mechanical description of physical systems. For example, they have made Bialynicki-Birula and Mycielski (BBM) find a new and stronger version of the Heisenberg uncertainty relation. The corresponding BBM inequality for single particle is given by

$$S_\rho+S_\gamma\ge D(1+\ln\pi). \tag{6.75}$$

Once the exact solutions of quantum systems are known, it is possible to study them numerically.

7 Conclusions

In this Chapter we have made use of previous results given in Chap. 3 to present the exact solutions of the harmonic oscillator in arbitrary dimensions. The recurrence relations for the radial functions have been established in terms of the Hamiltonian. In addition, we have shown how to realize the dynamic algebra su$(1, 1)$ by the factorization method. As an important generalization, we have also carried out the pseudoharmonic oscillator case and constructed the ladder operators directly from wavefunction by factorization method. The matrix elements of some related functions have been obtained analytically from the ladder operators $\hat{M}_{\pm}$. We find that this method represents a simple and elegant approach to obtain them. On the other hand, a useful recurrence relation among the wavefunction is derived on the basis of the recurrence relations on the associated Laguerre polynomials. Finally, we give a useful remark. We note that Oyewumi *et al.* have restudied pseudoharmonic potential in N-dimensions in 2008 and carried out some expectation values for $\langle r^{-2}\rangle$, $\langle r^2\rangle$, $\langle T\rangle$, $\langle V(r)\rangle$, $\langle H\rangle$ and $\langle p^2\rangle$ as well as virial theorem obtained by means of the Hellmann-Feynman theorems [102].

Chapter 7
Coulomb Potential

1 Introduction

The exact solutions of the non-relativistic and relativistic equations with a Coulomb field have been the subject both in quantum mechanics and in classical mechanics. The well-known exact solutions in almost all textbooks [1, 2] are important achievements at the beginning stage of quantum mechanics, which provided a strong evidence in favor of the quantum theory being correct.

The purpose of this Chapter is three-fold. The first is to study the analytical solutions of the D-dimensional Schrödinger equation with a Coulomb potential in arbitrary dimensions [26, 61–63, 78, 87, 99, 215], the relation between the radial equations of the D-dimensional hydrogen atom and harmonic oscillator [34]. The second is to realize the dynamic algebra su$(1, 1)$ for the radial Schrödinger Coulomb potential in terms of the Sturmian bases. The third is to study a generalized case, i.e., the Coulomb plus an inverse squared potential and then to analyze the variation of energy levels $E(n, l, D)$ on the dimension D [216]. As far as the potential energy term, we use results from scattering experiments to fix its form $1/r$. Indeed, since the results of Rutherford-type scattering experiments are independent of the spatial dimension, we can unambiguously conclude from the experimental data, that in arbitrary dimension D the potential must be of the form like $1/r$. This is of course consistent with the analysis of Refs. [217, 218] that atoms with the usual kinetic energy coupled to a modified potential of the form $1/r^{D-2}$ are not stable, where the exponent $(D-2)$ is due to the requirement that Gauss's law should be still valid in higher dimensions.

This Chapter is organized as follows. Section 2 is devoted to the study of the D-dimensional Schrödinger equation with a Coulomb potential. We establish the shift operators in Sect. 3. In Sect. 4 we illustrate the mapping between the Coulomb potential and harmonic oscillator radial functions. In Sect. 5 we show how to realize the dynamic algebra su$(1, 1)$ relying only on the radial Schrödinger equation. The generalized case, i.e., the Coulomb potential plus an inverse squared potential shall be investigated in Sect. 5. Some concluding remarks are given in Sect. 6.

S.-H. Dong, *Wave Equations in Higher Dimensions*,
DOI 10.1007/978-94-007-1917-0_7, © Springer Science+Business Media B.V. 2011

2 Exact Solutions

The D-dimensional Schrödinger equation ($\hbar = M = 1$) with a spherically symmetric potential $V(r)$ can be written as

$$-\frac{1}{2}\nabla^2 \Psi(\mathbf{x}) = [E - V(r)]\Psi(\mathbf{x}). \tag{7.1}$$

For central potentials, we may separate the angular variables from the radial one by taking the wavefunction $\Psi(\mathbf{x}) = R_l(r)Y^l_{l_{D-2}\dots l_1}(\hat{\mathbf{x}})$. Substitution of this into Eq. (7.1) allows us to obtain the D-dimensional radial Schrödinger equation with a spherically symmetric potential $V(r)$

$$\frac{1}{r^{D-1}}\frac{d}{dr}\left(r^{D-1}\frac{d}{dr}\right)R_l(r) + \left(2E - 2V(r) - \frac{l(l+D-2)}{r^2}\right)R_l(r) = 0, \tag{7.2}$$

where the potential $V(r)$ is taken as a Coulomb potential

$$V(r) = -\frac{Z\alpha}{r} = -\frac{\xi}{r}, \quad \xi = Z\alpha. \tag{7.3}$$

Upon taking a new variable

$$\rho = r\sqrt{-8E} \tag{7.4}$$

for the bound states, Eq. (7.2) is rearranged as

$$\frac{d^2R_l(\rho)}{d\rho^2} + \frac{D-1}{\rho}\frac{dR_l(\rho)}{d\rho} + \left\{-\frac{1}{4} + \frac{\tau}{\rho} - \frac{l(l+D-2)}{\rho^2}\right\}R_l(\rho) = 0, \tag{7.5}$$

where

$$\tau \equiv \xi\sqrt{\frac{1}{-2E}}. \tag{7.6}$$

From the behaviors of the radial function at the origin and at infinity, we define

$$R(\rho) = \rho^l e^{-\rho/2}F(\rho). \tag{7.7}$$

Substituting of Eq. (7.7) into Eq. (7.5), we find that $F(\rho)$ satisfies

$$\rho\frac{d^2F(\rho)}{d\rho^2} + (2l + D - 1 - \rho)\frac{dF(\rho)}{d\rho} + \left(\tau - l - \frac{D-1}{2}\right)F(\rho) = 0, \tag{7.8}$$

whose solutions are nothing but the confluent hypergeometric functions ${}_1F_1(a; b; \rho)$ with

$$a = l - \tau + \frac{D-1}{2}, \qquad b = 2l + D - 1. \tag{7.9}$$

Thus, the eigenfunctions can be expressed as

$$R(\rho) = N\rho^l e^{-\rho/2}{}_1F_1\left(l - \tau + \frac{D-1}{2}; 2l + D - 1; \rho\right), \tag{7.10}$$

where N is the normalization factor to be determined.

We now discuss the eigenvalues. From the consideration of the finiteness of the solutions at infinity, the general quantum condition is given by

$$\tau - l - \frac{D-1}{2} = n_r = 0, 1, 2, \dots. \tag{7.11}$$

Introduce a principal quantum number

$$n = n_r + l + 1. \tag{7.12}$$

As a result, we have

$$\xi\sqrt{\frac{1}{-2E}} = \tau = n + \frac{D-3}{2}. \tag{7.13}$$

From this we obtain the energy levels

$$E(n, D) = -\frac{\xi^2}{2[n + (D-3)/2]^2}. \tag{7.14}$$

Notice that the $E(n, D)$ is independent of the quantum number l. On the other hand, this result means that a stable hydrogen atom exists when the Coulomb-like potential is taken as the form Eq. (7.3) in higher dimensions.

For a large D, we have

$$E(n, D) \simeq -2\xi^2\{D^{-2} - 2(2n-3)D^{-3} + 3(2n-3)^2 D^{-4} - \cdots\}. \tag{7.15}$$

We now calculate the normalization factor N. Note that $n_r = \tau - l - (D-1)/2$ is a non-negative integer. We can express the confluent hypergeometric functions ${}_1F_1(-n_r; \beta+1; \rho)$ by the associated Laguerre polynomials $L_{n_r}^{\beta}(\rho)$. Based on formulas (5.21), we express the radial function as

$$R(\rho) = \mathcal{N}\rho^l e^{-\rho/2} L_{n-l-1}^{2l+D-2}(\rho). \tag{7.16}$$

The $\mathcal{N}$ can be obtained from the normalization condition

$$\int_0^\infty R(\rho)^2 r^{D-1} dr = 1. \tag{7.17}$$

Before proceeding to do so, we recall a generalized Coulomb-like integral $J_{n,\alpha}^{(\gamma)}$ given in Ref. [219]

$$\begin{aligned} J_{n,\alpha}^{(\gamma)} &= \int_0^\infty e^{-x} x^{\alpha+\gamma} [L_n^\alpha(x)]^2 dx \\ &= \frac{\Gamma(\alpha+n+1)}{n!} \sum_{k=0}^{n} \frac{(-1)^k \Gamma(n-k-\gamma)\Gamma(\alpha+k+\gamma+1)}{\Gamma(-k-\gamma)\Gamma(\alpha+k+1)k!(n-k)!}, \end{aligned} \tag{7.18}$$

with $\mathrm{Re}(\alpha+\gamma+1) > 0$. With the help of this integral formula, we are able to obtain

$$J_{n,\alpha}^{(1)} = \frac{(2n+\alpha+1)\Gamma(\alpha+n+1)}{n!}, \tag{7.19}$$

where two nonzero contributions ($k = n-1, n$) are made to Eq. (7.18) due to the Γ functions of the negative integers. It is shown from Eqs. (7.17)–(7.19) that the normalization factor is calculated as

$$\mathcal{N} = \left(\frac{4\xi}{2n+D-3}\right)^{\frac{D}{2}} \left[\frac{(n-l-1)!}{(2n+D-3)\Gamma(n+l+D-2)}\right]^{1/2}. \tag{7.20}$$

Before ending this section, we want to address the lower-dimensional case briefly due to recent interest in the lower-dimensional field theory. When $D = 1$, the eigenvalues (7.14) reduce to the one-dimensional case

$$E_n = -\frac{\xi^2}{2n^2}, \quad n = 1, 2, 3, \ldots, \tag{7.21}$$

where we have used shifted n in order to avoid the infinitely bound ground state. This is because the energy level E_0 goes to negative infinity if a principal quantum number $n = 0$. That is to say, the Coulomb potential $-\xi/|x|$ behaves like a negative $\delta(x)$ potential well.

The corresponding radial function is calculated as

$$R = C\rho e^{-\rho/2} L^1_{n-1}(\rho), \quad \rho = \frac{2\xi}{n}|x|, \tag{7.22}$$

where $C = \sqrt{\xi/n}$ is a normalization constant.

Similarly, when $D = 2$ the eigenvalues are given by

$$E_n = -\frac{\xi^2}{2(n-1/2)^2}, \quad n = 1, 2, 3, \ldots. \tag{7.23}$$

The corresponding eigenfunctions are obtained as

$$R(\rho) = \left(\frac{4\xi}{2n-1}\right)\sqrt{\frac{(n-|m|-1)!}{(2n-1)(n+|m|-1)!}}\,\rho^{|m|} e^{-\frac{\rho}{2}} L^{2|m|}_{n-|m|-1}(\rho), \tag{7.24}$$

where $\rho = 2\xi r/(n-1/2)$ and the angular momentum quantum number l is replaced by the traditional notation $|m|$.

3 Shift Operators

We now study the "ladder operators" for the wavefunction by the factorization method [3]. Generally speaking, we may apply the recurrence relations of the generalized Laguerre polynomials to obtain what appear to be the creation and annihilation operators for the radial wavefunction $R_{n,l}(\rho)$, but what makes us discouraged is we find that the variable ρ depends on n as shown in Eq. (7.4). In this case, we have to apply the creation operator acting on $R_{n,l}(\rho)$ to obtain

$$\hat{M}^+_n R_{n,l}(\rho_n) = \hat{C}_{n,l} R_{n+1,l}(\rho_n) \neq C_{n,l} R_{n+1,l}(\rho_{(n+1)}). \tag{7.25}$$

This means that we must apply a shift operator to change $\rho(n)$ to $\rho(n+1)$. Such a deep problem has been discussed by Aebersold *et al.* [220].

Consequently, we are going to establish the l raising and lowering operators since the variable ρ does not depend on the quantum number l. That is, we attempt to look for the shift operators which generate a group of potentials but with the same energy. This is the so-called potential group approach.

For this purpose, we begin by acting the differential operator $d/d\rho$ on the radial wavefunction $R_{n,l}(\rho)$:

$$\frac{d}{d\rho}R_{n,l}(\rho)=\left(\frac{l}{\rho}-\frac{1}{2}\right)R_{n,l}(\rho)+\mathcal{N}_{n,l}\rho^l e^{-\rho/2}\frac{d}{d\rho}L_{n-l-1}^{2l+D-2}(\rho). \tag{7.26}$$

It should be aware that the n in $L_n^\alpha(y)$ corresponds to n_r.

Before proceeding, let us first recall an important relation for the derivative of the associated Laguerre functions [192]

$$\frac{d}{dy}L_n^\alpha(y)=-\frac{1}{(\alpha+1)}[yL_{n-1}^{\alpha+2}(y)+nL_n^\alpha(y)]. \tag{7.27}$$

Substituting this into (7.26) allows us to obtain the following relation

$$\left\{\frac{d}{d\rho}-\frac{l}{\rho}+\frac{1}{2}\left(\frac{2n+D-3}{2l+D-1}\right)\right\}R_{n,l}(\rho)=-\frac{1}{2l+D-1}\frac{\mathcal{N}_{n,l}}{\mathcal{N}_{n,(l+1)}}R_{n,(l+1)}. \tag{7.28}$$

As a result, we may define the raising operator as follows:

$$\hat{M}_l^+=-\frac{d}{d\rho}+\frac{l}{\rho}-\frac{1}{2}\left(\frac{2n+D-3}{2l+D-1}\right), \tag{7.29}$$

with the following property

$$\hat{M}_l^+R_{n,l}(\rho)=m_l^+R_{n,(l+1)}(\rho), \tag{7.30}$$

with

$$m_l^+=\frac{1}{2l+D-1}\sqrt{(n-l-1)(n+l+D-2)}. \tag{7.31}$$

We now proceed to find the corresponding lowering operator. First, we should keep in mind that we need obtain a relation between $\frac{d}{d\rho}L_n^\alpha(\rho)$ and $L_{n+1}^{\alpha-2}(\rho)$ since this implies a relation between $\frac{d}{d\rho}R_{n,l}(\rho)$ and the radial wavefunction $R_{n,(l-1)}(\rho)$. To achieve this task we start with the relation [192]

$$y\frac{d}{dy}L_n^\alpha(y)=nL_n^\alpha(y)-(n+\alpha)L_{n-1}^\alpha(y), \tag{7.32}$$

which, when taking into account the relation [192]

$$(n+1)L_{n+1}^\alpha(y)-(2n+\alpha+1-y)L_n^\alpha(y)+(n+\alpha)L_{n-1}^\alpha(y)=0, \tag{7.33}$$

can be transformed into

$$y\frac{d}{dy}L_n^\alpha(y)=(-n-\alpha-1+y)L_n^\alpha(y)+(n+1)L_{n+1}^\alpha(y). \tag{7.34}$$

On the other hand, the relation

$$L_n^{\alpha-1}(y) = L_n^{\alpha}(y) - L_{n-1}^{\alpha}(y), \tag{7.35}$$

together with Eq. (7.33), allows us to set up the following relation[1]

$$\frac{(\alpha-1)}{(n+\alpha)} L_{n+1}^{\alpha}(y) = \frac{(\alpha+y-1)}{(\alpha+n)} L_n^{\alpha}(y) + L_{n+1}^{\alpha-2}(y). \tag{7.41}$$

This relation in turn can be substituted into Eq. (7.34) to give

$$\begin{aligned}(\alpha-1)\frac{d}{dy}L_n^{\alpha}(y) = &\left[(\alpha+n) - \frac{\alpha(\alpha-1)}{y}\right] L_n^{\alpha}(y) \\ &+ \frac{(n+1)(n+\alpha)}{y} L_{n+1}^{\alpha-2}(y).\end{aligned} \tag{7.42}$$

Finally, when this equation is substituted into (7.26), this allows us to obtain the following relation

$$\begin{aligned}&\left[\frac{d}{d\rho} + \frac{l+D-2}{\rho} - \frac{1}{2}\left(\frac{2n+D-3}{2l+D-3}\right)\right] R_{n,l}(\rho) \\ &= \frac{(n-l)(n+l+D-3)}{2l+D-3} \frac{\mathcal{N}_{n,l}}{\mathcal{N}_{n,(l-1)}} R_{n,(l-1)}(\rho),\end{aligned} \tag{7.43}$$

[1] To show this formula explicitly, we want to derive it in more detail. It is shown from Eqs. (7.33) and (7.35) that

$$\begin{aligned}&(n+1)L_{n+1}^{\alpha}(y) - 2(n+\alpha)L_n^{\alpha}(y) + (\alpha+y-1)L_n^{\alpha}(y) \\ &+ (n+\alpha)[L_n^{\alpha}(y) - L_n^{\alpha-1}(y)] = 0,\end{aligned} \tag{7.36}$$

which can be further modified to

$$\begin{aligned}&(n+1)L_{n+1}^{\alpha}(y) - (n+\alpha)L_n^{\alpha}(y) \\ &- (n+\alpha)L_n^{\alpha-1}(y) + (\alpha+y-1)L_n^{\alpha}(y) = 0,\end{aligned} \tag{7.37}$$

from which, together with Eq. (7.35) again, we have

$$\begin{aligned}&(n+1)L_{n+1}^{\alpha}(y) - (n+\alpha)[L_{n+1}^{\alpha}(y) - L_{n+1}^{\alpha-1}(y)] \\ &- (n+\alpha)L_n^{\alpha-1}(y) + (\alpha+y-1)L_n^{\alpha}(y) = 0.\end{aligned} \tag{7.38}$$

Moreover, this equation can be rewritten as

$$\begin{aligned}&(\alpha-1)L_{n+1}^{\alpha}(y) + (n+\alpha)[L_n^{\alpha-1}(y) - L_{n+1}^{\alpha-1}(y)] \\ &- (\alpha+y-1)L_n^{\alpha}(y) = 0.\end{aligned} \tag{7.39}$$

Using Eq. (7.35) once again, we have

$$(\alpha-1)L_{n+1}^{\alpha}(y) - (n+\alpha)L_{n+1}^{\alpha-2}(y) - (\alpha+y-1)L_n^{\alpha}(y) = 0, \tag{7.40}$$

which is nothing but Eq. (7.41).

from which we are able to define the lowering operator as follows:

$$\hat{M}_l^- = \frac{d}{d\rho} + \frac{l+D-2}{\rho} - \frac{1}{2}\left(\frac{2n+D-3}{2l+D-3}\right), \tag{7.44}$$

with the property

$$\hat{M}_l^- R_{nl}(\rho) = m_l^- R_{n(l-1)}(\rho), \tag{7.45}$$

where

$$m_l^- = \frac{1}{(2l+D-3)}\sqrt{(n-l)(n+l+D-3)}. \tag{7.46}$$

Essentially, these results (7.29) and (7.44) coincide with those given in [87].

4 Mapping Between the Coulomb and Harmonic Oscillator Radial Functions

In this section we are going to establish a mapping between the Coulomb and harmonic oscillator radial functions in arbitrary dimensions [34]. As shown above, the Coulomb radial functions satisfy the following differential equation

$$\frac{d^2R_l(y)}{dy^2} + \frac{D-1}{y}\frac{dR_l(y)}{dy} + \left\{-\frac{1}{4} + \frac{\tau}{y} - \frac{l(l+D-2)}{y^2}\right\}R_l(y) = 0, \tag{7.47}$$

where $y = \rho = 2r\xi/\tau$ and $\tau = n + (D-3)/2$ as defined in Eq. (7.13). The exact solutions are given by

$$\begin{aligned}
&R(y) = \mathcal{N} y^l e^{-y/2} L_{n-l-1}^{2l+D-2}(y),\\
&\mathcal{N} = \frac{(4\xi)^{D/2}}{(2n+D-3)^{(D+1)/2}}\left[\frac{(n-l-1)!}{\Gamma(n+l+D-2)}\right]^{1/2},\\
&E_n = -\frac{\xi^2}{2\tau^2} = -\frac{\xi^2}{2[n+(D-3)/2]^2},\\
&n = n_r + l + 1.
\end{aligned} \tag{7.48}$$

On the other hand, the radial equation of the harmonic oscillator in arbitrary dimensions can be written out

$$\frac{d^2R_L(Y)}{dY^2} + \frac{d-1}{Y}\frac{dR_L(Y)}{dY} + \left\{-Y^2 - \frac{L(L+d-2)}{Y^2} + \epsilon\right\}R_L(Y) = 0, \tag{7.49}$$

where $Y = R$, $\epsilon = 2N + d$ and $N \geq L$. The corresponding solutions are given by

$$\begin{aligned}
&R_L(Y) = N e^{-Y^2/2} Y^L L_{(N-L)/2}^{L-1+d/2}(Y^2),\\
&N = \sqrt{\frac{2\Gamma[(N-L)/2+1]}{\Gamma[(N+L+D)/2]}},\\
&E_N = \frac{\epsilon}{2} = N + \frac{d}{2}.
\end{aligned} \tag{7.50}$$

The map taking Eq. (7.48) into Eq. (7.50) is $y = Y^2$. In particular, for integers D, d, N, n, L and l, we find that the solutions (7.48) for $R(y, D, n, l)$ can be related to the solutions (7.50) $R(Y, d, N, L)$ by

$$R(y, D, n, l) = c_0 R(Y, 2D-2, 2n-2, 2l), \tag{7.51}$$

where

$$c_0 = \frac{1}{2}\sqrt{\frac{(2\xi)^d}{\tau^{d+1}}}. \tag{7.52}$$

The identity (7.51) establishes the following relations

$$d = 2D-2, \qquad N = 2n-2, \qquad L = 2l. \tag{7.53}$$

Therefore, we observe that the Coulomb problem in three dimensions is in one-to-one correspondence with half the states of the four-dimensional harmonic oscillator for even values of the quantum numbers N and L.

In affect, there exists a further degree of freedom in the map $y = Y^2$, i.e.,

$$R(y, D, n, l) = c_0 R(Y, 2D-2-2\kappa, 2n-2+\kappa, 2l+\kappa), \quad \kappa \in \mathbf{Z}, \tag{7.54}$$

from which we have the following identity

$$d = 2D-2-2\kappa, \qquad N = 2n-2+\kappa, \qquad L = 2l+\kappa. \tag{7.55}$$

This is a general feature of this map that the spectrum of the D-dimensional Coulomb problem is related to half the spectrum of the d-dimensional harmonic oscillator for any even integer d.

5 Realization of Dynamic Group SU(1, 1)

In this section we are ready to realize the dynamic group SU(1, 1) for the radial Schrödinger equation with the Coulomb potential [194]. For simplicity, we write out the generalized D-dimensional Sturm basis

$$S^D_{n_r l}(r) = 2^{\frac{D-1}{2}}\sqrt{\frac{n_r!}{\Gamma(n_r+2l+D-1)}}(2r)^l e^{-r} L^{2l+D-2}_{n_r}(2r). \tag{7.56}$$

Making use of the relations for the Laguerre functions (6.35) and formula (C.29), we may obtain the raising and lowering operators for the quantum number n_r

$$M_\pm = \pm r\left(r\frac{d}{dr} - r + M_0 \pm \frac{D-1}{2}\right), \tag{7.57}$$

where M_0 is given by

$$M_0 = -\frac{1}{2}r\left\{\frac{d^2}{dr^2} + \frac{D-1}{r}\frac{d}{dr} - \frac{l(l+D-2)}{r^2} - 1\right\}. \tag{7.58}$$

It is found that $M_\pm$ and M_0 satisfies the su(1, 1) algebra commutation relations

$$[M_0, M_\pm] = \pm M_\pm, \qquad [M_+, M_-] = -2M_0, \tag{7.59}$$

with the following properties

$$\begin{aligned} M_+ S^D_{n_r l}(r) &= \sqrt{(n_r+1)(n_r+2l+D-1)}\, S^D_{(n_r+1)l}(r), \\ M_- S^D_{n_r l}(r) &= \sqrt{n_r(n_r+2l+D-2)}\, S^D_{(n_r-1)l}(r), \\ M_0 S^D_{n_r l}(r) &= [n_r + l + (D-1)/2] S^D_{n_r l}(r). \end{aligned} \tag{7.60}$$

Based on $M_\pm = M_1 \pm i M_2$, we have

$$M_1 = M_0 - r, \qquad M_2 = -i\left(r\frac{d}{dr} + \frac{D-1}{2}\right). \tag{7.61}$$

The Casimir operator for this group is calculated by

$$C = M_0^2 - \frac{1}{2}(M_+M_- + M_-M_+), \tag{7.62}$$

with eigenvalue $k(k-1)$. After calculating we find the Casimir eigenvalue is

$$CS^D_{n_r l}(r) = \left(\frac{D^2}{4} + D(l-1) - 2l + l^2 + \frac{3}{4}\right) S^D_{n_r l}(r), \tag{7.63}$$

from which we get two possible solutions $k = (2l + D - 1)/2$ or $k = (3 - D - 2L)/2$. However, we are interested only in the positive discrete representations of su(1, 1), $D^+(k)$. Thus, we take $k = (2l + D - 1)/2$.

To establish the connection between the Sturmian functions and the radial Coulomb functions in arbitrary dimensions D, let us write the D-dimensional radial Schrödinger equation for the hydrogen atom as

$$HR^D_{n_r l} = ER^D_{n_r l}, \tag{7.64}$$

where

$$H = -\frac{1}{2}\left\{\frac{d^2}{dr^2} + \frac{D-1}{r}\frac{d}{dr} - \frac{l(l+D-2)}{r^2}\right\} - \frac{\xi}{r}. \tag{7.65}$$

If we multiply Eq. (7.64) by r, then we may write the above equation as

$$(\bar{H} - \xi)R^D_{n_r l} = 0, \tag{7.66}$$

where $\bar{H}$ is named the modified Hamiltonian

$$\bar{H} = \frac{1}{2}(M_0 + M_1) - E(M_0 - M_1). \tag{7.67}$$

In terms of the tilting transformation [137]

$$R^D_{n_r l} = Ce^{i\theta M_2} S^D_{n_r l}, \tag{7.68}$$

where the normalization constant C can be calculated by Eq. (7.17). In the calculation, we consider the relation $r = M_0 - M_1$ given in Eq. (7.61).

Multiplying Eq. (7.66) from the left by $e^{-i\theta M_2}$ yields

$$(\mathcal{H}-\xi)S_{n_r l}^D = 0, \tag{7.69}$$

where

$$\begin{aligned}\mathcal{H} &= e^{-i\theta M_2}\bar{H}e^{i\theta M_2}\\ &= \frac{1}{2}e^{\theta}(M_0+M_1) - Ee^{-\theta}(M_0-M_1)\\ &= [(1/2+E)\cosh\theta + (1/2-E)\sinh\theta]M_1\\ &\quad + [(1/2+E)\sinh\theta + (1/2-E)\cosh\theta]M_0.\end{aligned} \tag{7.70}$$

The final expression was obtained by performing a similarity transformation on both the compact generator M_0 and the noncompact generator M_1. During the calculation, we have used the Hausdroff-Campbell-Baker formula

$$e^{-i\alpha M_2}(M_0 \pm M_1)e^{i\alpha M_2} = e^{\pm\alpha}(M_0 \pm M_1). \tag{7.71}$$

The choice $\theta = \frac{1}{2}\ln(-2E)$ makes the coefficient of M_1 vanish. (Note that the expression for θ is actually n-dependent due to its definition.) Thus, we have

$$\mathcal{H} = \sqrt{-2E}M_0. \tag{7.72}$$

How to calculate the normalization constant C and the radial function? For this purpose, we require the following condition due to different metrics for two basses

$$1 = \int_0^\infty |R_{n_r l}^D|^2 r^{D-1}dr = \langle S_{n_r l}^D r|S_{n_r l}^D\rangle = Ce^{-\theta}\langle R_{n_r l}^D|(M_0-M_1)|S_{n_r l}^D\rangle, \tag{7.73}$$

from which we have $C = \sqrt{\xi/\kappa^2}$, $\kappa = n + (D-3)/2$.

Now, let us obtain the radial Coulomb functions. In terms of Eq. (7.68) we are able to write its expression as follows:

$$\begin{aligned}R_{n_r l}^D(r) &= Ce^{\theta(r\frac{d}{dr}+\frac{D-1}{2})}S_{n_r l}^D(r)\\ &= \sqrt{\frac{\xi^D}{\kappa^{D+1}}}S_{n_r l}^D(\rho)\\ &= \sqrt{\frac{(2\xi)^D}{2\kappa^{D+1}}}\sqrt{\frac{(n-l-1)!}{(n+l+D-3)!}}\rho^l e^{-\rho/2}L_{n-l-1}^{2l+D-2}(\rho),\end{aligned} \tag{7.74}$$

where we have used the following relations

$$n = n_r + l + 1, \qquad \rho = \frac{2\xi r}{\kappa}, \qquad e^{\ln\theta r\frac{d}{dr}}g(r) = g(\theta r). \tag{7.75}$$

6 Generalization to Kratzer Potential

In this section let us study a generalized Coulomb potential named Kratzer potential. It is the sum of a Coulomb and an inverse squared potential as discussed in three dimensions by Landau and Lifshitz [2].

As mentioned above, the radial equation of the D-dimensional Schrödinger equation with a spherically symmetric potential $V(r)$ can be written as

$$\frac{d^2}{dr^2}R_l(r)+\frac{D-1}{r}\frac{d}{dr}R_l(r)+\left\{2E-2V(r)-\frac{l(l+D-2)}{r^2}\right\}R_l(r)=0, \quad (7.76)$$

where the $V(r)$ is taken as a Kratzer type potential

$$V(r)=\frac{A}{r^2}-\frac{B}{r}. \quad (7.77)$$

Upon taking a new variable $\rho=r\sqrt{-8E}$ for the bound states, Eq. (7.76) can be rearranged as

$$\frac{d^2R_l(\rho)}{d\rho^2}+\frac{D-1}{\rho^2}\frac{dR_l(\rho)}{d\rho}+\left\{-\frac{1}{4}+\frac{\tau}{\rho}-\frac{2A+l(l+D-2)}{\rho^2}\right\}R_l(\rho)=0, \quad (7.78)$$

where

$$\tau\equiv B\sqrt{\frac{1}{-2E}}. \quad (7.79)$$

From the behaviors of the radial function at the origin and at infinity, we define

$$R(\rho)=\rho^{\lambda}e^{-\rho/2}F(\rho), \quad (7.80)$$

where

$$\lambda=\frac{2-D+\sqrt{8A+\kappa^2}}{2}, \qquad \kappa\equiv|2l+D-2|. \quad (7.81)$$

The constraint condition is taken as $8A+\kappa^2\geq 0$. Substituting of Eq. (7.80) into Eq. (7.78), we find that $F(\rho)$ satisfies

$$\rho\frac{d^2F(\rho)}{d\rho^2}+(2\lambda+D-1-\rho)\frac{dF(\rho)}{d\rho}+\left(\tau-\lambda-\frac{D-1}{2}\right)F(\rho)=0, \quad (7.82)$$

whose solutions are nothing but the ${}_1F_1[\lambda-\tau+(D-1)/2;2\lambda+D-1;\rho]$.

Thus, the eigenfunctions can be expressed as

$$R(\rho)=N\rho^{\lambda}e^{-\rho/2}{}_1F_1\left(\lambda-\tau+\frac{D-1}{2};2\lambda+D-1;\rho\right), \quad (7.83)$$

where N is the normalization factor to be determined.

We now discuss the eigenvalues. From consideration of the finiteness of the solutions at infinity, the general quantum condition is obtained from Eq. (7.83)

$$\tau-\lambda-\frac{D-1}{2}=n'=0,1,2,\ldots. \quad (7.84)$$

Introducing a principal quantum number

$$n=n'+\kappa/2-D/2+2=n'+l+1, \quad (7.85)$$

from which, together with Eqs. (7.79), (7.84) and (7.85), we have

$$B\sqrt{\frac{1}{-2E}}=\tau=n-l-1+\lambda+\frac{D-1}{2}>0, \quad (7.86)$$

which leads to

$$E(n,l,D) = -\frac{2B^2}{(2n-2l-1+\sqrt{8A+\kappa^2})^2}. \tag{7.87}$$

For a large D, we have

$$E \simeq -2B^2\{D^{-2} - 2(2n-3)D^{-3} + [3(2n-3)^2 - 8A - (2l-2)^2]D^{-4} - \cdots\}, \tag{7.88}$$

which implies that the energy E is almost independent of the quantum number l for a large D, but the quantum number l devotes a small contribution to the energy $E(n,l,D)$ for a small D.

For a small A, we have

$$\begin{aligned} E \simeq -2B^2\bigg\{&(2n+D-3)^{-2} - \frac{8A}{\kappa}(2n+D-3)^{-3} \\ &+ \frac{16A^2}{\kappa^3}(2n+D-3)^{-3} + \frac{48A^2}{\kappa^2}(2n+D-3)^{-4} - \cdots\bigg\}. \end{aligned} \tag{7.89}$$

We now calculate the normalization factor. Note $n' = \tau - \lambda - (D-1)/2$ is a non-negative integer. In terms of the formula (5.21), we can express the radial function as

$$R(\rho) = \mathcal{N}\rho^{\lambda} e^{-\rho/2} L_{n-l-1}^{2\lambda+D-2}(\rho), \tag{7.90}$$

where the $\mathcal{N}$ can be obtained from the normalization condition

$$\int_0^{\infty} R(\rho)^2 r^{D-1} dr = 1 \tag{7.91}$$

as follows:

$$\begin{aligned} \mathcal{N} &= \left(\frac{4B}{2n-2l+2\lambda+D-3}\right)^{\frac{D}{2}} \\ &\times \left[\frac{(n-l-1)!}{(2n-2l+2\lambda+D-3)\Gamma(n-l+2\lambda+D-2)}\right]^{1/2}. \end{aligned} \tag{7.92}$$

In the calculation, the formulas (7.18) and (7.19) are used.

We now briefly address the lower-dimensional case. When $D = 1$, the eigenvalues (7.87) reduce to

$$E_n = -\frac{2B^2}{(2n-1+\sqrt{1+8A})^2}. \tag{7.93}$$

The corresponding radial function becomes

$$R(\rho) = \frac{\sqrt{2B}}{2n+2s_1-2}\left[\frac{(n-1)!}{\Gamma(n+2s_1-1)}\right]^{1/2} \rho^{s_1} e^{-\rho/2} L_{n-1}^{2s_1-1}(\rho), \tag{7.94}$$

with

$$s_1 = \frac{1+\sqrt{1+8A}}{2}. \tag{7.95}$$

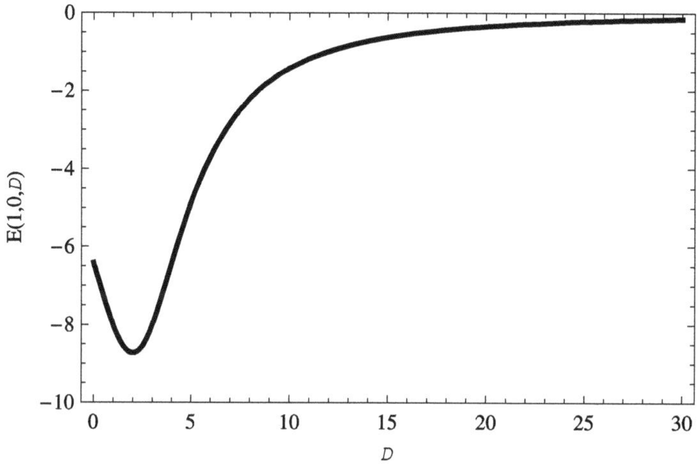

Fig. 7.1 The energy spectrum $E(1,0,D)$ decreases with the increasing dimension $D \in (0,2]$, but increases with the dimension $D \geq 2$. This is a common property of the energy levels $E(n,0,D)$ regardless of the quantum number n. The parameters $A=1$ and $B=8$ are taken

Similarly, when $D=2$ the eigenvalues are given by

$$E_{n,|m|} = -\frac{2B^2}{(2n-2|m|-1+2\sqrt{2A+m^2})^2}, \tag{7.96}$$

where the angular momentum quantum number l is replaced by the traditional notation $|m|$. The corresponding eigenfunctions are obtained as

$$R(\rho) = \frac{4B}{2(n-|m|+s_2)-1}\sqrt{\frac{(n-|m|-1)!}{[2(n-|m|+s_2)-1]\Gamma(n-|m|+2s_2)}} \times \rho^{s_2} e^{-\rho/2} L_{n-|m|-1}^{2s_2}(\rho), \tag{7.97}$$

with

$$s_2 = \sqrt{2A+m^2}. \tag{7.98}$$

Finally, let us elucidate the properties of energy $E(n,l,D)$ as shown in Figs. 7.1, 7.2, 7.3. We take $A=1$ and $B=8$ for definiteness. It is shown in Fig. 7.1 that the energy $E(1,0,D)$ decreases with the dimension D for $D \in (0,2]$, but increases with it for $D \geq 2$, so do the energy levels $E(n,0,D)$. This is a common variation for energy levels $E(n,0,D)$ regardless of the principal quantum number n. This kind of property can be explained well by the first derivative of the energy with respect to dimension D

$$\frac{\partial E(n,l,D)}{\partial D} = \frac{4B^2\kappa}{\sqrt{8A+\kappa^2}(2n-2l-1+\sqrt{8A+\kappa^2})^3}. \tag{7.99}$$

This implies that $\partial E(n,l,D)/\partial D = 0$ due to $\kappa \equiv 2l+D-2=0$ when $l=0$ and $D=2$. That is to say, there exists a turning point at $D=2$ for $l=0$. It is

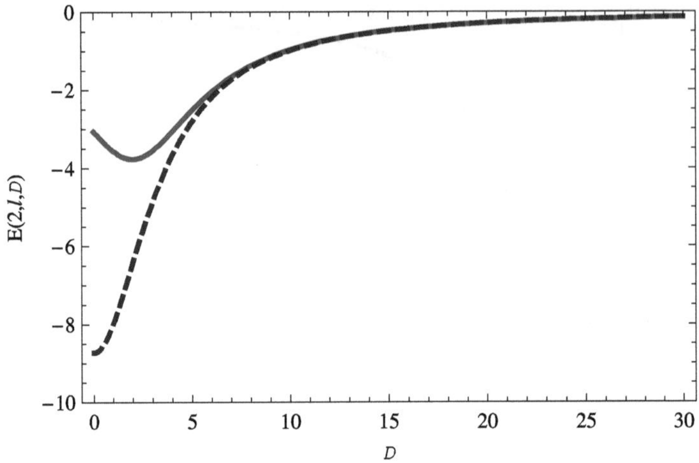

Fig. 7.2 The variation of energy $E(2, 0, D)$ (*red solid line*) on the dimension D is very similar to $E(1, 0, D)$. The $E(2, 1, D)$ (*blue dashed line*) increases with the increasing dimension D. Specially, note that the $E(2, 1, D)$ almost overlaps $E(2, 0, D)$ for a large D

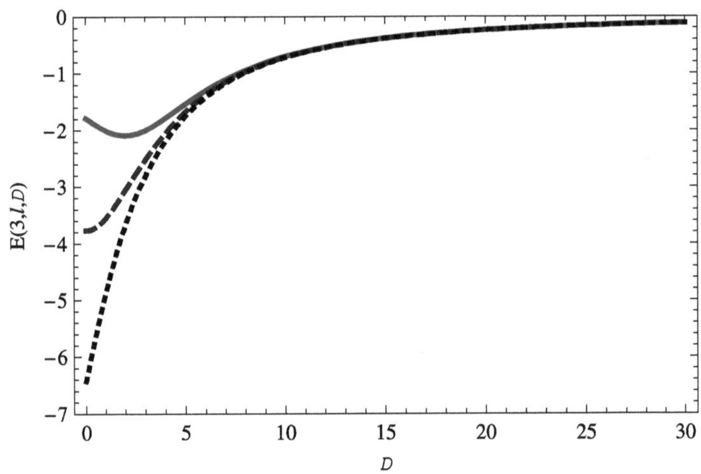

Fig. 7.3 The variations of energy $E(3, l, D)$ on the dimension D are very similar to $E(2, l, D)$. The *red*, *blue dashed* and *black dotted* lines correspond to the different angular momentum quantum numbers $l = 0, 1, 2$, respectively

found from Eq. (7.99) that $\partial E(n, l, D)/\partial D \leq 0$ for $D \in (0, 2)$. On the contrary, $\partial E(n, 0, D)/\partial D \geq 0$ for $D > 2$. This has been shown clearly in Figs. 7.1, 7.2, 7.3 for $E(n, 0, D)$. When $l \neq 0$, the κ is always positive regardless of dimension D. Thus, the $\partial E(n, l, D)/\partial D$ is positive, too. This means that the energy $E(n, l, D)$ ($l \neq 0$) monotonically increases with the increasing dimension D as shown in Figs. 7.2 and 7.3.

Second, we study the variations of energy $E(n,l,D)$ on the quantum number l. It is shown in Figs. 7.2 and 7.3 that the energy $E(n,l,D)$ is inversely proportional to quantum number l. For a large D, however, the energy $E(n,l,D)$ is almost independent of the quantum number l as shown in Figs. 7.2 and 7.3. This can be explained well by series expansion for $1/D$ as given in Eq. (7.15). When D is very large, the third term involving D^{-4} can be ignored. Thus, the energy $E(n,l,D)$ is independent of quantum number l. However, when D is not too large, note that the quantum number l will make a small contribution to energy $E(n,l,D)$ as shown in Figs. 7.2 and 7.3. In fact, the relation between $E(n,l,D)$ and D can be explained well by the first derivative of energy $E(n,l,D)$ with respect to quantum number l

$$\frac{\partial E(n,l,D)}{\partial l} = \frac{8B^2(\kappa - \sqrt{8A+\kappa^2})}{\sqrt{8A+\kappa^2}(2n-2l-1+\sqrt{8A+\kappa^2})^3}. \tag{7.100}$$

This implies that $\partial E(n,l,D)/\partial l < 0$ regardless of the quantum number l and dimension D. If $A=0$, it is shown from Eq. (7.100) that $\partial E(n,l,D)/\partial l = 0$, which means that the energy $E(n,l,D)$ is independent of quantum number l.

Third, we investigate the variations of energy levels $E(n,l,D)$ on the principal quantum number n. We observe that the energy $E(n,l,D)$ monotonically increases with the increasing n. This can be explained by the first derivative of the energy $E(n,l,D)$ with respect to the principal quantum number n

$$\frac{\partial E(n,l,D)}{\partial n} = \frac{8B^2}{(2n-2l-1+\sqrt{8A+\kappa^2})^3}. \tag{7.101}$$

It should be pointed out that the energy levels $E(n,l,D)$ will gradually tend to zero for a large dimension D, so do the energy levels $E(n,l,D)$ for a large n.

Before ending this part, we outline a generalized Coulomb potential in arbitrary dimensions [38]. In that work, they proposed a unified approach to treat the Coulomb and harmonic oscillator potentials in D dimensions. The generalized Coulomb potential is defined as

$$\begin{aligned} v(r) = &-\frac{1}{r^2}\left(l+\frac{D-3}{2}\right)\left(l+\frac{D-1}{2}\right) \\ &+\left(\beta-\frac{1}{2}\right)\left(\beta-\frac{3}{2}\right)\frac{\gamma}{4h(r)[h(r)+\alpha]} \\ &-\frac{a}{h(r)+\alpha}-\frac{3\gamma}{16[h(r)+\alpha]^2}+\frac{5\alpha\gamma}{16[h(r)+\alpha]^3}, \end{aligned} \tag{7.102}$$

where $h(r)$ is defined in terms of its inverse function

$$r = r(h) = \gamma^{-1/2}\left[\alpha \tanh^{-1}\left(\sqrt{\frac{h}{h+\alpha}}\right)+\sqrt{h(h+\alpha)}\right]. \tag{7.103}$$

The exact solutions of this system are given by

$$\begin{aligned} E_n &= -\frac{\hbar^2\gamma}{8\mu}\rho_n^2, \\ \psi_n &= \gamma^{\frac{1}{4}}\rho_n^{\frac{\beta+1}{2}}\sqrt{\frac{\Gamma(n+1)}{(2n+\beta+\rho_n\alpha)\Gamma(n+\beta)}} \\ &\quad \times [h(r)+\alpha]^{\frac{1}{4}}[h(r)]^{\frac{2\beta-1}{4}}e^{-\frac{\rho_n h(r)}{2}}L_n^{(\beta-1)}(\rho_n h(r)), \end{aligned} \tag{7.104}$$

where

$$\rho_n = \frac{2}{\alpha}\left[\sqrt{(n+\beta/2)^2+\frac{a\alpha}{\gamma}}-(n+\beta/2)\right]. \tag{7.105}$$

Here, we give a useful remark on Eq. (7.102). It is clear to see that this potential carries angular momentum dependence: its first term merely compensates the centrifugal term arising from the kinetic term of the Hamiltonian. Its second term also has r^{-2}-like singularity, and thus it cancels the angular momentum dependent term in two important limiting cases that recover the D-dimensional Coulomb and the harmonic oscillator potentials. The third term represents an asymptotically Coulomb-like interaction, while the remaining two terms behave like r^{-2} and r^{-3} for large values of the variable r. In addition, it should be noted that half century ago Tangherlini [215] proposed a generalized Keplerian problem when he studied the hydrogen atom in n dimensions, i.e.,

$$V(n,r) = -\frac{e^2}{(n-2)r^{n-2}}, \tag{7.106}$$

which implies that it cannot be acceptable for $n=2$. This is the reason why the Coulomb potential is taken as the logarithmic form $\ln r$ in two dimensions.

7 Concluding Remarks

In this Chapter we have studied the analytical solutions of the D-dimensional Schrödinger equation with a Coulomb potential in arbitrary dimensions. The shift operators for the Coulomb potential have been established by factorization method. These operators constructed a set of potentials with the same energy level, i.e., keep the energy level fixed but change the potential shape. Additionally, the mapping between the radial equations of the D-dimensional hydrogen atom and harmonic oscillator has been shown. In addition to these, we have realized the dynamical group SU(1, 1) using the Sturm basis. Finally, we have investigated the variations of energy $E(n,l,D)$ on the dimension D for the Kratzer potential and reviewed the generalized Coulomb potential.

Chapter 8
Wavefunction Ansatz Method

1 Introduction

It is well known that the solutions of the fundamental dynamic equations play a very important role in physics. The exact solutions of wave equations are possible only for a few potentials. We have to use some approximation methods to obtain their solutions. Until now, many efforts have been made to solve the stationary Schrödinger equation with the anharmonic potentials containing negative powers of the radial coordinate [221–243]. Interest in them stems from the fact that the study of the Schrödinger equation with these potentials provides us for insight into the physical problem. However, most of those works have been mainly carried out in the lower-dimensional space. The purpose of this Chapter is, by applying a suitable ansatz to the wavefunction, to analyze the D-dimensional radial Schrödinger equation with anharmonic potentials such as the sextic potential $V(r) = ar^6 + br^4 + cr^2$, the singular integer power potentials $V(r) = ar^2 + br^{-2} + cr^{-4} + dr^{-6}$, the singular fraction power potentials $V(r) = ar^{-1/2} + br^{-3/2}$ and others.

This Chapter is organized as follows. The Schrödinger equation with these anharmonic potentials will be treated by the wavefunction ansatz method in Sects. 2–4. The concluding remarks will be given in Sect. 5.

2 Sextic Potential

For simplicity the atomic units $\hbar = \mu = 1$ are employed if not explicitly stated otherwise. Since the Schrödinger equation with spherically symmetric central fields

$$\left\{-\frac{1}{2}\nabla^2 + V(r)\right\}\psi(\mathbf{r}) = E\psi(\mathbf{r}) \tag{8.1}$$

keeps invariant in spatial rotation, the solutions of radial Schrödinger equation can be expressed as

$$\psi(\mathbf{r}) = r^{-(D-1)/2}U(r)Y^{l}_{l_{D-1}\ldots l_1}(\hat{\mathbf{x}}), \tag{8.2}$$

S.-H. Dong, *Wave Equations in Higher Dimensions*,
DOI 10.1007/978-94-007-1917-0_8, © Springer Science+Business Media B.V. 2011

where the $Y^l_{l_{D-1}\ldots l_1}(\hat{\mathbf{x}})$ is the generalized spherical functions. Substitution of this into the Schrödinger equation allows us to obtain the radial Schrödinger equation

$$\left\{\frac{d^2}{dr^2} - \frac{l(l+D-2)+(D-1)(D-3)/4}{r^2}\right\}U(r) = -2[E-V(r)]U(r), \quad (8.3)$$

which can be rearranged as

$$\left\{\frac{d^2}{dr^2} - \frac{\kappa^2-1/4}{r^2}\right\}U(r) = -2[E-V(r)]U(r), \quad (8.4)$$

where κ is given in Eq. (5.4).

Now, we are going to study the sextic potential

$$V(r) = ar^6 + br^4 + cr^2, \quad a > 0. \quad (8.5)$$

For the solutions of Eq. (8.4) with this potential, we take an ansatz for the radial wavefunction

$$U(r) = \exp[p(\alpha,\beta,r)]\sum_{n=0} a_n r^{2n+\kappa+1/2}, \quad (8.6)$$

where

$$p(\alpha,\beta,r) = \frac{1}{2}\beta r^2 + \frac{1}{4}\alpha r^4. \quad (8.7)$$

Substituting Eq. (8.6) into Eq. (8.4) and equating the coefficient of $r^{2n+\kappa+5/2}$ to zero, one can obtain

$$A_n a_n + B_{n+1}a_{n+1} + C_{n+2}a_{n+2} = 0, \quad (8.8)$$

where

$$\begin{aligned} A_n &= \beta^2 + 2(2+\kappa+2n)\alpha - 2c, \\ B_n &= 2E + 2(1+\kappa+2n)\beta, \\ C_n &= 4n(\kappa+n), \end{aligned} \quad (8.9)$$

and two constraint conditions

$$\alpha^2 = 2a, \qquad 2\alpha\beta = 2b. \quad (8.10)$$

It is shown from Eq. (8.10) that the values of parameters for $p(\alpha,\beta,r)$ can be evaluated as

$$\alpha = \pm\sqrt{2a}, \qquad \beta = \frac{b}{\alpha}. \quad (8.11)$$

To retain the well-behaved solutions at the origin and at infinity, we choose the physically acceptable solutions as follows:

$$\alpha = -\sqrt{2a}, \qquad \beta = -\frac{b}{\sqrt{2a}}. \quad (8.12)$$

On the other hand, we know that the physical solutions of (8.8) and (8.9), with real E, require that $A_{n-1}C_n > 0$ for all relevant n, because the matrix equivalent can then be reduced to a real symmetric form by the substitution [234]

$$d_n = \sqrt{\frac{A_{n-1}}{C_n}\frac{A_{n-2}}{C_{n-1}}\cdots\frac{A_0}{C_1}}d'_n, \tag{8.13}$$

where d_n represents the basis vectors of the matrix. However, $C_n > 0$ for all n, while A_n becomes positive for sufficiently large n, unless the series is truncated by the condition $A_p = 0$. Therefore, we have

$$2c + 2\sqrt{2a}(2+\kappa+2p) - \frac{b^2}{2a} = 0. \tag{8.14}$$

This is a restriction on the parameters of the potential and κ. As our previous work [234], A_n, B_n and C_n must satisfy the determinant relation for a nontrivial solution

$$\det\begin{vmatrix} B_0 & C_1 & \cdots & \cdots & \cdots & 0 \\ A_0 & B_1 & C_2 & \cdots & \cdots & 0 \\ \vdots & \vdots & \vdots & \ddots & \vdots & \vdots \\ 0 & 0 & 0 & 0 & A_{p-1} & B_p \end{vmatrix} = 0. \tag{8.15}$$

To interpret this method, we will give the exact solutions for $p = 0, 1$ as follows:

(1): when $p = 0$, it is found from Eq. (8.15) that $B_0 = 0$, which, together with the values of α and β, leads to

$$E_0 = \frac{b(1+\kappa)}{\sqrt{2a}}. \tag{8.16}$$

The restriction on the parameters of the potential and κ can be obtained from Eq. (8.14)

$$\kappa = \frac{b^2 - 4ac}{4a\sqrt{2a}} - 2, \tag{8.17}$$

from which, together with κ we are able to obtain an important constraint on the parameters of potential and the angular momentum quantum number l for a given dimension D. That is to say, the choice of those parameters of potential can be taken arbitrarily. The wavefunction for $p = 0$ can be written as

$$\psi^{(0)}(r) = a_0 r^l \exp\left[-\frac{b}{2\sqrt{2a}}r^2 - \frac{\sqrt{2a}}{4}r^4\right], \tag{8.18}$$

where a_0 is the normalization constant.

(2): when $p = 1$, we can obtain the following relation from Eq. (8.15)

$$B_0B_1 - A_0C_1 = 0, \tag{8.19}$$

from which, we have

$$E_1 = \frac{b(2+\kappa)}{\sqrt{2a}} \pm \frac{1}{\sqrt{2a}}\sqrt{b^2(2+\kappa) - 4ac(1+\kappa) - 4\sqrt{2}a^{3/2}(2+3\kappa+\kappa^2)}. \tag{8.20}$$

Likewise, the restriction on the parameters of the potential and κ can be obtained from Eq. (8.14)

$$\kappa = \frac{b^2 - 4ac}{4a\sqrt{2a}} - 4. \tag{8.21}$$

The eigenfunction for $p = 1$ can be read as

$$\psi^{(1)}(r) = (a_0 + a_1 r^2) r^l \exp\left[-\frac{b}{2\sqrt{2a}} r^2 - \frac{\sqrt{2a}}{4} r^4\right], \tag{8.22}$$

where the a_0 and a_1 are the expansion constants. Following this way, we can generate a class of exact solutions through setting $p = 0, 1, 2, \ldots$, etc. Generally, if $a_p \neq 0$, but $a_{p+1} = a_{p+2} = \cdots = 0$, we have $A_p = 0$, which leads to a restriction on the parameters of the potential and κ. The eigenvalue can be obtained from Eq. (19). The eigenfunction becomes

$$\psi^{(p)}(r) = (a_0 + a_1 r^2 + \cdots + a_p r^{2p}) r^l \exp\left[-\frac{b}{2\sqrt{2a}} r^2 - \frac{\sqrt{2a}}{4} r^4\right], \tag{8.23}$$

where a_i $(i = 0, 1, 2, \ldots, p)$ are the expansion constants.

3 Singular One Fraction Power Potential

This potential can be taken as

$$V(r) = \frac{a}{r^{1/2}} + \frac{b}{r^{3/2}}. \tag{8.24}$$

Take the following ansatz for the wavefunction

$$U(r) = \exp[\alpha r + 2\beta r^{1/2}] \sum_{n=0} a_n r^{n/2+\kappa+1/2}. \tag{8.25}$$

On substituting this into Eq. (8.4) and setting the coefficient of $r^{n/2+\kappa-1/2}$ to zero, we have

$$A_n a_n + B_{n+1} a_{n+1} + C_{n+2} a_{n+2} = 0, \tag{8.26}$$

where

$$\begin{aligned} A_n &= \beta^2 + \alpha(n + 2\kappa + 1), \\ B_n &= \beta(n + 2\kappa + 1/2) - 2b, \\ C_n &= n(\kappa + n/4), \end{aligned} \tag{8.27}$$

and

$$\alpha^2 + 2E = 0, \qquad \alpha\beta = a, \tag{8.28}$$

from which we take the physically acceptable solutions

$$\alpha = -\sqrt{-2E}, \qquad \beta = -\frac{a}{\sqrt{-2E}}. \tag{8.29}$$

Similarly, taking $A_p = 0$ leads to

$$E_p^\kappa = -\frac{1}{2}\left(\frac{a^2}{2\kappa + 1 + p}\right)^{2/3}. \tag{8.30}$$

(1): when $p = 0$, it is easy to obtain

$$E_0^\kappa = -\frac{1}{2}\left(\frac{a^2}{2\kappa + 1}\right)^{2/3}. \tag{8.31}$$

Besides, it is shown from Eq. (8.15) that $B_0 = 0$, which leads to a constraint condition between the parameters of the potential and κ,

$$(\kappa + 1/4)[a(2\kappa + 1)]^{1/3} + b = 0. \tag{8.32}$$

The eigenfunction for $p = 0$ can be written as

$$\psi^{(0)}(r) = \exp\left[-\sqrt{-2E_0^\kappa}\, r - \frac{2a}{\sqrt{-2E_0^\kappa}} r^{1/2}\right] a_0 r^l. \tag{8.33}$$

(2): when $p = 1$, the eigenvalue can be obtained from Eq. (8.30) as

$$E_1^\kappa = -\frac{1}{2}\left(\frac{a^2}{2\kappa + 2}\right)^{2/3}. \tag{8.34}$$

On the other hand, based on Eq. (8.15) we have $B_0 B_1 = A_0 C_1$, which leads to

$$[\beta(2\kappa + 1/2) - 2b][\beta(2\kappa + 3/2) - 2b] = [\beta^2 + \alpha(1 + 2\kappa)](\kappa + 1/4), \tag{8.35}$$

where

$$\beta = -[2a(1+\kappa)]^{1/3}, \qquad \alpha = -\left(\frac{a^2}{2 + 2\kappa}\right)^{1/3}. \tag{8.36}$$

The eigenfunction for $p = 1$ now becomes

$$\psi^{(1)}(r) = \exp\left[-\sqrt{-2E_1^\kappa}\, r - \frac{2a}{\sqrt{-2E_1^\kappa}} r^{1/2}\right](a_0 + a_1 r^{1/2}) r^l, \tag{8.37}$$

where the a_0 and a_1 are expansion constants.

For a given p, the wavefunction can be written as

$$\psi^{(p)}(r) = \exp\left[-\sqrt{-E_p^\kappa}\, r - \frac{a}{\sqrt{-E_p^\kappa}} r^{1/2}\right](a_0 + a_1 r^{1/2} + \cdots + a_p r^{p/2}) r^l, \tag{8.38}$$

where a_i $(i = 0, 1, 2, \ldots, p)$ are the expansion constants.

4 Mixture Potential

This potential has the form

$$V(r) = ar^2 + br + \frac{c}{r}, \quad a > 0. \tag{8.39}$$

Take the ansatz as follows:

$$U(r) = \exp[p(\alpha, \beta, r)] \sum_{n=0} a_n r^{n+\kappa+1/2}, \tag{8.40}$$

where

$$p(\alpha, \beta, r) = \beta r + \frac{1}{2}\alpha r^2. \tag{8.41}$$

Similarly, we can obtain the following set of equations on substituting Eq. (8.40) into Eq. (8.4) and equating the coefficients of $r^{\kappa+n+1/2}$ to zero,

$$A_n a_n + B_{n+1} a_{n+1} + C_{n+2} a_{n+2} = 0, \tag{8.42}$$

where

$$\begin{aligned} A_n &= 2E + 2\alpha(1 + n + \kappa) + \beta^2, \\ B_n &= -2c + \beta(2n + 2\kappa + 1), \\ C_n &= n(n + 2\kappa), \end{aligned} \tag{8.43}$$

and

$$\alpha^2 = 2a, \qquad \alpha\beta = b. \tag{8.44}$$

Similar to above choice, the physically acceptable solutions are chosen as

$$\alpha = -\sqrt{2a}, \qquad \beta = -\frac{b}{\sqrt{2a}}. \tag{8.45}$$

Let us now consider the case $a_p \neq 0$, but $a_{p+1} = a_{p+2} = \cdots = 0$. From Eq. (8.15) we have $A_p = 0$, from which we can obtain the eigenvalue

$$E_p^\kappa = \sqrt{2a}(1 + \kappa + p) - \frac{b^2}{4a}. \tag{8.46}$$

For example, the exact solutions for $p = 0, 1$ can be illustrated below.

(1): when $p = 0$, we have

$$E_0^\kappa = \sqrt{2a}(1 + \kappa) - \frac{b^2}{4a}. \tag{8.47}$$

On the other hand, it is shown from Eq. (8.15) that $B_0 = 0$, which leads to the following restriction on the parameters of the potential and κ,

$$2c\sqrt{2a} + b(1 + 2\kappa) = 0. \tag{8.48}$$

The eigenfunction for $p = 0$ can be written as

$$\psi^{(0)}(r) = a_0 r^l \exp\left[-\frac{br + ar^2}{\sqrt{2a}}\right], \tag{8.49}$$

where the a_0 are the expansion constant.

(2): when $p=1$, the eigenvalue becomes

$$E_1^{\kappa}=\sqrt{2a}(\kappa+2)-\frac{b^2}{4a}. \tag{8.50}$$

Moreover, it is shown from Eq. (8.15) that $B_0B_1=A_0C_1$, which leads to a restriction on the parameters of the potential and κ

$$\left(2c+\frac{(1+2\kappa)b}{\sqrt{2a}}\right)\left(2c+\frac{(3+2\kappa)b}{\sqrt{2a}}\right)=2\sqrt{2a}(1+2\kappa). \tag{8.51}$$

The eigenfunction for $p=1$ can be read as

$$\psi^{(1)}(r)=(a_0+a_1r)r^l\exp\left[-\frac{br+ar^2}{\sqrt{2a}}\right], \tag{8.52}$$

where the a_0 and a_1 are the expansion constants. Similarly, if $a_p\neq 0$, but $a_{p+1}=a_{p+2}=\cdots=0$, we have $A_p=0$, from which we can obtain the eigenvalue. For a given p, the restriction can be determined by Eq. (14). The eigenfunction can be read as

$$\psi^{(p)}(r)=(a_0+a_1r+\cdots+a_pr^p)r^l\exp\left[-\frac{br+ar^2}{\sqrt{2a}}\right], \tag{8.53}$$

where the coefficients a_i $(i=0,1,2,\ldots,p)$ are the expansion constants.

5 Non-polynomial Potential

This potential has the form

$$V(r)=r^2+\frac{\lambda r^2}{1+gr^2}. \tag{8.54}$$

Take the following ansatz

$$U(r)=\exp\left[\frac{1}{2}\alpha r^2\right]\sum_{n=0}a_nr^{2n+\kappa+1/2}. \tag{8.55}$$

On substituting this into Eq. (8.4) and setting the coefficient of $r^{2n+\kappa+5/2}$ to zero, one can obtain the following recurrence relation

$$A_na_n+B_{n+1}a_{n+1}+C_{n+2}a_{n+2}=0, \tag{8.56}$$

where

$$\begin{aligned}A_n&=-2\lambda+2g[E+\alpha(1+\kappa+2n)],\\B_n&=2E+2\alpha(2n+\kappa+1)+4ng(\kappa+n),\\C_n&=4n(n+\kappa),\end{aligned} \tag{8.57}$$

and

$$\alpha^2=2. \tag{8.58}$$

With the same reason, we choose $\alpha = -\sqrt{2}$. Likewise, if $a_p \neq 0$, but $a_{p+1} = a_{p+2} = \cdots = 0$, we can obtain $A_p = 0$, which leads to the following relation

$$E_p^\kappa = \frac{\lambda}{g} + \sqrt{2}(\kappa + 1 + 2p). \tag{8.59}$$

To expound this method, we will give the exact solutions for $p = 0, 1$ as follows.

(1): when $p = 0$, it is easy to obtain

$$E_0^\kappa = \frac{\lambda}{g} + \sqrt{2}(\kappa + 1). \tag{8.60}$$

On the other hand, it is shown from Eq. (8.15) that $B_0 = 0$, which means $\lambda = 0$. This refers to harmonic oscillator case. The eigenfunction for $p = 0$ can be simply written as

$$\psi^{(0)}(r) = a_0 r^l \exp\left[-\frac{r^2}{\sqrt{2}}\right], \tag{8.61}$$

where a_0 is the expansion constant.

(2): when $p = 1$, the eigenvalue can be obtained as

$$E_1^\kappa = \frac{\lambda}{g} + \sqrt{2}(\kappa + 3). \tag{8.62}$$

In this case, it is shown from $B_0B_1 - A_0C_1 = 0$ that a constraint on the potential parameters and κ is given by

$$2\sqrt{2}g + 2g^2(1+\kappa) + \lambda = 0. \tag{8.63}$$

The eigenfunction for $p = 1$ can be read as

$$\psi^{(1)}(r) = (a_0 + a_1 r^2) r^l \exp\left[-\frac{1}{\sqrt{2}} r^2\right], \tag{8.64}$$

where the a_0 and a_1 are the expansion constants. Generally, if $a_p \neq 0$, but $a_{p+1} = a_{p+2} = \cdots = 0$, we can obtain $A_p = 0$. The eigenfunction can be written as

$$\psi^{(p)}(r) = (a_0 + a_1 r^2 + \cdots + a_p r^{2p}) r^l \exp\left[-\frac{1}{\sqrt{2}} r^2\right], \tag{8.65}$$

where a_i $(i = 0, 1, 2, \ldots, p)$ are the expansion constants.

6 Screened Coulomb Potential

For this potential

$$V(r) = \frac{a}{r} + \frac{b}{r+\lambda}, \quad a < -b, \tag{8.66}$$

take the following ansatz

$$U(r) = e^{\beta r} \sum_{n=0} a_n (r+\lambda) r^{n+\kappa+1/2}. \tag{8.67}$$

Substituting this into Eq. (8.4) and setting the coefficient of $r^{n+\kappa+1/2}$ to zero, we have

$$A_n a_n + B_{n+1} a_{n+1} + C_{n+2} a_{n+2} = 0, \tag{8.68}$$

where

$$\begin{aligned} A_n &= -2a - 2b + (3 + 2n + 2\kappa)\beta, \\ B_n &= 2(1+\beta\lambda)(n+\kappa+1/2) - 2a\lambda + n(n+2\kappa), \\ C_n &= \lambda n(n+2\kappa), \end{aligned} \tag{8.69}$$

and

$$\beta^2 + 2E = 0. \tag{8.70}$$

Based on this Eq. (8.70) we take

$$\beta = -\sqrt{-2E}. \tag{8.71}$$

This is required by the physically acceptable solution. On the other hand, if $a_p \neq 0$, but $a_{p+1} = a_{p+2} = a_{p+3} = \cdots = 0$, it is easy to obtain $A_p = 0$, i.e.,

$$E_p^\kappa = -\frac{(a+b)^2}{2(3/2+\kappa+p)^2}. \tag{8.72}$$

To show this method, we present exact solutions for $p = 0$ and 1 below.

(1): when $p = 0$, it is found from Eq. (8.72) that

$$E_0^\kappa = -\frac{(a+b)^2}{2(3/2+\kappa)^2}. \tag{8.73}$$

The restriction on the parameters of the potential and κ can be obtained from $B_0 = 0$ as

$$(\kappa + 1/2)(\kappa + 3/2) = \lambda(2a\kappa + b\kappa + 2a + b/2). \tag{8.74}$$

The eigenfunction for $p = 0$ now becomes

$$\psi^{(0)}(r) = a_0 r^l \exp\left[-\sqrt{-2E_0^\kappa}\, r\right](r+\lambda), \tag{8.75}$$

where a_0 is the expansion constant.

(2): when $p = 1$, one can obtain the eigenvalue as follows:

$$E_1^\kappa = -\frac{(a+b)^2}{2(5/2+\kappa)^2}. \tag{8.76}$$

The restriction on the parameters of the potential and κ can be obtained from $B_0 B_1 = A_0 C_1$. The eigenfunction for $p = 1$ becomes

$$\psi^{(1)}(r) = (a_0 + a_1 r) r^l \exp\left[-\sqrt{-2E_1^\kappa}\, r\right](r+\lambda), \tag{8.77}$$

where the a_0 and a_1 are expansion constants. For a given p, if $a_p \neq 0$, but $a_{p+1} = a_{p+2} = \cdots = 0$, we have $A_p = 0$. The eigenfunction can be written as

$$\psi_r^{(p)} = (a_0 + a_1 r + \cdots + a_p r^p) r^l \exp[-\sqrt{-2E_p^\kappa} r](r + \lambda), \tag{8.78}$$

where a_i $(i = 0, 1, 2, \ldots, p)$ are the expansion constants.

7 Morse Potential

Due to its mathematical advantages, the harmonic oscillator model has been widely used to describe the interaction force of the diatomic molecule. Nevertheless, it is well known that the real molecular vibrations are anharmonic. Among many molecular potentials, the Morse potential as an ideal and typical anharmonic potential permits an exactly mathematical treatment and has been the subject of interest since 1929 [244]. In particular, the Morse potential will reduce to the harmonic oscillator in the harmonic limit.

Choose the separated atoms limit as the zero-point energy. The Morse potential has the following form

$$V(r) = V_0(e^{-2\beta r} - 2e^{-\beta r}), \tag{8.79}$$

where $V_0 > 0$ corresponds to its depth, β is related to the range of the potential, and r gives the relative distance from the equilibrium position of the atoms.

Following approach [245] and our recent work [246], we can take the ansatz for the wavefunction

$$R(r) = r^\tau e^{-br} G(r), \quad b = \sqrt{-2E}, \ E < 0. \tag{8.80}$$

Substitution of this into (8.4) leads to

$$\begin{aligned} &r^\tau \frac{d^2G(r)}{dr^2} + (2\tau r^{\tau-1} - 2br^\tau)\frac{dG(r)}{dr} + \Bigg[\tau(\tau-1)r^{\tau-2} - 2b\tau r^{\tau-1} \\ &\quad + (b^2 + 2E - 2V(r))r^\tau - \left(\kappa^2 - \frac{1}{4}\right)r^{\tau-2}\Bigg]G(r) = 0, \end{aligned} \tag{8.81}$$

which is re-arranged as

$$\begin{aligned} &[\tau(\tau-1) - (\kappa^2 - 1/4)]G(r) + \left(2\tau\frac{dG(r)}{dr} - 2b\tau G(r)\right)r \\ &\quad + \left(\frac{d^2G(r)}{dr^2} - 2b\frac{dG(r)}{dr} - 2V(r)\,G(r)\right)r^2 = 0. \end{aligned} \tag{8.82}$$

From the behavior of the wavefunction at the origin, it is shown from Eqs. (8.81) and (8.82) that

$$\tau = \kappa + \frac{1}{2} = l + \frac{D-1}{2}, \tag{8.83}$$

where another solution $\tau = 1/2 - \kappa$ is ignored in physics.

As shown in [246], we expand the Morse potential about the origin as follows:

$$V(r) = 2V_0 \sum_{l=0}^{\infty} c_l r^l, \tag{8.84}$$

where we have used an important formula

$$\begin{aligned} e^{-2\beta r} - 2e^{-\beta r} &= (e^{-\beta r} - 1)^2 - 1 \\ &= 2\sum_{l=0}^{\infty} (-1)^l (2^{l-1} - 1)\frac{\beta^l}{l!} r^l \\ &= 2\sum_{l=0}^{\infty} c_l r^l. \end{aligned} \tag{8.85}$$

We take the standard series for $G(r)$ with the form

$$G(r) = \sum_{k=0}^{\infty} \gamma_k r^k, \quad \gamma_0 \neq 0. \tag{8.86}$$

Substituting of this, together with Eq. (8.85), into Eq. (8.82) and setting the coefficients of the powers of r^n $(n = l + k + 2)$ to zero, one can obtain

$$\sum_{k=0}^{\infty} [k(2\kappa + k)]\gamma_k r^k - \sum_{k=0}^{\infty} 2b(k + \kappa + 1/2)\gamma_k r^{k+1} - 4V_0 \sum_{k=0}^{\infty}\sum_{l=0}^{\infty} c_l \gamma_k r^{l+k+2} = 0, \tag{8.87}$$

from which we have

$$\begin{aligned} \gamma_1 &= b\gamma_0, \\ \gamma_2 &= \frac{b(2\kappa + 3)\gamma_1 + 4V_0\gamma_0 c_0}{4 + 4\kappa}, \\ \gamma_3 &= \frac{b(2\kappa + 5)\gamma_2 + 4V_0(\gamma_1 c_0 + \gamma_0 c_1)}{9 + 6\kappa}. \end{aligned} \tag{8.88}$$

Similarly continuing to use Eq. (8.87), we can finally obtain the expansion coefficients γ_n as

$$\gamma_n = \frac{b(2\kappa + 2n - 1)\gamma_{n-1} + 4V_0 S_{n,k}}{n(2\kappa + n)}, \tag{8.89}$$

with

$$S_{n,k} = \sum_{l=0, l\geq 2}^{n-2=N} c_l \gamma_{N-l}. \tag{8.90}$$

Accordingly, we can finally obtain the analytical solutions of the Schrödinger equation with the Morse potential in arbitrary dimensions as

$$R(r) = r^{1/2+\kappa} e^{-br} \sum_{n=0}^{\infty} \gamma_n r^n, \quad \gamma_0 \neq 0. \tag{8.91}$$

Thus, for a given Morse potential, we can always find the suitable eigenvalue to make the wavefunction $R(r)$ convergent when r tends to infinity.

8 Conclusions

In this Chapter we have carried out the solutions of the D-dimensional radial Schrödinger equation with some anharmonic potentials applying an ansatz to the wavefunction and obtained the restrictions on the parameters of the potential and κ. These potentials include the sextic potential, singular one fraction power potential, mixture potential, non-polynomial potential, screened Coulomb potential and Morse potential. Before ending this Chapter, we give a useful remark on this method. As far as all insoluble physical quantum systems, we have to treat them by different approximate methods. By using present wavefunction ansatz method, one is able to only obtain some so-called eigenfunctions with some constraints on the potential parameters and quantum numbers.

Chapter 9
The Levinson Theorem for Schrödinger Equation

1 Introduction

The Levinson theorem [109] as an important theorem in quantum scattering theory establishes the relation between the total number n_l of bound states and the phase shift $\delta_l(0)$ of the scattering states at zero momentum. For the Schrödinger equation with a spherically symmetric potential $V(r)$, the short-range central potential $V(r)$ satisfies the boundary conditions

$$\begin{aligned} &\int_0^1 r|V(r)|dr < \infty, \\ &\int_1^\infty r^2|V(r)|dr < \infty. \end{aligned} \tag{9.1}$$

They are required for the nice behavior of the wavefunction at the origin and the analytic property of the Jost function, respectively.

Since 1949 the Levinson theorem has been proved by different methods and generalized to different equations and potentials [184, 247–296]. With the recent interest in higher-dimensional field theory, we attempt to establish the Levinson theorem for the D-dimensional Schrödinger equation by the Sturm-Liouville theorem.

This Chapter is organized as follows. We study the scattering states and bound states in Sects. 2 and 3, respectively. The Sturm-Liouville theorem is carried out in Sect. 4. The non-relativistic Levinson theorem in D dimensions is established in Sect. 5. The general case will be discussed in Sect. 6. In Sect. 7 we present the comparison theorem. We summarize our conclusions in Sect. 8.

2 Scattering States and Phase Shifts

Let us consider the D-dimensional Schrödinger equation

$$\left(-\frac{\hbar^2}{2M}\nabla_D^2 + V(r)\right)\Psi(\mathbf{r}) = E\Psi(\mathbf{r}), \tag{9.2}$$

S.-H. Dong, *Wave Equations in Higher Dimensions*,
DOI 10.1007/978-94-007-1917-0_9, © Springer Science+Business Media B.V. 2011

which keeps invariant in spatial rotation. Therefore, choose the wavefunction as

$$\Psi(\mathbf{r}) = r^{-(D-1)/2} R_l(r) Y^l_{l_{D-1}\dots l_1}(\hat{\mathbf{x}}). \tag{9.3}$$

By separating the angular variables from the wavefunction we obtain the radial Schrödinger equation as

$$\left\{\frac{d^2}{dr^2} - \frac{(l-1+D/2)^2 - 1/4}{r^2}\right\} R_l(r) = -\frac{2M}{\hbar^2}\{E - V(r)\} R_l(r), \tag{9.4}$$

which is a real equation so that we only discuss its real solutions.

For simplicity, we first study Eq. (9.4) with a cutoff potential

$$V(r) = 0, \quad \text{when } r \geq r_0, \tag{9.5}$$

where r_0 is a sufficiently large radius. The general case will be studied in Sect. 6.

Similar to previous works [290, 291], we introduce a parameter λ for $V(r)$

$$V(r,\lambda) = \lambda V(r), \qquad V(r,1) = V(r). \tag{9.6}$$

Thus, Eq. (9.4) can be modified as

$$\left\{\frac{\partial^2}{\partial r^2} - \frac{\eta^2 - 1/4}{r^2}\right\} R_l(r,\lambda) = -\frac{2M}{\hbar^2}[E - V(r,\lambda)] R_l(r,\lambda), \tag{9.7}$$

where $\eta = |l - 1 + D/2|$ is the same as κ given in Eq. (5.4) for difference. As you see below, we use κ to denote the energy.

We now solve this equation in two regions and match the logarithmic derivative of the radial function at r_0:

$$A_l(E,\lambda) \equiv \left[\frac{1}{R_l(r,\lambda)} \frac{\partial R_l(r,\lambda)}{\partial r}\right]_{r=r_0^-} = \left[\frac{1}{R_l(r,\lambda)} \frac{\partial R_l(r,\lambda)}{\partial r}\right]_{r=r_0^+}. \tag{9.8}$$

From (9.1) there exists a convergent solution to Eq. (9.7) at the region $[0, r_0]$. When $V(r,0) = 0$ this solution is given by

$$R_l(r,0) = (\pi k r/2)^{1/2} J_\eta(kr), \tag{9.9}$$

when $E > 0$ and $k = (2ME)^{1/2}/\hbar$,

$$R_l(r,0) = e^{-i\eta\pi/2} (\pi \kappa r/2)^{1/2} J_\eta(i\kappa r), \tag{9.10}$$

when $E \leq 0$ and $\kappa = (-2ME)^{1/2}/\hbar$. $J_\nu(x)$ is the Bessel function and $R_l(r,0)$ in Eqs. (9.9) and (9.10) is a real function. A multiplied factor on $R_l(r,0)$ is unimportant.

In the region (r_0, ∞), we have $V(r,\lambda) = 0$. When $E > 0$ the combination of two oscillatory solutions to Eq. (9.7) can always satisfy Eq. (9.8) so that there exists a continuous spectrum

$$\begin{aligned} R_l(r,\lambda) &= (\pi k r/2)^{1/2} [\cos\delta_l(k,\lambda) J_\eta(kr) - \sin\delta_l(k,\lambda) N_\eta(kr)] \\ &\sim \cos\left[kr - \frac{(2\eta-1)\pi}{4} + \delta_l(k,\lambda)\right], \quad \text{when } r \to \infty, \end{aligned} \tag{9.11}$$

where $N_\nu(kr)$ is the Neumann function. Although $V(r,\lambda)$ does not depend on λ in the region (r_0,∞), through Eq. (9.8), $R_l(r,\lambda)$ and the phase shift $\delta_l(k,\lambda)$ depend on λ. In fact, from Eq. (9.8) we can obtain

$$\tan\delta_l(k,\lambda)=\frac{J_\eta(kr_0)}{N_\eta(kr_0)}\frac{A_l(E,\lambda)-k\frac{J'_\eta(kr_0)}{J_\eta(kr_0)}-\frac{1}{2r_0}}{A_l(E,\lambda)-k\frac{N'_\eta(kr_0)}{N_\eta(kr_0)}-\frac{1}{2r_0}}, \tag{9.12}$$

$$\delta_l(k)\equiv\delta_l(k,1), \tag{9.13}$$

where the prime denotes the derivative of the Bessel function, the Neumann function, and later the Hankel function with respect to their arguments. It is found from Eq. (9.12) that the $\delta_l(k,\lambda)$ is determined up to a multiple of π due to the period of the tangent function. As usual, we may use the convention that

$$\delta_l(k)=0, \quad \text{when } V(r)=0, \tag{9.14}$$

which implies that $\delta_l(\infty)=0$ as defined in [1].

3 Bound States

Because when $E\le 0$ there is only one convergent solution to Eq. (9.7) in the region $r>r_0$, Eq. (9.7) is not always satisfied

$$R_l(r,\lambda)=e^{i(\eta+1)\pi/2}(\pi\kappa r/2)^{1/2}H_\eta^{(1)}(i\kappa r)\sim e^{-\kappa r}, \quad \text{when } r\to\infty, \tag{9.15}$$

where $H_\nu^{(1)}(x)$ is the Hankel function of the first kind. Actually, $R_l(r,\lambda)$ in Eq. (9.15) does not depend on λ. The matching condition (9.8) may be satisfied only for some discrete energy E, where a bound state appears. Therefore, there exists a discrete spectrum for $E\le 0$.

It is worth paying attention to the solutions with $E=0$. If $A_l(0,1)$ (zero momentum and $\lambda=1$) is equal to $(1-2\eta)/(2r_0)$, it matches a solution of zero energy

$$R_l(r,1)=r^{(1-2\eta)/2}, \quad r_0<r<\infty, \tag{9.16}$$

from which we know the solution describes a bound state for $l>2-D/2$ and a half bound state for $l\le 2-D/2$.

4 The Sturm-Liouville Theorem

Since Eq. (9.7) is a Sturm-Liouville type equation, then it must satisfy the Sturm-Liouville theorem. For this problem, it is known that the logarithmic derivative of wavefunction is monotonic with respect to the energy [297]. Due to this property, the Sturm-Liouville theorem has become a powerful tool in proving the Levinson theorem.

Denote by $\overline{R}_l(r,\lambda)$ the solution to Eq. (9.7) for the energy $\overline{E}$

$$\left\{\frac{\partial^2}{\partial r^2}-\frac{\eta^2-1/4}{r^2}\right\}\overline{R}_l(r,\lambda)=-\frac{2M}{\hbar^2}[\overline{E}-V(r,\lambda)]\overline{R}_l(r,\lambda). \tag{9.17}$$

Multiplying Eq. (9.7) and Eq. (9.17) by $\overline{R}_l(r,\lambda)$ and $R_l(r,\lambda)$, respectively and calculating their difference, we have

$$\begin{aligned}&\frac{\partial}{\partial r}\left\{R_l(r,\lambda)\frac{\partial\overline{R}_l(r,\lambda)}{\partial r}-\overline{R}_l(r,\lambda)\frac{\partial R_l(r,\lambda)}{\partial r}\right\}\\&\quad=-\frac{2M}{\hbar^2}(\overline{E}-E)\overline{R}_l(r,\lambda)R_l(r,\lambda).\end{aligned} \tag{9.18}$$

From the boundary condition, both solutions $R_l(r,\lambda)$ and $\overline{R}_l(r,\lambda)$ vanish at the origin. Integrating Eq. (9.18) from 0 to r_0, we obtain

$$\begin{aligned}&\frac{1}{\overline{E}-E}\left[R_l(r,\lambda)\frac{\partial\overline{R}_l(r,\lambda)}{\partial r}-\overline{R}_l(r,\lambda)\frac{\partial R_l(r,\lambda)}{\partial r}\right]_{r=r_0^-}\\&\quad=-\frac{2M}{\hbar^2}\int_0^{r_0}\overline{R}_l(r,\lambda)R_l(r,\lambda)dr.\end{aligned} \tag{9.19}$$

By taking the limit $\overline{E}\to E$, we have

$$\begin{aligned}\frac{\partial A_l(E,\lambda)}{\partial E}&=\frac{\partial}{\partial E}\left[\frac{1}{R_l(r,\lambda)}\frac{\partial R_l(r,\lambda)}{\partial r}\right]_{r=r_0^-}\\&=-\frac{2M}{\hbar^2R_l(r_0,\lambda)^2}\int_0^{r_0}R_l(r,\lambda)^2dr<0.\end{aligned} \tag{9.20}$$

When $E=\hbar^2k^2/(2M)$ is larger than zero and tends to zero, we have

$$A_l(E,\lambda)=A_l(0,\lambda)-c^2k^2+\cdots,\quad c^2\le 0. \tag{9.21}$$

Similarly, from the boundary condition that $R_l(r,\lambda)$ tends to zero at infinity, we have

$$\frac{\partial}{\partial E}\left[\frac{1}{R_l(r,\lambda)}\frac{\partial R_l(r,\lambda)}{\partial r}\right]_{r=r_0^+}=\frac{2M}{\hbar^2R_l(r_0,\lambda)^2}\int_{r_0}^{\infty}R_l(r,\lambda)^2dr>0. \tag{9.22}$$

This is another form of the Sturm-Liouville theorem [297]. As E increases, the logarithmic derivative of the radial function at r_0^- decreases monotonically, but that at r_0^+ for $E\le 0$ increases monotonically. When $E\le 0$ because both sides of Eq. (9.8) are monotonic as energy changes, the variety of the $A_l(0,\lambda)$ determines the number of bound states as λ changes.

5 The Levinson Theorem

In this section we apply the Sturm-Liouville theorem to show that both phase shifts and the number of bound states are related with the variety of the logarithmic derivative $A_l(0,\lambda)$.

We first study relation between the n_l and the $A_l(0,\lambda)$ when the potential changes. From the Sturm-Liouville theorem, when $E \le 0$ the logarithmic derivative of the radial function at r_0 is monotonic with respect to E. From Eq. (9.15) we obtain the logarithmic derivative

$$\left(\frac{1}{R_l(r,\lambda)}\frac{\partial R_l(r,\lambda)}{\partial r}\right)_{r=r_0^+} = \frac{i\kappa H_\eta^{(1)}(i\kappa r_0)'}{H_\eta^{(1)}(i\kappa r_0)} - \frac{1}{2r_0}$$
$$= \begin{cases} \frac{1-2\eta}{2r_0}, & \text{when } E \sim 0, \\ -\infty, & \text{when } E \to -\infty, \end{cases} \tag{9.23}$$

which does not depend on λ. On the other hand, when $\lambda = 0$ we obtain from Eq. (9.10)

$$A_l(E,0) = \left(\frac{1}{R_l(r,0)}\frac{\partial R_l(r,0)}{\partial r}\right)_{r=r_0^-} = \frac{i\kappa J_\eta'(i\kappa r_0)}{J_\eta(i\kappa r_0)} - \frac{1}{2r_0}$$
$$= \begin{cases} \frac{2\eta+1}{2r_0}, & \text{when } E \sim 0, \\ \infty, & \text{when } E \to -\infty. \end{cases} \tag{9.24}$$

It is evident from Eqs. (9.23) and (9.24) that there is no overlap between two variant ranges of two logarithmic derivatives when $\lambda = 0$ and as the energy increases from $-\infty$ to 0. Thus, there is no bound state except for the case of $l = 0$ and $D = 2$, where a half bound state occurs at $E = 0$.

Second, we study the relation between the $\delta_l(0,\lambda)$ and the $A_l(0,\lambda)$ when the potential changes. The $\delta_l(0,\lambda)$ is the limit of the $\delta_l(k,\lambda)$ as k tends to zero. Therefore, we are interested in the $\delta_l(k,\lambda)$ at a sufficiently small momentum k, i.e., $k \ll 1/r_0$. In this case we obtain from Eq. (9.12)

$$\tan\delta_l(k,\lambda) = \frac{-\pi(kr_0)^{2\eta}}{2^{2\eta}\Gamma(\eta+1)\Gamma(\eta)} \cdot \frac{A_l(l,\lambda) - \frac{2\eta+1}{2r_0}}{A_l(0,\lambda) - c^2k^2 - \frac{1-2\eta}{2r_0}\left[1 - \frac{(kr_0)^2}{(2\eta-1)(\eta-1)}\right]}, \tag{9.25}$$

for $l > 2 - D/2$,

$$\tan\delta_l(k,\lambda) = \frac{-\pi(kr_0)^2}{4}\frac{A_l(0,\lambda) - \frac{3}{2r_0}}{A_l(0,\lambda) - c^2k^2 + \frac{1}{2r_0}[1 + 2(kr_0)^2\ln(kr_0)]} \tag{9.26}$$

for $l = 2 - D/2$ ($D = 4$ and $l = 0$ or $D = 2$ and $l = 1$),

$$\tan\delta_l(k,\lambda) = -(kr_0)\frac{A_l(0,\lambda) - \frac{1}{r_0}}{A_l(0,\lambda) - c^2k^2 + k^2r_0}, \tag{9.27}$$

for $l = (3-D)/2$ ($D = 3$ and $l = 0$), and

$$\tan\delta_l(k,\lambda) = \frac{\pi}{2\ln(kr_0)}\frac{A_l(0,\lambda) - c^2k^2 - \frac{1}{2r_0}[1 - (kr_0)^2]}{A_l(0,\lambda) - c^2k^2 - \frac{1}{2r_0}\left[1 + \frac{2}{\ln(kr_0)}\right]}, \tag{9.28}$$

for $l = 1 - D/2$ ($D = 2$ and $l = 0$).

Likewise, we can also obtain from Eq. (9.12) that the $\delta_l(k,\lambda)$ increases monotonically as the $A_l(E,\lambda)$ decreases

$$\left.\frac{\partial\delta_l(k,\lambda)}{\partial A_l(E,\lambda)}\right|_k = \frac{-8r_0\cos^2\delta_l(k,\lambda)}{\pi\{[2r_0A_l(E,\lambda)-1]N_\eta(kr_0)-2kr_0N'_\eta(kr_0)\}^2} \le 0, \tag{9.29}$$

where $k=(2ME)^{1/2}/\hbar$.

Since Eq. (9.7) is analogous to that of two-dimensional case [291], we suggest the readers refer to Ref. [291]. By repeating the proof in [290, 291], we can obtain the Levinson theorem for the D-dimensional Schrödinger equation in non-critical cases

$$\delta_l(0) = n_l\pi. \tag{9.30}$$

Similarly, in critical cases $l=2-D/2$ and $l=(3-D)/2$, the Levinson theorem must be modified as

$$\delta_l(0) = \begin{cases} (n_m+1)\pi, & \text{when } l=1,\ D=2,\ \text{or } l=0,\ D=4, \\ (n_m+1/2)\pi, & \text{when } l=0,\ D=3, \end{cases} \tag{9.31}$$

when a half bound state occurs. The Levinson theorems given in Eqs. (9.30) and (9.31) coincide with the results obtained in two and three dimensions.

6 Discussions on General Case

We now discuss the general case where the potential $V(r)$ has a tail at $r>r_0$. Let r_0 be so large that only the leading term in $V(r)$ is concerned in the region $r>r_0$:

$$V(r) \sim \frac{\hbar^2}{2M}br^{-n}, \quad \text{when } r\to\infty, \tag{9.32}$$

where b is a non-vanishing constant and n is a positive constant, not necessarily to be an integer. From (9.1) n should be larger than 3. However, we are also interested in the modification of the Levinson theorem for $n=2$, for which Newton [249–253] found two counterexamples where the Levinson theorem is violated. Substituting Eq. (9.32) into Eq. (9.7) and changing the variable r to ξ

$$\xi = \begin{cases} kr = r\sqrt{2ME}/\hbar, & \text{when } E>0, \\ \kappa r = r\sqrt{-2ME}/\hbar, & \text{when } E\le 0, \end{cases} \tag{9.33}$$

we get the radial equation at the region (r_0,∞)

$$\frac{\partial^2 R_l(\xi,\lambda)}{\partial\xi^2} + \left\{1-\frac{b}{\xi^n}k^{n-2}-\frac{4\eta^2-1}{4\xi^2}\right\}R_l(\xi,\lambda)=0, \tag{9.34}$$

for $E>0$, and

$$\frac{\partial^2 R_l(\xi,\lambda)}{\partial\xi^2} + \left\{-1-\frac{b}{\xi^n}\kappa^{n-2}-\frac{4\eta^2-1}{4\xi^2}\right\}R_l(\xi,\lambda)=0, \tag{9.35}$$

for $E\le 0$.

When $n = 2$, if we define

$$\nu^2 = \eta^2 + b, \tag{9.36}$$

then Eq. (9.34) becomes

$$\frac{\partial^2 R_l(r,\lambda)}{\partial r^2} + \left\{\frac{2ME}{\hbar^2} - \frac{\nu^2 - 1/4}{r^2}\right\} R_l(r,\lambda) = 0, \quad r \geq r_0. \tag{9.37}$$

If $\nu^2 < 0$, there are infinite number of bound states. Here, we will not discuss this case as well as the case with $\nu = 0$. When $\nu^2 > 0$, we take $\nu > 0$. Some formulas given in the previous sections will be changed by replacing the parameter η with ν ($\eta \neq \nu$). Equation (9.23) becomes

$$\begin{aligned}\left(\frac{1}{R_l(r,\lambda)}\frac{\partial R_l(r,\lambda)}{\partial r}\right)_{r=r_0^+} &= \frac{i\kappa H_\nu^{(1)}(i\kappa r_0)'}{H_\nu^{(1)}(i\kappa r_0)} - \frac{1}{2r_0} \\ &= \begin{cases} \frac{1-2\nu}{2r_0}, & \text{when } E \sim 0, \\ -\kappa \sim -\infty, & \text{when } E \to -\infty. \end{cases}\end{aligned} \tag{9.38}$$

The scattering solution (9.11) in the region (r_0, ∞) is modified as

$$\begin{aligned} R_l(r,\lambda) &= \sqrt{\frac{\pi k r}{2}}\{\cos\eta_l(k,\lambda) J_\nu(kr) - \sin\eta_l(k,\lambda) N_\nu(kr)\} \\ &\sim \sin\left(kr - \frac{\nu\pi}{2} + \frac{\pi}{4} + \eta_l(k,\lambda)\right), \quad \text{when } r \to \infty. \end{aligned} \tag{9.39}$$

The $\delta_l(k)$ can be thus calculated from $\eta_l(k,1)$

$$\delta_l(k) = \eta_l(k,1) + (\eta - \nu)\pi/2. \tag{9.40}$$

The $\eta_l(k,\lambda)$ satisfies

$$\tan\eta_l(k,\lambda) = \frac{J_\nu(kr_0)}{N_\nu(kr_0)} \frac{A_l(E,\lambda) - k\frac{J_\nu'(kr_0)}{J_\nu(kr_0)} - \frac{1}{2r_0}}{A_l(E,\lambda) - k\frac{N_\nu'(kr_0)}{N_\nu(kr_0)} - \frac{1}{2r_0}}, \tag{9.41}$$

and it increases monotonically as the $A_l(E,\lambda)$ decreases:

$$\left.\frac{\partial\eta_l(k,\lambda)}{\partial A_l(E,\lambda)}\right|_k = \frac{-8r_0\cos^2\eta_l(k,\lambda)}{\pi\{[2r_0 A_l(E,\lambda) - 1]N_\nu(kr_0) - 2kr_0 N_\nu'(kr_0)\}^2} \leq 0. \tag{9.42}$$

For a sufficiently small k, the asymptotic formulas for $\tan\eta_l(k,\lambda)$ can be obtained from the formulas of $\tan\delta_l(k,\lambda)$ given in Eqs. (9.25)–(9.28) by replacing l with $\nu + 1 - D/2$, except for the cases of $0 < \nu < 1/2$ and $1/2 < \nu < 1$. For the latter cases, we have

$$\begin{aligned}\tan\eta_l(k,\lambda) = &-\frac{\pi(kr_0)^{2\nu}}{2^{2\nu}\nu\Gamma(\nu)^2} \\ &\cdot \frac{A_l(0,\lambda) - \frac{\nu+1/2}{r_0}}{A_l(0,\lambda) - c^2k^2 - \frac{1-2\nu}{2r_0} + \frac{2\pi\cot(\nu\pi)}{r_0\Gamma(\nu)^2}\left(\frac{kr_0}{2}\right)^{2\nu}}. \end{aligned} \tag{9.43}$$

Repeating the proof for the Levinson theorem (9.30) and (9.31), we obtain the modified Levinson theorem for the non-critical cases

$$\delta_l(0) = \frac{\pi}{2}(2n_l + \eta - \nu). \tag{9.44}$$

For the critical case where $A_l(0, 1) = (-\nu + 1/2)/r_0$, the modified Levinson theorem (9.44) holds for $\nu > 1$, but it should be revised for $0 < \nu \leq 1$ as

$$\delta_l(0) = \frac{\pi}{2}(2n_l + \eta + \nu). \tag{9.45}$$

When $n > 2$, for any arbitrarily given small ϵ, one can always find a sufficiently large r_0 such that $|V(r)| < \epsilon/r^2$ in the region (r_0, ∞). Since $\nu^2 = \eta^2 + \epsilon \sim \eta^2$, the Levinson theorems (9.30) and (9.31) still hold in this case. This conclusion can also be shown from Eqs. (9.34) and (9.35). In comparison with the centrifugal term, the term with a factor k^{n-2} (or κ^{n-2}) is too small to affect the phase shift at a sufficiently small k and the variant range of the logarithmic derivative $[\partial R_l(r, \lambda)/\partial r]/R_l(r, \lambda)$ at r_0^+. Therefore, the proof given in the previous sections is still valid for this case and the Levinson theorems (9.30) and (9.31) still hold.

Finally, we check whether two counterexamples in three dimensions ($D = 3$) pointed by Newton (see pp. 438–439 in [249–253]), where the Levinson theorems (9.30) and (9.31) are violated, satisfy the modified Levinson theorems (9.44) and (9.45).

Example 1

$$V(r) = \frac{2c^2}{(1 + cr)^2} \to \frac{2}{r^2}, \quad \text{when } r \to \infty. \tag{9.46}$$

The phase shift of the S wave is $\delta_0(0) = -\pi/2$, and there is no bound state for the S wave, $n_0 = 0$. The Levinson theorem (9.30) is violated. However, from Eq. (9.36) one has $\nu = 3/2$. The modified Levinson theorem (9.44) still holds.

Example 2

$$V(r) = -6r\frac{2c^2 - r^3}{(c^2 + r^3)^2} \to \frac{6}{r^2}, \quad \text{when } r \to \infty, \tag{9.47}$$

where c is a constant. The solution of the S wave is

$$\begin{aligned} R(k, r) &= \frac{\sin kr}{k} - \frac{3r(\sin kr - kr\cos kr)}{k^3(c^2 + r^3)} \\ &\to c^2 r/(c^2 + r^3), \quad \text{when } k \to 0. \end{aligned} \tag{9.48}$$

The phase shift $\delta_0(0)$ at zero momentum is equal to zero, but there is a bound state with $E = 0$. The Levinson theorem (9.30) is violated. However, from Eq. (9.36) we have $\nu = 5/2$. The modified Levinson theorem (9.44) still holds.

7 Comparison Theorem

In this section we are going to study the comparison theorem for the D-dimensional radial Schrödinger equation. The comparison theorem of quantum mechanics states that if two real potentials are ordered $V_1(r) \leq V_2(r)$, then each corresponding pair of eigenvalues is ordered $E_1 \leq E_2$. Considering the relation between the Levinson theorem and the comparison theorem, it is necessary to show how we get the comparison theorem. In fact, the derivation is very closely related to the Sturm-Liouville theorem as shown above except for the difference in disappearance of the central potential.

For a pair of radial functions $\{R_1(r), R_2(r)\}$ in a second-order linear differential equation, it is shown from Eq. (9.4) that the radial equations can be written as

$$\frac{d^2}{dr^2}R_1(r) - \frac{C}{r^2}R_1(r) = -\frac{2M}{\hbar^2}[E_1 - V_1(r)]R_1(r), \tag{9.49}$$

$$\frac{d^2}{dr^2}R_2(r) - \frac{C}{r^2}R_2(r) = -\frac{2M}{\hbar^2}[E_2 - V_2(r)]R_2(r), \tag{9.50}$$

where

$$C = \frac{(2l + D - 1)(2l + D - 3)}{4}. \tag{9.51}$$

With these two equations, on multiplying Eq. (9.49) by $R_2(r)$ and Eq. (9.50) by $R_1(r)$, and subtracting one from the other, integrating this over the variable $r \in [0, \infty)$ leads to the following equation

$$(E_1 - E_2)\int_0^\infty R_1(r)R_2(r)dr = [V_1(r) - V_2(r)]\int_0^\infty R_1(r)R_2(r)dr, \tag{9.52}$$

which leads to the comparison theorem

$$V_1(r) \leq V_2(r) \Rightarrow E_1 \leq E_2. \tag{9.53}$$

It should be noted that the wavefunction $R(0) = R(\infty) = 0$ are used. Since the Schrödinger equation belongs to the Sturm-Liouville type differential equation, then this theorem is also called Sturm-Liouville's comparison theorem.

8 Conclusions

In this Chapter we have discussed scattering states and bound states of the D-dimensional radial Schrödinger equation with cutoff potential. The Sturm-Liouville theorem has been used to establish the non-relativistic Levinson theorem in arbitrary dimensions D. The general case has also been discussed. The comparison theorem has also been presented briefly.

Chapter 10
Generalized Hypervirial Theorem

1 Introduction

There has been a long history of attempts to calculate the matrix elements and the recurrence relations among them for some important wavefunctions such as the Coulomb-like potential, harmonic oscillator, Kratzer oscillator and others [298–324] because of their wide applications. For example, in order to simplify the calculations of the matrix elements for the Coulomb-like potential, some methods like the relation among the Laguerre polynomials, the Dirac's "q-number", the generalized hypergeometric function, the group theoretical approach, the Schrödinger radial ladder operators, the hypervirial theorem and the various sum rules were used [298–312]. The recurrence relations for the Dirac equation with the Coulomb-like potential were also discussed [320, 321]. Recently, we have studied the Klein-Gordon equation case [322]. On the other hand, the recurrence relations for the harmonic oscillator, Kratzer oscillator and Morse potential were studied by the generalized expression of the second hypervirial theorem [317, 318]. Moreover, the two-center matrix elements and the recurrence relations for the Kratzer oscillator were investigated [323], which has also been derived by means of a hypervirial-like theorem procedure [324]. It should be pointed out that almost all contributions appearing in the literature have been made in three dimensions.

Due to the recent interest in the higher dimensional field theory [13–15, 53, 325] and the fact that one can easily obtain the results in lower dimensions from the general higher dimensional results, the purpose of this Chapter is to derive general Blanchard's and Kramers' recurrence relations for arbitrary central potentials in arbitrary dimensions D. These relations are applied to study quantum systems like the Coulomb-like potential,[1] isotropic harmonic oscillator and Kratzer oscillator. In

[1] It is worth addressing that the "Coulomb-like" potential in almost all contributions mentioned above and others [326, 327] has the form $1/r$. Even though the real Coulomb-like potential in two dimensions is taken as a logarithmic form $\ln r$, its exact solutions have not been obtained except for the approximate solutions [328].

S.-H. Dong, *Wave Equations in Higher Dimensions*,
DOI 10.1007/978-94-007-1917-0_10, © Springer Science+Business Media B.V. 2011

addition, we also present the results in two dimensions because of recent interest in lower dimensional field theory and condensed matter physics.

This Chapter is organized as follows. In Sect. 2 we first present a generalized second hypervirial formula in dimensions D and then obtain the general Blanchard's and Kramers' recurrence relations. In Sect. 3 we shall apply them to obtain the corresponding Blanchard's and Kramers' recurrence relations for those certain central potentials. We also present the corresponding results in two dimensions. Finally, some concluding remarks are given in Sect. 4.

2 Generalized Blanchard's and Kramers' Recurrence Relations in Arbitrary Dimensions D

We begin by considering the Schrödinger equation in arbitrary dimensions D ($\hbar = M = 1$)

$$H\psi(\mathbf{r}) = -\frac{1}{2}\nabla^2\psi(\mathbf{r}) + V(r)\psi(\mathbf{r}) = E\psi(\mathbf{r}), \tag{10.1}$$

where $\mathbf{r}$ is a D-dimensional position vector with the Cartesian components x_1, x_2, $\ldots$, x_D. As shown previously, we know that one is able to express wavefunction as the product of a radial wavefunction $R_l(r)$ and the generalized spherical harmonics $Y^l_{l_{D-2}\ldots l_1}(\hat{\mathbf{x}})$

$$\Psi(\mathbf{r}) = R_l(r)Y^l_{l_{D-2}\ldots l_1}(\hat{\mathbf{x}}). \tag{10.2}$$

Substitution of this into Eq. (10.1) allows us to obtain the D-dimensional radial Schrödinger equation

$$\left\{\frac{d^2}{dr^2} + \frac{D-1}{r}\frac{d}{dr} - \frac{l(l+D-2)}{r^2}\right\}R_l(r) + 2[E - V(r)]R_l(r) = 0, \tag{10.3}$$

from which we may define the radial Hamiltonian as

$$H_i = -\frac{1}{2}\frac{d^2}{dr^2} - \frac{D-1}{2r}\frac{d}{dr} + \frac{l_i(l_i+D-2)}{2r^2} + V(r), \quad i = 1, 2, \tag{10.4}$$

with the properties

$$H_2|n_2l_2\rangle = E_{n_2l_2}|n_2l_2\rangle, \qquad \langle n_1l_1|H_1 = E_{n_1l_1}\langle n_1l_1|, \tag{10.5}$$

where we have used the Dirac notation $|n_il_i\rangle \equiv R_{n_il_i}(r)$.

Before further proceeding, we introduce an arbitrary function $f(r) = f$, which is independent of the potential $V(r)$ and assumed to have continuous second, third and fourth derivatives with respect to all r. From Eq. (10.4) we have

$$H_1f - fH_2 = -\frac{1}{2}f'' - f'\frac{d}{dr} - \frac{D-1}{2r}f' + \frac{(l_1-l_2)(-2+D+l_1+l_2)}{2r^2}f, \tag{10.6}$$

where the prime denotes the first derivative of the f with respect to the variable r. In the calculation, we have acted the above operator on the radial wavefunction $R_l(r)$.

Similarly, we may obtain the following expression

$$
\begin{aligned}
&H_1(H_1 f - f H_2) - (H_1 f - f H_2)H_2 \\
&= \frac{1}{4} f'''' + f''' \frac{d}{dr} + f'' \frac{d^2}{dr^2} + \frac{D-1}{2r} f''' + \frac{D-1}{r} f'' \frac{d}{dr} \\
&+ \frac{(D-1)(D-3) + (l_1 - l_2)[4 - 2D - 2(l_1 + l_2)]}{4r^2} f'' \\
&- \frac{(l_1 - l_2)(-2 + D + l_1 + l_2)}{r^2} f' \frac{d}{dr} + V'(r) f' \\
&+ \frac{(l_1 - l_2)(-2 + D + l_1 + l_2)}{r^3} f \frac{d}{dr} + \frac{\alpha}{4r^3} f' + \frac{\tau}{4r^4} f,
\end{aligned} \tag{10.7}
$$

with

$$
\begin{aligned}
\alpha &= 2l_2(D-5)(l_2 + D - 2) \\
&- (D-3)[-1 + D + 2Dl_1 + 2(-2 + l_1)l_1],
\end{aligned} \tag{10.8}
$$

$$
\begin{aligned}
\tau &= (l_1 - l_2)(-2 + D + l_1 + l_2)[(2 + l_1)(-4 + D + l_1) \\
&- (-2 + D)l_2 - l_2^2].
\end{aligned} \tag{10.9}
$$

Substitution of the following identity

$$
\frac{d^2}{dr^2} = -\frac{D-1}{r}\frac{d}{dr} + \frac{l_2(l_2 + D - 2)}{r^2} + 2V(r) - 2H_2
$$

into Eq. (10.7) leads to

$$
\begin{aligned}
&H_1(H_1 f - f H_2) - (H_1 f - f H_2)H_2 \\
&= \frac{1}{4} f'''' + \frac{D-1}{2r} f''' + \frac{\tau}{4r^4} f + 2V(r) f'' - 2f'' H_2 + V'(r) f' \\
&+ \frac{3 + D^2 - 2(-2 + l_1)l_1 - 2D(2 + l_1 - 3l_2) + 6(-2 + l_2)l_2}{4r^2} f'' \\
&+ \frac{\alpha}{4r^3} f' + B\left(f, l_1, l_2, \frac{d}{dr}\right)
\end{aligned} \tag{10.10}
$$

with

$$
\begin{aligned}
B\left(f, l_1, l_2, \frac{d}{dr}\right) &= f''' \frac{d}{dr} - \frac{(l_1 - l_2)(-2 + D + l_1 + l_2)}{r^2} f' \frac{d}{dr} \\
&+ \frac{(l_1 - l_2)(-2 + D + l_1 + l_2)}{r^3} f \frac{d}{dr}.
\end{aligned} \tag{10.11}
$$

Specially, if taking $f = r^\kappa$ (κ is assumed as a non-negative integer), we have

$$
B\left(r^\kappa, m_1, m_2, \frac{d}{dr}\right) = (\kappa - 1)[\kappa(\kappa - 2) - (l_1 - l_2)(D - 2 + l_1 + l_2)]r^{\kappa - 3} \frac{d}{dr}. \tag{10.12}
$$

To remove the operator d/dr appearing in Eq. (10.12), by considering $f = r^{\kappa - 2}$ and Eq. (10.6), we have

$$r^{\kappa-3}\frac{d}{dr} = \frac{1}{\kappa-2}(r^{\kappa-2}H_2 - H_1 r^{\kappa-2}) + \frac{1}{2}\left[\frac{(l_1-l_2)(D-2+l_1+l_2)}{\kappa-2} - \kappa - D + 4\right] r^{\kappa-4}. \quad (10.13)$$

Combining Eqs. (10.12) and (10.13) into Eq. (10.10) and taking $f = r^\kappa$ result in the following generalized second hypervirial for an arbitrary central potential $V(r)$

$$\begin{aligned} H_1(H_1 r^\kappa - r^\kappa H_2) &- (H_1 r^\kappa - r^\kappa H_2)H_2 \\ &= -\kappa(\kappa-1)(H_1 r^{\kappa-2} + r^{\kappa-2}H_2) \\ &\quad + \frac{\kappa-1}{\kappa-2}(l_1-l_2)(D-2+l_1+l_2)(H_1 r^{\kappa-2} - r^{\kappa-2}H_2) \\ &\quad + \kappa r^{\kappa-1}V'(r) + 2\kappa(\kappa-1)r^{\kappa-2}V(r) + \eta r^{\kappa-4} \end{aligned} \quad (10.14)$$

with

$$\eta = \frac{-\kappa[(l_1-l_2)^2 - (\kappa-2)^2](D-\kappa+l_1+l_2)(-4+D+\kappa+l_1+l_2)}{4(\kappa-2)}. \quad (10.15)$$

Based on Eqs. (10.5) and (10.14), we may obtain a useful general Blanchard's recurrence relation for arbitrary central potential wavefunction in arbitrary dimensions D

$$\begin{aligned} (E_{n_1l_1} &- E_{n_2l_2})^2\langle n_1l_1|r^\kappa|n_2l_2\rangle \\ &= \eta\langle n_1l_1|r^{\kappa-4}|n_2l_2\rangle \\ &\quad + \left\{\frac{\kappa-1}{\kappa-2}(l_1-l_2)(D-2+l_1+l_2)(E_{n_1l_1} - E_{n_2l_2}) \right. \\ &\quad \left. - \kappa(\kappa-1)(E_{n_1l_1} + E_{n_2l_2})\right\}\langle n_1l_1|r^{\kappa-2}|n_2l_2\rangle \\ &\quad + \kappa\langle n_1l_1|V'(r)r^{\kappa-1}|n_2l_2\rangle \\ &\quad + 2\kappa(\kappa-1)\langle n_1l_1|V(r)r^{\kappa-2}|n_2l_2\rangle. \end{aligned} \quad (10.16)$$

For diagonal case $n_i = n$ and $l_i = l$ $(i = 1, 2)$, we obtain a simple expression for the Kramers' recurrence relation

$$\begin{aligned} &\frac{1}{4}(\kappa-2)(D-\kappa+2l)(D-4+\kappa+2l)\langle nl|r^{\kappa-4}|nl\rangle \\ &\quad - 2(\kappa-1)E_{nl}\langle nl|r^{\kappa-2}|nl\rangle + \langle nl|V'(r)r^{\kappa-1}|nl\rangle \\ &\quad + 2(\kappa-1)\langle nl|V(r)r^{\kappa-2}|nl\rangle = 0. \end{aligned} \quad (10.17)$$

For $\kappa = 0, 2$, from Eq. (10.16) we obtain a very useful Pasternack-Sternheimer selection rule in arbitrary dimensions D

$$\frac{1}{2}(l_1-l_2)(D-2+l_1+l_2)\langle n_1l_1|r^{-2}|n_2l_2\rangle = (E_{n_1l_1} - E_{n_2l_2})\langle n_l l_1|n_2l_2\rangle, \quad (10.18)$$

which means that this selection rule is independent of the central potential $V(r)$.

Moreover, it is worth studying the recurrence relation among the off-diagonal matrix elements of $V'(r)$. For $\kappa = 1$, it is shown from Eq. (10.16) that

$$\langle n_1l_1|V'(r)|n_2l_2\rangle = (E_{n_1l_1} - E_{n_2l_2})^2\langle n_1l_1|r|n_2l_2\rangle - \frac{[(l_1-l_2)^2-1](l_1+l_2+D-1)(l_1+l_2+D-3)}{4} \cdot \langle n_1l_1|r^{-3}|n_2l_2\rangle. \tag{10.19}$$

3 Applications to Certain Central Potentials

3.1 Coulomb-like Potential Case

Let us study the Coulomb-like potential $V(r) = -\xi/r$ in dimensions D. The eigenvalues are given in [325]

$$E_{n_i} = -\frac{\xi^2}{2[n_i + (D-3)/2]^2}. \tag{10.20}$$

By replacing κ by $\kappa + 2$ in Eq. (10.16), we may obtain the general Blanchard's recurrence relation for this potential

$$\frac{\kappa\xi^4}{2}\left(\frac{1}{[n_2+(D-3)/2]^2} - \frac{1}{[n_1+(D-3)/2]^2}\right)^2 \langle n_1l_1|r^{\kappa+2}|n_2l_2\rangle + \sum_{j=0}^{2} D_j\langle n_1l_1|r^{\kappa-j}|n_2l_2\rangle = 0 \tag{10.21}$$

where

$$D_0 = \xi^2(\kappa+1)\left\{(l_2-l_1)(D-2+l_1+l_2) \times \left[\frac{1}{[n_2+(D-3)/2]^2} - \frac{1}{[n_1+(D-3)/2]^2}\right] - \kappa(\kappa+2)\left[\frac{1}{[n_2+(D-3)/2]^2} + \frac{1}{[n_1+(D-3)/2]^2}\right]\right\}, \tag{10.22}$$

$$D_1 = 2\kappa\xi(\kappa+2)(2\kappa+1),$$
$$D_2 = \frac{1}{2}(\kappa+2)[(l_1-l_2)^2-\kappa^2][(l_1+l_2+D-2)^2-\kappa^2]. \tag{10.23}$$

For diagonal case, we may obtain the Kramers' recurrence relation

$$\frac{1}{4}\kappa[\kappa^2-(D+2l-2)^2]\langle nl|r^{\kappa-2}|nl\rangle + \xi(2\kappa+1)\langle nl|r^{\kappa-1}|nl\rangle - \frac{\xi^2(1+\kappa)}{[n+(D-3)/2]^2}\langle nl|r^{\kappa}|nl\rangle = 0. \tag{10.24}$$

Let us discuss some special cases. First, we study the case $\kappa = |l_1 - l_2|$ or $\kappa = l_1 + l_2 + D - 2$. As a result, it is found from Eq. (10.23) that the coefficient $D_2 = 0$. The corresponding general Blanchard's recurrence relation is simplified as

$$\frac{\kappa'\xi^4}{2}\left\{\frac{1}{[n_2+(D-3)/2]^2}-\frac{1}{[n_1+(D-3)/2]^2}\right\}^2\langle n_1l_1|r^{\kappa'+2}|n_2l_2\rangle$$
$$+\sum_{i=0}^{1}D_i\langle n_1l_1|r^{\kappa'-i}|n_2l_2\rangle=0, \tag{10.25}$$

where the parameter κ appearing in the coefficients D_0 and D_1 is replaced by $\kappa'=|l_1-l_2|$ or $\kappa'=l_1+l_2+D-2$.

Second, we consider the case $n_1=n_2=n$, but $l_1\neq l_2$. If so, we have

$$\frac{\xi^2}{[n+(D-3)/2]^2}\kappa(\kappa+1)\langle nl_1|r^{\kappa}|nl_2\rangle$$
$$=\xi\kappa(1+2\kappa)\langle nl_1|r^{\kappa-1}|nl_2\rangle$$
$$+\frac{[(l_1-l_2)^2-\kappa^2][(l_1+l_2+D-2)^2-\kappa^2]}{4}\langle nl_1|r^{\kappa-2}|nl_2\rangle, \tag{10.26}$$

from which we may obtain a special result for $\kappa=0$

$$\langle nl_1|\frac{1}{r^2}|nl_2\rangle=0. \tag{10.27}$$

This can also be obtained from Eq. (10.18). On the other hand, let us consider the following two interesting cases for $\kappa=|l_1-l_2|$ and $\kappa=l_1+l_2+D-2$. It is shown from Eq. (10.26) that

$$\frac{\langle nl_1|r^{|l_1-l_2|}|nl_2\rangle}{\langle nl_1|r^{|l_1-l_2|-1}|nl_2\rangle}=\frac{[n+(D-3)/2]^2}{\xi}\cdot\frac{1+2|l_1-l_2|}{1+|l_1-l_2|}, \tag{10.28}$$

for $\kappa=|l_1-l_2|$, and

$$\frac{\langle nl_1|r^{l_1+l_2+D-2}|nl_2\rangle}{\langle nl_1|r^{l_1+l_2+D-3}|nl_2\rangle}=\frac{[n+(D-3)/2]^2}{\xi}\cdot\frac{2(l_1+l_2+D)-3}{l_1+l_2+D-1}, \tag{10.29}$$

for $\kappa=l_1+l_2+D-2$.

Third, from Eq. (10.19) we have for the Coulomb-like potential $V(r)=-\xi/r$

$$\xi\langle n_1l_1|r^{-2}|n_2l_2\rangle$$
$$=\frac{\xi^4}{4}\left\{\frac{1}{[n_2+(D-3)/2]^2}-\frac{1}{[n_1+(D-3)/2]^2}\right\}^2\langle n_1l_1|r|n_2l_2\rangle$$
$$-\frac{[(l_1-l_2)^2-1](l_1+l_2+D-1)(l_1+l_2+D-3)}{4}\langle n_1l_1|r^{-3}|n_2l_2\rangle, \tag{10.30}$$

from which, together with Eq. (10.27), we have

$$\langle nl_1|r^{-3}|nl_2\rangle=0. \tag{10.31}$$

Fourth, we study two special cases for the general Kramers' recurrence relation (10.24), i.e.,

$$\langle nl|r^{-1}|nl\rangle = \frac{\xi}{[n+(D-3)/2]^2}, \qquad \text{for } \kappa = 0, \tag{10.32}$$

$$\frac{\langle nl|r^{2l+D-2}|nl\rangle}{\langle nl|r^{2l+D-3}|nl\rangle} = \frac{[n+(D-3)/2]^2}{\xi} \cdot \frac{4l+2D-3}{2l+D-1},$$
$$\text{for } \kappa = 2l+D-2, \tag{10.33}$$

which can also be obtained from Eqs. (10.28) and (10.29) under the condition $l_1 = l_2 = l$.

3.2 Harmonic Oscillator

Let us study the isotropic harmonic oscillator $V(r) = \omega^2 r^2/2$ with $M = 1$ in dimensions D. The eigenvalues are given in [326, 327]

$$E_n = \omega(D/2+n), \quad n = 2n'+l, \; n' = 0, 1, 2, \ldots. \tag{10.34}$$

We find from Eq. (10.16) that the Blanchard's recurrence relation for the harmonic oscillator is given by

$$\begin{aligned}
&\omega^2[(n_1-n_2)^2-\kappa^2]\langle n_1l_1|r^\kappa|n_2l_2\rangle \\
&\quad = \eta\langle n_1l_1|r^{\kappa-4}|n_2l_2\rangle \\
&\qquad + \left[\frac{\kappa-1}{\kappa-2}\omega(n_1-n_2)(l_1-l_2)(D-2+l_1+l_2) \right.\\
&\qquad \left. -\,\omega\kappa(\kappa-1)(n_1+n_2+D)\right]\langle n_1l_1|r^{\kappa-2}|n_2l_2\rangle,
\end{aligned} \tag{10.35}$$

from which we obtain the corresponding Kramers' recurrence relation

$$\begin{aligned}
&\frac{\kappa-2}{4}(D-\kappa+2l)(-4+D+\kappa+2l)\langle nl|r^{\kappa-4}|nl\rangle \\
&\quad = 2\omega(\kappa-1)(n+D/2)\langle nl|r^{\kappa-2}|nl\rangle - \kappa\omega^2\langle nl|r^\kappa|nl\rangle,
\end{aligned} \tag{10.36}$$

from which, we obtain the following two identities

$$\langle nl|r|nl\rangle = \frac{(D-1+2l)(D-3+2l)}{4\omega^2}\langle nl|r^{-3}|nl\rangle, \quad \text{for } \kappa = 1, \tag{10.37}$$

$$\langle nl|r^2|nl\rangle = \frac{n+D/2}{\omega}, \quad \text{for } \kappa = 2. \tag{10.38}$$

For $\kappa = 1$, from Eq. (10.35) we obtain a more general identity for the off-diagonal case

$$\begin{aligned}
&\omega^2[(n_1-n_2)^2-1]\langle n_1l_1|r|n_1l_1\rangle \\
&\quad = \frac{[(l_1-l_2)^2-1](l_1+l_2+D-1)(l_1+l_2+D-3)}{4}\langle n_1l_1|r^{-3}|n_1l_1\rangle.
\end{aligned} \tag{10.39}$$

3.3 Kratzer Oscillator

Let us investigate the Kratzer oscillator [6]

$$V(r) = V_0\left(\frac{b^2}{r^2} - \frac{2b}{r}\right). \tag{10.40}$$

The exact solutions in D dimensions are given by [322]

$$E_{nl} = -\frac{8b^2V_0^2}{[2n - 2l - 1 + \sqrt{8b^2V_0 + (2l + D - 2)^2}]^2}. \tag{10.41}$$

It is shown from Eq. (10.16) that the Blanchard's recurrence relation is obtained as

$$\begin{aligned}
&(E_{n_1l_1} - E_{n_2l_2})^2\langle n_1l_1|r^\kappa|n_2l_2\rangle \\
&\quad = [2V_0b^2\kappa(\kappa - 2) + \eta]\langle n_1l_1|r^{\kappa-4}|n_2l_2\rangle \\
&\qquad + \left[\frac{\kappa - 1}{\kappa - 2}(l_1 - l_2)(D - 2 + l_1 + l_2)(E_{n_1l_1} - E_{n_2l_2})\right. \\
&\qquad \left. - \kappa(\kappa - 1)(E_{n_1l_1} + E_{n_2l_2})\right]\langle n_1l_1|r^{\kappa-2}|n_2l_2\rangle \\
&\qquad + 2V_0b\kappa(3 - 2\kappa)\langle n_1l_1|r^{\kappa-3}|n_2l_2\rangle.
\end{aligned} \tag{10.42}$$

On the other hand, we may obtain the corresponding general Kramers' recurrence relation

$$\begin{aligned}
&(\kappa - 2)\left[2V_0b^2 + \frac{(2l + D - \kappa)(2l + D + \kappa - 4)}{4}\right]\langle nl|r^{\kappa-4}|nl\rangle \\
&\quad = 2(\kappa - 1)E_{nl}\langle nl|r^{\kappa-2}|nl\rangle + 2V_0b(2\kappa - 3)\langle nl|r^{\kappa-3}|nl\rangle.
\end{aligned} \tag{10.43}$$

In addition, for $\kappa = 1$ we may obtain other interesting and important particular results from Eq. (10.42), i.e.,

$$\begin{aligned}
&(E_{n_1l_1} - E_{n_2l_2})^2\langle n_1l_1|r|n_2l_2\rangle \\
&\quad = 2V_0b\langle n_1l_1|r^{-2}|n_2l_2\rangle \\
&\qquad + \left\{\frac{[(l_1 - l_2)^2 - 1](D - 1 + l_1 + l_2)(-3 + D + l_1 + l_2)}{4} - 2V_0b^2\right\} \\
&\qquad \cdot \langle n_1l_1|r^{-3}|n_2l_2\rangle
\end{aligned} \tag{10.44}$$

from which, we can obtain the following identity

$$\langle nl|\frac{1}{r^2}|nl\rangle = \left\{b + \frac{(2l + D - 1)(2l + D - 3)}{8bV_0}\right\}\langle nl|\frac{1}{r^3}|nl\rangle, \tag{10.45}$$

which can also be derived from Eq. (10.43) by setting $\kappa = 1$. We now study Eq. (10.43) in detail. It is interesting to find that some useful recurrence relations among the diagonal matrix elements can be obtained from Eq. (10.43)

$$E_{nl} = \begin{cases} -V_0 b\langle nl|r^{-1}|nl\rangle, \\ \frac{[2V_0 b^2 + \frac{(2l+D-3)(2l+D-1)}{4}]\langle nl|r^{-1}|nl\rangle - 6bV_0}{4\langle nl|r|nl\rangle}, \\ \frac{[2V_0 b^2 + \frac{(2l+D-4)(2l+D)}{4}] - 5bV_0\langle nl|r|nl\rangle}{3\langle nl|r^2|nl\rangle}, \end{cases} \tag{10.46}$$

for $\kappa = 2, 3$ and 4, respectively.

In addition, we may obtain the recurrence relation and identity for the Morse potential $V(r) = D(e^{-2\beta r} - 2e^{-\beta r})$, as shown in Refs. [317, 318]. From Eq. (10.19), we are able to obtain the following recurrence relation among off-diagonal matrix elements

$$\begin{aligned} &2\beta D\langle n_1 l_1|e^{-\beta r} - e^{-2\beta r}|n_2 l_2\rangle \\ &\quad = (E_{n_1 l_1} - E_{n_2 l_2})^2 \langle n_1 l_1|r|n_2 l_2\rangle \\ &\quad\quad - \frac{[(l_1 - l_2)^2 - 1](l_1 + l_2 + D - 1)(l_1 + l_2 + D - 3)}{4}\langle n_1 l_1|r^{-3}|n_2 l_2\rangle \end{aligned} \tag{10.47}$$

and the identity between two particular diagonal matrix elements

$$\langle nl|e^{-\beta r} - e^{-2\beta r}|nl\rangle = \frac{(2l + D - 1)(2l + D - 3)}{8\beta D}\langle nl|r^{-3}|nl\rangle. \tag{10.48}$$

Specially, when $D = 3$ and $l = 0$ we have

$$\langle nl|e^{-\beta r}|nl\rangle = \langle nl|e^{-2\beta r}|nl\rangle. \tag{10.49}$$

This coincides with the result given in [329]. It should be noted that the exact solutions of this system have not been obtained up to now. Nevertheless, we can predict the results (10.47)–(10.49) in theory. Particularly, we find that the formulas (10.48) and (10.49) are independent of energy levels of this quantum system in dimensions D.

Before ending this section, we present the results in two dimensions. This can be easily realized by setting $D = 2$ and through replacing l_1, l_2 by m_1, m_2, respectively. For the Blanchard's recurrence relation, we have

$$\begin{aligned} &\frac{1}{\kappa - 1}[(E_{n_1 m_1} - E_{n_2 m_2})^2 \langle n_1 m_1|r^\kappa|n_2 m_2\rangle - \lambda\langle n_1 m_1|r^{\kappa-4}|n_2 m_2\rangle \\ &\quad - \kappa\langle n_1 m_1|V'(r) r^{\kappa-1}|n_2 m_2\rangle] \\ &\quad = \left[\frac{m_1^2 - m_2^2}{\kappa - 2}(E_{n_1 m_1} - E_{n_2 m_2}) - \kappa(E_{n_1 m_1} + E_{n_2 m_2})\right] \\ &\quad \cdot \langle n_1 m_1|r^{\kappa-2}|n_2 m_2\rangle + 2\kappa\langle n_1 m_1|V(r) r^{\kappa-2}|n_2 m_2\rangle, \end{aligned} \tag{10.50}$$

where

$$\lambda = \frac{-\kappa[(m_1 + m_2)^2 - (\kappa - 2)^2][(m_1 - m_2)^2 - (\kappa - 2)^2]}{4(\kappa - 2)}. \tag{10.51}$$

For the Kramers' recurrence relation, however, we have

$$-\frac{1}{4}(\kappa-2)[(\kappa-2)^2-4m^2]\langle nm|r^{\kappa-4}|nm\rangle$$
$$=2(\kappa-1)E_{nm}\langle nm|r^{\kappa-2}|nm\rangle-\langle nm|V'(r)r^{\kappa-1}|nm\rangle$$
$$-2(\kappa-1)\langle nm|V(r)r^{\kappa-2}|nm\rangle. \qquad (10.52)$$

Similarly, we may obtain similar recurrence relations for $\kappa = 0, 2$ and $\kappa = 1$ and the corresponding general Blanchard's and Kramers' recurrence relations for those physical potentials. However, we do not present them here for simplicity.

4 Concluding Remarks

In this Chapter based on the Hamiltonian identity we have presented a useful generalized second hypervirial for arbitrary central potential wavefunction in dimensions D and shown that this formula is very powerful in deriving the general Blanchard's and Kramers' recurrence relations. Interestingly, we have found that the generalized Pasternack-Sternheimer selection rule is independent of $V(r)$ for $f = r^\kappa$ with $\kappa = 0, 2$. We have applied the proposed general Blanchard's and Kramers' recurrence relations to study the quantum systems for three certain central potentials. Some interesting and useful results have been obtained simply. It should be pointed out that the present approach can be extended to consider $f \neq r^\kappa$ off-diagonal matrix elements for arbitrary central potential wavefunction. For example, we have established the recurrence relations between the exponential functions and the powers of the radial function for the Morse potential. Finally, we have briefly presented the general Blanchard's and Kramers' recurrence relations in two dimensions.

Before ending this Chapter, we give some useful remarks here. First, it should be noted that Eqs. (10.32) and (10.38) are two well known virial relations. Second, in terms of Eqs. (10.50) and (10.52), it is possible to obtain the general Blanchard's and Kramers' recurrence relations among the matrix elements $\langle n_1m_1|\ln r\ r^{\kappa-2}|n_2m_2\rangle$ for the logarithmic potential $\ln r$ in two dimensions. The merit of this method is that we need not know the exact solutions of the studied quantum system, but we may predict some useful results in theory. Third, due to the specificity of the Klein-Gordon (KG) equation, i.e., the energy levels are involved in the potential $V(r)$, which arises from the KG equation $[E - V(r)]^2\Phi(\mathbf{r}) = (m^2c^4 - \hbar^2c^2\nabla^2)\Phi(\mathbf{r})$, it seems that the present approach is unsuitable for this equation. Nevertheless, we have applied the so-called Kramers' approach to obtain the recurrence relation for the Coulomb-like potential case [322]. Fourth, it is possible to use this method to study the Dirac equation with the Coulomb-like potential in dimensions D.

Chapter 11
Exact and Proper Quantization Rules and Langer Modification

1 Introduction

A fundamental interest in quantum mechanics is to obtain the right result without invoking the full mathematics of the Schrödinger equation. Since last decade there has been a great revival of interest in semiclassical methods for obtaining approximate solutions to the Schrödinger equation. Among them, the WKB approximation and its generalization have attracted much attention to many authors [330–333] since this method is proven to be useful in obtaining an approximate solution to the Schrödinger equation with solvable potentials.

Since the radial Schrödinger equation in three dimensions can be written in a similar form to that of one-dimensional case it is not surprising to lead us to apply one-dimensional quantization rule to study energy levels of quantum system in three dimensions. However, this is not always correct and sometimes becomes invalid. For example, when one utilizes the first-order WKB integral to derive the eigenvalues of hydrogen atom and harmonic oscillator, the quantity $\ell(\ell+1)$ has to be replaced by $(\ell+1/2)^2$ in quantization rule [334]. Such study was first advocated by Young and Uhlenbeck [335]. In 1937 this problem was re-considered and explained well by Langer [336]. It should be noted that replacing $\ell(\ell+1)$ by $(\ell+1/2)^2$ is valid only for the first-order integral considered, but the Langer modification is no longer valid when the second-order integral is included [337, 338]. On the other hand, the algebraic procedure to adjust Langer modification for higher-order integral, the discussions related to its physical nature and other studies have been carried out [339–347].

Recently, the modified WKB method proposed by Friedrich and Trost [342] has been found to avoid the Langer modification. This can be realized by introducing a non-integral Maslov index. Furthermore, an exact quantization rule presented by Ma and his coauthors has been shown to be powerful in calculating the energy levels of some exactly solvable quantum systems [348–351]. In fact, this exact quantization rule was relied in some sense on the previous work by Cao and his collaborators [352–355]. This method has become one of several important formalisms to deal with solvable quantum systems, but the integrals, in particular the calculations of the

S.-H. Dong, *Wave Equations in Higher Dimensions*,
DOI 10.1007/978-94-007-1917-0_11, © Springer Science+Business Media B.V. 2011

quantum correction term become rather complicated. To overcome this difficulty, we have improved it and found a proper quantization rule [356, 357]. By this rule the energy spectra of all solvable systems can be determined from its ground state energy only. The trick and simplicity of the rule come from its meaning—whenever the number of the nodes of $\phi(x)$ or the number of the nodes of the wavefunction $\psi(x)$ increases by one, the momentum integral $\int_{x_A}^{x_B} k(x)dx$ will increase by π. This proper quantization rule has ended the history of semiclassical quantization rules and opened a new formalism to carry out all solvable potentials.

The purpose of this Chapter is following. We shall first give a brief review of the fundamental development of the quantization rule including the WKB method, the exact quantization rule and proper quantization rule. After that we shall establish the relation between the proper quantization rule, the Maslov index and the Langer modification. As illustrations we shall choose a few solvable potentials and study them via these quantization rules.

This Chapter is organized as follows. In Sect. 2 we briefly review the WKB method since it is closely related to recently proposed exact and proper quantization rules. We shall review the exact quantization rule in Sect. 3. As an illustration, we present its application to asymmetric trigonometric Rosen-Morse potential in Sect. 4. Section 5 is devoted to extension of the exact quantization rule, i.e., the proper quantization rule. In Sect. 6 the performance of the proper quantization rule is demonstrated in four different situations, the harmonic oscillator, modified Rosen-Morse potential, Coulombic ring-shaped noncentral Hartmann system, the Manning-Rosen effective potential. In Sect. 7 the evaluation of the Langer modification and Maslov index in D dimensions are carried out. The results for most exactly solvable potentials are presented in Tables 11.1, 11.2, 11.3. In Sect. 8 we illustrate the calculations of the logarithmic derivatives of wavefunction. Finally, in Sect. 9 we will summarize our conclusions.

2 WKB Approximation

The success of quantum theory in atomic domain prompted physicists to apply the Bohr atomic model to complex atoms. It was soon obvious that although the Bohr model is basically correct, it has many minor flaws. Some flaws in the details of the Bohr-Sommerfeld-Wilson (BSW) quantization hypothesis were pointed out by Einstein [358] in 1917, and subsequently corrected by Brillouin [359] in 1926 and by Keller [360] in 1958. Schrödinger's wave equation of quantum mechanics was published in 1926, and in the same year Wentzel, Kramers and Brillouin developed the semiclassical approximation now known as the Wentzel-Kramers-Brillouin (WKB) [361–363] approximation. It is of importance because it exhibits the connection with the older quantization rules of Bohr and Sommerfeld. Important contributions were also made by Langer [336] in 1937 and by Maslov [364] in 1972. The modern form of the semiempirical hypothesis, which elucidates the quantum mechanical formulation of level energies, is known as the Maslov-indexed Einstein-Brillouin-Keller (EBK) quantization [365, 366].

Table 11.1 Some useful integral formulae

$\int_{r_1}^{r_2} \frac{1}{r}\sqrt{(r-r_1)(r_2-r)}dr = \frac{\pi}{2}(r_1+r_2) - \pi\sqrt{r_1 r_2}$
$\int_{r_1}^{r_2} \sqrt{(r-r_1)(r_2-r)}dr = \frac{\pi}{8}(r_2-r_1)^2$
$\int_{r_1}^{r_2} \sqrt{(r-r_1)/(r_2-r)}dr = \int_{r_1}^{r_2} \sqrt{(r_2-r)/(r-r_1)}dr = \frac{\pi}{2}(r_2-r_1)$
$\int_{r_1}^{r_2} r\sqrt{(r-r_1)/(r_2-r)}dr = \frac{\pi}{8}(r_2-r_1)(r_1+3r_2)$
$\int_{r_1}^{r_2} r\sqrt{(r_2-r)/(r-r_1)}dr = \frac{\pi}{8}(r_2-r_1)(3r_1+r_2)$
$\int_{r_1}^{r_2} \frac{dr}{1+r^2}\sqrt{(r-r_1)(r_2-r)} = -\pi + \frac{\pi}{2}\sqrt{1-r_1r_2+\sqrt{(1+r_1^2)(1+r_2^2)}}$
$\int_{r_1}^{r_2} \frac{dr}{(a+br)\sqrt{(r-r_1)(r_2-r)}} = \frac{\pi}{\sqrt{(a+br_1)(a+br_2)}}$
$\int_{r_1}^{r_2} \frac{dr}{1-r^2}\sqrt{(r-r_1)(r_2-r)} = \frac{\pi}{2}[2-\sqrt{(1-r_1)(1-r_2)} - \sqrt{(1+r_1)(1+r_2)}]$

Table 11.2 Abbreviated symbols and exactly solvable potentials

Abbreviated symbols	Solvable potentials	Formulas
HO	harmonic oscillator	$\frac{1}{2}M\omega^2x^2$
MP	Morse potential	$U_0(e^{-2x/a} - U_1e^{-x/a})$
GMP	generalized Morse potential	$U_0[1-b(e^{r/a}-1)^{-1}]^2$
SRMP	symmetric Rosen-Morse potential	$-U_0\,\mathrm{sech}^2(x/a)$
ARMP	asymmetric Rosen-Morse potential	$-U_0\,\mathrm{sech}^2(x/a) + U_1\tanh(x/a)$
PT-I	Pöschl-Teller I	$\frac{\hbar^2}{2Ma^2}[\frac{\mu(\mu-1)}{\sin^2(x/a)} + \frac{\lambda(\lambda-1)}{\cos^2(x/a)}]$
PT-II	Pöschl-Teller II	$\frac{\hbar^2}{2Ma^2}[\frac{\mu(\mu-1)}{\sinh^2(x/a)} - \frac{\lambda(\lambda+1)}{\cosh^2(x/a)}]$
EP	Eckart potential	$U_0\,\mathrm{csch}^2(r/a) - U_1\coth(r/a)$
HP	Hulthén potential	$-\frac{U_0}{e^{r/a}-1}$
STRM	symmetric trigonometric Rosen-Morse	$U_0\cot^2(\pi x/a)$
ATRM	asymmetric trigonometric Rosen-Morse	$U_0\cot^2(\pi x/a) + U_1\cot(\pi x/a)$
HO3D	harmonic oscillator in $3D$	$\frac{1}{2}M\omega^2r^2 + \frac{\ell(\ell+1)\hbar^2}{2Mr^2}$
HA3D	hydrogen atom in $3D$	$-\frac{e^2}{r} + \frac{\hbar^2\ell(\ell+1)}{2Mr^2}$
HOD	harmonic oscillator in D dimensions	$\frac{1}{2}M\omega^2r^2 + \frac{\hbar^2\ell'(\ell'+1)}{2Mr^2}$
HAD	hydrogen atom in D dimensions	$-\frac{e^2}{r} + \frac{\hbar^2\ell'(\ell'+1)}{2Mr^2}$

The well known conventional quantization rule in the EBK method can be expressed in the form

$$\int_{x_1}^{x_2} k(x)dx = \left(n + \frac{\mu}{4}\right)\pi, \tag{11.1}$$

where μ is the Maslov index. It is found that when $\mu = 0$, Eq. (11.1) reduces to the Bohr-Sommerfeld form. The most popular WKB approximation is the special case of Eq. (11.1) with $\mu = 2$, i.e.,

$$\int_{x_1}^{x_2} k(x)dx = \left(n + \frac{1}{2}\right)\pi. \tag{11.2}$$

Table 11.3 Solvable potentials, Maslov index μ and eigenvalues E_n

Solvable potentials	Maslov index μ	Eigenvalues E_n
HO	2	$\hbar\omega(n+\frac{1}{2})$
MP	2	$-\frac{U_0}{4}[U_1-\frac{(2n+1)}{a\sqrt{U_0}}]$
GMP	$4a(C-b\sqrt{2MU_0}/\hbar)$	$U_0-\frac{\hbar^2}{2M}[\frac{MU_0b(b+2)}{\hbar^2(C+n/a)}-\frac{C+n/a}{2}]^2$, $C=[1+\sqrt{1+8MU_0a^2b^2/\hbar^2}]/(2a)$
SRMP	$4a(\sqrt{2MU_0}/\hbar-C)$	$-\frac{\hbar^2(C-n/a)^2}{2M}$, $C=[\sqrt{1+8MU_0a^2/\hbar^2}-1]/(2a)$
ARMP	$4a(\sqrt{2MU_0}/\hbar-C)$	$-[\frac{\hbar^2(C-n/a)^2}{2M}+\frac{MU_1^2}{2\hbar^2(C-n/a)^2}]$, $C=[\sqrt{1+8MU_0a^2/\hbar^2}-1]/(2a)$
PT-I	$2(\mu+\lambda)$ $-4(\sqrt{\mu(\mu-1)}+\sqrt{\lambda(\lambda-1)})$	$\frac{\hbar^2(\mu+\lambda+2n)^2}{2Ma^2}$
PT-II	$2(\mu-\lambda)$ $-4(\sqrt{\mu(\mu-1)}-\sqrt{\lambda(\lambda+1)})$	$-\frac{\hbar^2(\lambda-\mu-2n)^2}{2Ma^2}$
EP	$4a(C-\sqrt{2MU_0}/\hbar)$	$-[\frac{MU_1^2}{2\hbar^2(C+n/a)^2}+\frac{\hbar^2(C+n/a)^2}{2M}]$, $C=[\sqrt{1+8MU_0a^2/\hbar^2}+1]/(2a)$
HP	4	$-U_0[\frac{a\sqrt{2MU_0}}{2\hbar(n+1)}+\frac{\hbar(n+1)}{2a\sqrt{2MU_0}}]^2$
STRM	$\frac{4a}{\pi}(C-\sqrt{2MU_0}/\hbar)$	$\frac{\hbar^2}{2Ma^2}(aC+n\pi)^2-U_0$, $C=\frac{\pi}{2a}[\sqrt{1+\frac{8MU_0a^2}{\pi^2\hbar^2}}+1]$
ATRM	$\frac{4a}{\pi}(C-\sqrt{2MU_0}/\hbar)$	$\frac{\hbar^2(aC+n\pi)^2}{2Ma^2}-\frac{Ma^2U_1^2}{2\hbar^2(aC+n\pi)^2}-U_0$, $C=\frac{\pi}{2a}[\sqrt{1+\frac{8MU_0a^2}{\pi^2\hbar^2}}+1]$
HO3D	$3+2\ell-2\sqrt{\ell(\ell+1)}$	$\hbar\omega(n+3/2)$
HA3D	$4(\ell+1-\sqrt{\ell(\ell+1)})$	$-\frac{Me^4}{2n^2\hbar^2}$
HOD	$D+2(\ell-\sqrt{\ell'(\ell'+1)})$	$\hbar\omega(n+D/2)$
HAD	$4(\ell'+1-\sqrt{\ell'(\ell'+1)})$	$-\frac{Me^4}{2(n+\frac{D-3}{2})^2\hbar^2}$

For a diverse class of problems and a variety of potentials, the WKB quantization rule has proven to be very useful in finding an approximate solution of the one-dimensional Schrödinger equation. However, except for the harmonic oscillator and the Morse potential, the WKB quantization rule fails to reproduce exactly analytic results for other solvable potentials. Further, in the WKB analysis of the radial Schrödinger equation, exact result can not be obtained unless the classical Hamiltonian is slightly modified. The Langer modification, which prescribes replacing the angular momentum factor $\ell(\ell+1)$ in the effective potential by $(\ell+1/2)^{1/2}$ [335, 336, 367], is seen as a standard ingredient of WKB theory for the quantum systems with radial symmetry [344, 368]. However, it emerges that the algebraic procedure for adjusting the Langer-type corrections for higher-order approximations is quite difficult and cumbersome [339].

Frequently alternative methods of improving the conventional WKB approximation are proposed, such as supersymmetry quantum mechanics [369], phase loss method [341–343] and periodic orbit theory [370, 371]. Among various versions of the modified quantum conditions [372, 373], one of the promising methods is the exact quantization rule approach [348, 349, 354], which allows one to determine eigenvalues of known analytically solvable potentials without ever having to solve the Schrödinger equation. As mentioned above, this method has become one of several important formalisms to deal with solvable quantum systems, but the integrals, in particular the calculations of the quantum correction term become rather complicated. To overcome this difficulty we have proposed a proper quantization rule.

3 Exact Quantization Rule

Here, we give a brief and necessarily sketchy review of the exact quantization rule method which is necessary for the subsequent sections. For more elaborate discussions, the reader is referred to Refs. [348, 349].

The one-dimensional Schrödinger equation is given by

$$\frac{d^2}{dx^2}\psi(x) = -\frac{2M}{\hbar^2}[E - V(x)]\psi(x), \tag{11.3}$$

where the potential $V(x)$ is a piecewise continuous real function of x satisfying

$$\begin{aligned} &V(x) < E, \quad && x_A < x < x_B, \\ &V(x) = E, \quad && x = x_A \text{ or } x = x_B, \\ &V(x) > E, \quad && x \in (-\infty, x_A) \text{ or } x \in (x_B, \infty), \end{aligned} \tag{11.4}$$

where x_A and x_B are two turning points determined by $E = V(x)$.

The Schrödinger equation is equivalent to a non-linear Riccati equation

$$-\frac{d}{dx}\phi(x) = \frac{2M}{\hbar^2}[E - V(x)] + \phi(x)^2, \tag{11.5}$$

where $\phi(x) = \psi'(x)/\psi(x)$ is the logarithmic derivative of wavefunction $\psi(x)$. The exact quantization rule for one-dimensional Schrödinger equation proposed and studied well in [348, 349] is given by

$$\begin{aligned} &\int_{x_A}^{x_B} k_n(x)dx = (n+1)\pi + \int_{x_A}^{x_B} k'(x)\frac{\phi(x)}{\phi'(x)}dx, \\ &k_n(x) = \sqrt{2M[E_n - V(x)]}/\hbar, \qquad E \geq V(x). \end{aligned} \tag{11.6}$$

The first term $(n+1)\pi$ is the contribution from the nodes of the logarithmic derivative of wavefunction, and the second is called the quantum correction. It is observed that, for all well-known exactly solvable quantum systems, this quantum correction is independent of the number of nodes of wavefunction. Hence, using the subscription 0 to denote the ground state, it is enough to consider the ground state in calculating the quantum correction

$$Q_0 = \int_{x_A}^{x_B} k_0'(x) \frac{\phi_0(x)}{\phi_0'(x)} dx. \tag{11.7}$$

Let us turn to the three-dimensional Schrödinger equation with a spherically symmetric potential. After separation of the angular part of the wavefunction

$$\psi(\mathbf{r}) = r^{-1} R(r) Y_m^{\ell}(\theta, \varphi), \tag{11.8}$$

the radial Schrödinger equation becomes

$$\begin{aligned} \frac{d^2}{dr^2} R''(r) &= -\frac{2M}{\hbar^2}[E - V_{\text{eff.}}(r)]R(r), \\ V_{\text{eff.}}(r) &= V(r) + \frac{\hbar^2 \ell(\ell+1)}{2Mr^2}. \end{aligned} \tag{11.9}$$

Since Eq. (11.9) is similar to Eq. (11.3), its energy levels can be calculated by the matching conditions of the logarithmic derivatives, where the logarithmic derivative is defined as $\phi(r) = R'(r)/R(r)$. It indicates that the exact quantization rule (11.6) still holds in this case simply by replacing the variable x with r, which is led to

$$\int_{r_A}^{r_B} k_n(r) dr = (n+1)\pi + \int_{r_A}^{r_B} k'(r) \frac{\phi(r)}{\phi'(r)} dr, \tag{11.10}$$

$$k_n(r) = \sqrt{2M[E_n - V_{\text{eff.}}(r)]}/\hbar, \qquad E \geq V_{\text{eff.}}(r), \tag{11.11}$$

where r_A and r_B are two turning points determined by $E = V_{\text{eff.}}(r)$. It should be mentioned that some useful integral formulae are given in Table 11.1 for convenience.

4 Application to Trigonometric Rosen-Morse Potential

In this section, we will show how to employ the exact quantization rule approach to calculate the energy spectra of exactly solvable potentials. Consider, for instance, the asymmetric trigonometric Rosen-Morse potential [5, 374] given by

$$V(x) = U_0 \cot^2(\pi x/a) + U_1 \cot(\pi x/a), \tag{11.12}$$

where $U_0 > 0$, $x \in [0, a]$. Introduce a new variable

$$y = -\cot\left(\frac{\pi x}{a}\right), \qquad y \in (-\infty, \infty). \tag{11.13}$$

The turning points y_A and y_B are determined by solving $V(x) = E_n$, where

$$\begin{aligned} E_n &= U_0 y^2 - U_1 y, \\ y_A + y_B &= \frac{U_1}{U_0}, \\ y_A y_B &= -\frac{E_n}{U_0}. \end{aligned} \tag{11.14}$$

The momentum $k_n(x)$ between them is given by

$$\begin{aligned} k_n(x) &= \frac{\sqrt{2M}}{\hbar}\sqrt{E_n - U_0 y^2 + U_1 y}, \\ k_n'(y) &= -\frac{\sqrt{2M}\,U_0}{\hbar}\frac{y - U_1/(2U_0)}{\sqrt{E_n - U_0 y^2 + U_1 y}}. \end{aligned} \tag{11.15}$$

Ma *et al.* [350] found that the solution of the ground state is given by

$$E_0 = \frac{\hbar^2 C^2}{2M} - U_0 - \frac{MU_1^2}{2\hbar^2 C^2}, \qquad \phi_0 = -Cy + B, \tag{11.16}$$

where

$$C = \frac{\pi}{2a}\left(1 + \sqrt{1 + \frac{8Ma^2 U_0}{\pi^2\hbar^2}}\right), \qquad B = \frac{MU_1}{C\hbar^2}. \tag{11.17}$$

The integral of the momentum $k_n(x)$ is calculated to be

$$\begin{aligned} &\int_{-x_B}^{x_B} k_n(x)dx \\ &= \frac{a\sqrt{2MU_0}/\hbar}{\pi}\int_{-y_B}^{y_B}\frac{\sqrt{(y_B - y)(y - y_A)}}{(y^2+1)}dy \\ &= -a\sqrt{2MU_0}/\hbar + \frac{a\sqrt{2MU_0}/\hbar}{\pi}\int_{-y_B}^{y_B}\frac{y(y_A + y_B) - y_A y_B + 1}{(1+y^2)\sqrt{(y_B - y)(y - y_A)}}dy \\ &= -a\sqrt{2MU_0}/\hbar + \frac{a\sqrt{2M}}{\hbar}\left|\mathrm{Re}\left\{\sqrt{E_n + U_0 - iU_1}\right\}\right|. \end{aligned} \tag{11.18}$$

The quantum correction Q_0 in the exact quantization rule can be calculated as follows:

$$\begin{aligned} &\int_{x_A}^{x_B} k_0'(x)\frac{\phi_0(x)}{\phi_0'(x)}dx \\ &= -\frac{a\sqrt{2MU_0}}{\pi\hbar}\int_{y_A}^{y_B}\frac{(y - B/C)[y - U_1/(2U_0)]}{(1+y^2)\sqrt{(y_B - y)(y - y_A)}}dy \\ &= -\frac{a\sqrt{2MU_0}}{\pi\hbar}\left\{\int_{y_A}^{y_B}\frac{dy}{\sqrt{(y_B - y)(y - y_A)}}\right. \\ &\quad \left. - \int_{y_A}^{y_B}\frac{y[B/C + U_1/(2U_0)] + [1 - BU_1/(2CU_0)]}{(1+y^2)\sqrt{(y_B - y)(y - y_A)}}dy\right\} \\ &= -\frac{a\sqrt{2MU_0}}{\pi\hbar}\left\{\pi + \pi\,\mathrm{Re}\left[\frac{[U_1/(2U_0) + i](1 - iB/C)}{\sqrt{-E_0/U_0 + iU_1/U_0 - 1}}\right]\right\} \\ &= -\frac{a\sqrt{2MU_0}}{\pi\hbar}\left(\pi + \frac{a\sqrt{2MU_0}\pi}{\hbar C}\right) \\ &= -a\sqrt{2MU_0}/\hbar + aC - \pi. \end{aligned} \tag{11.19}$$

Substituting Eqs. (11.18) and (11.19) into Eq. (11.6) gives

$$\frac{a\sqrt{2M}}{\hbar}\left|\mathrm{Re}\left\{\sqrt{E_n+U_0-iU_1}\right\}\right|=aC+n\pi, \tag{11.20}$$

from which we obtain the eigenvalues as

$$E_n=\frac{\hbar^2(aC+n\pi)^2}{2Ma^2}-\frac{Ma^2U_1^2}{2\hbar^2(aC+n\pi)^2}-U_0, \tag{11.21}$$

where $n=0,1,2,\ldots$. When $U_1=0$, this result reduces to the symmetric case. This result is consistent with that in Refs. [5, 375]. We find that the calculation of the quantum correction term becomes rather complicated. As what follows, we shall show how to simplify these tedious and complicated calculation with the help of the proper quantization rule [356, 357].

5 Proper Quantization Rule

Let us start with the original formulation of the exact quantization rule. Notice that if one takes $n=0$ in Eq. (11.6), then one may find that

$$\begin{aligned}\int_{x_{0A}}^{x_{0B}} k_0(x)dx &= \pi+\int_{x_{0A}}^{x_{0B}} k_0'(x)\frac{\phi_0(x)}{\phi_0'(x)}dx,\\ k_0(x)&=\sqrt{2M[E_0-V(x)]}/\hbar.\end{aligned} \tag{11.22}$$

Then the quantum correction term will become

$$\int_{x_{0A}}^{x_{0B}} k_0'(x)\frac{\phi_0(x)}{\phi_0'(x)}dx=\int_{x_{0A}}^{x_{0B}} k_0(x)dx-\pi. \tag{11.23}$$

Substituting this equation into Eq. (11.6), we obtain

$$\int_{x_A}^{x_B} k_n(x)dx-\int_{x_{0A}}^{x_{0B}} k_0(x)dx=(N-1)\pi=n\pi. \tag{11.24}$$

Similarly, after this important transformation for the quantization rule in three dimensions, Eq. (11.10) can also be written in the same form

$$\int_{r_A}^{r_B} k_n(r)dr-\int_{r_{0A}}^{r_{0B}} k_0(r)dr=n\pi. \tag{11.25}$$

Equations (11.24) and (11.25) are called the proper quantization rules. It may be found that two integrals involved in the proper quantization rule have the completely same mathematical form in the present version of the formulation. Therefore, when applying it to calculate the energy levels, one needs to calculate only the first integral with respect to the momentum $k_n(x)$ or $k_n(r)$. Then, replacing energy levels E_n in the expression with the ground state energy E_0, one immediately obtains the result for the second integral of $k_0(x)$ or $k_0(r)$. From now on, the complicated calculations encountered previously [348, 349, 351, 354, 376–380] will be greatly simplified.

6 Illustrations of Proper Quantization Rule

In this section, the nice performance of the proper quantization rule will be demonstrated in following different situations such as the modified Rosen-Morse potential, a Coulombic ring-shaped noncentral Hartmann system and the Manning-Rosen effective potential.

6.1 Energy Spectra for Modified Rosen-Morse Potential

The modified Rosen-Morse potential is given by [381, 382]

$$V(x) = -\frac{U_0 - U_1 \sinh(x/a)}{\cosh^2(x/a)}. \tag{11.26}$$

With $y = \sinh(x/a)$, the potential can be re-expressed as

$$V(y) = -\frac{U_0 - U_1 y}{1 + y^2}. \tag{11.27}$$

In our previous study [382], the ground state energy was found by solving the non-linear Riccati equation (11.5)

$$E_0 = -\frac{\hbar^2 (G_0 - 1)^2}{8Ma^2}, \tag{11.28}$$

where

$$G_0^2 = \frac{1}{2} + \frac{4Ma^2 U_0}{\hbar^2} + \left\{ \left(\frac{1}{2} + \frac{4Ma^2 U_0}{\hbar^2} \right)^2 + \left(\frac{4Ma^2 U_1}{\hbar^2} \right)^2 \right\}^{1/2}. \tag{11.29}$$

By solving $V(x_A) = V(x_B) = E_n$, we have two turning points y_A and y_B satisfying

$$\begin{aligned} y_A &= \sinh(x_A/a) = \frac{-U_1 - \sqrt{U_1^2 - 4E_n(U_0 + E_n)}}{2E_n}, \\ y_B &= \sinh(x_B/a) = \frac{-U_1 + \sqrt{U_1^2 - 4E_n(U_0 + E_n)}}{2E_n}, \\ y_A + y_B &= \frac{U_1}{E_n}, \qquad y_A y_B = 1 + \frac{U_0}{E_n}. \end{aligned} \tag{11.30}$$

The momentum $k_n(x)$ between two turning points can then be written as

$$k_n(x) = \frac{\sqrt{-2ME_n}}{\hbar\sqrt{1 + y^2}} \sqrt{(y_B - y)(y - y_A)}. \tag{11.31}$$

Now, let us calculate the first integral in Eq. (11.24):

$$\int_{x_A}^{x_B} k_n(x) dx = \int_{x_A}^{x_B} \frac{1}{\hbar} \sqrt{2M(E_n - V(x))} dx$$

$$= a\sqrt{-2ME_n}/\hbar \int_{y_A}^{y_B} \frac{\sqrt{(y_B - y)(y - y_A)}}{1 + y^2} dy$$

$$= -\pi a\sqrt{-2ME_n}/\hbar + \frac{a\pi\sqrt{M}}{\hbar}\sqrt{U_0 + \sqrt{U_0^2 + U_1^2}}. \quad (11.32)$$

Then, replacing E_n in Eq. (11.32) with E_0 given in Eq. (11.28), we obtain

$$\int_{x_{0A}}^{x_{0B}} k_0(x)dx = \frac{\pi}{2}\left[1 - G_0 + \frac{2a\sqrt{M}}{\hbar}\sqrt{U_0 + \sqrt{U_0^2 + U_1^2}}\right]. \quad (11.33)$$

Substituting Eqs. (11.32) and (11.33) into Eq. (11.24) leads to

$$-\pi\left[\frac{a\sqrt{-2E_n M}}{\hbar} - \frac{G_0 - 1}{2}\right] = n\pi, \quad (11.34)$$

from which we get the eigenvalues

$$E_n = -\frac{\hbar^2(G_0 - 2n - 1)^2}{8Ma^2}, \quad (11.35)$$

where $n = 0, 1, 2, \ldots, [(G_0 - 1)/2]$. This result is the same as that in [382]. However, the procedure in practical calculations for energy spectra for asymmetric trigonometric Rosen-Morse potential have been greatly simplified [356, 382].

6.2 Energy Spectra for the Coulombic Ring-Shaped Hartmann Potential

In this subsection, we wish to present that the bound states of an electron in a noncentral but separable potential can be handled as well within this proper quantization rule frame. Consider, for example, the Hartmann potential, which is the Coulomb potential surrounded by a ring-shaped inverse square potential. This potential was originally considered by Hartmann [383, 384] to study ring-shaped molecules.

In spherical coordinates (r, θ, φ), the Hartmann potential is defined by

$$V(r, \theta) = \eta\sigma^2\varepsilon_0\left(\frac{2a}{r} - \frac{\eta a^2}{r^2\sin^2\theta}\right), \quad (11.36)$$

where a is the Bohr radius, $a = \hbar^2/Me^2$, ε_0 is the ground state energy of the hydrogen atom, $\varepsilon_0 = -Me^4/2\hbar^2$, η and σ are dimensionless positive parameters which range from about 1 to 10 in theoretical chemistry applications.

In terms of the atomic units ($2M = \hbar = 1$), the ring-shaped Hartmann potential in Eq. (11.36) can be expressed as

$$V(r, \theta) = \frac{V_1}{r} + \frac{V_2}{r^2\sin^2\theta}, \quad (11.37)$$

with $V_1 = -\eta\sigma^2 e^2$ and $V_2 = \eta^2\sigma^2$.

In order to evaluate bound-state energies of the ring-shaped Hartmann potential, the total wavefunction can be written as

$$\Psi(r,\theta,\varphi)=\frac{1}{r}R(r)Y(\theta)e^{im\varphi}. \tag{11.38}$$

Separating the variables in analogy with the usual treatment to a spherical potential, the Schrödinger equation in spherical coordinates for a particle in the presence of the Hartmann potential (11.37) is reduced to two ordinary differential equations

$$\frac{d^2R(r)}{dr^2}+\left(E-\frac{V_1}{r}-\frac{\ell(\ell+1)}{r^2}\right)R(r)=0, \tag{11.39}$$

$$\frac{d^2Y(\theta)}{d\theta^2}+\cot\theta\frac{dY(\theta)}{d\theta}+\left\{\ell(\ell+1)-\frac{m^2}{\sin^2\theta}-\frac{V_2}{\sin^2\theta}\right\}Y(\theta)=0. \tag{11.40}$$

It is obvious that Eq. (11.39) is the same as that we obtain in solving the problem of an electron in a Coulomb-like field. Through a mapping function $\theta=w(x)$, Eq. (11.40) can be transformed to the form

$$\begin{aligned}&\frac{d^2Y(\theta)}{dx^2}+\left(-\frac{w''}{w'}+w'\cot w\right)\frac{dY(\theta)}{dx}\\&\quad+\left\{\ell(\ell+1)-\frac{m^2}{\sin^2 w}-\frac{V_2}{\sin^2 w}\right\}Y(\theta)=0,\end{aligned} \tag{11.41}$$

which seems a Schrödinger-like equation if

$$-\frac{w''}{w'}+w'\cot w=0, \tag{11.42}$$

that leads to

$$\tan(w/2)=e^x,\qquad \sin^2 w=\operatorname{sech}^2 x,\qquad x\in(-\infty,\infty). \tag{11.43}$$

Using Eqs. (11.42) and (11.43) in the expression for Eq. (11.41) we arrive at

$$\frac{d^2Y(\theta)}{dx^2}+[\ell(\ell+1)\operatorname{sech}^2 x]Y(\theta)=(m^2+V_2)Y(\theta). \tag{11.44}$$

We are now ready to consider the treatment of two Eqs. (11.39) and (11.44) through the proper quantization rule approach, respectively.

First, we will derive the solutions for Eq. (11.44) with the effective potential

$$V_{\text{eff.}}(x)=-\ell(\ell+1)\operatorname{sech}^2 x. \tag{11.45}$$

Let

$$y=\tanh(x),\qquad y_A=\tanh(x_A),\qquad y_B=\tanh(x_B), \tag{11.46}$$

where x_A and x_B are two turning points satisfying $V(x_A)=V(x_B)=E_\ell$. This gives

$$\begin{aligned}&E_\ell=-\ell(\ell+1)(1-y^2),\\&y_A+y_B=0,\\&y_Ay_B=-\left(1+\frac{E_\ell}{\ell(\ell+1)}\right).\end{aligned} \tag{11.47}$$

From the Riccati equation (11.5) with the potential in Eq. (11.45), one may obtain $E_0 = -\ell^2$.

Then, the integral of the momentum $k_n(x)$ in Eq. (11.24)

$$\int_{x_A}^{x_B} k_n(x)dx = \pi\left(\sqrt{\ell(\ell+1)} - \sqrt{-E_\ell}\right). \tag{11.48}$$

We can obtain the second integral in Eq. (11.24) from the ground state

$$\int_{x_{0A}}^{x_{0B}} k_0(x)dx = \pi\left(\sqrt{\ell(\ell+1)} - \ell\right). \tag{11.49}$$

In conjunction with Eqs. (11.48) and (11.49), the proper quantization rule in Eq. (11.24) becomes

$$\pi\left(\sqrt{\ell(\ell+1)} - \sqrt{-E_\ell}\right) = n_\ell\pi + \pi\left(\sqrt{\ell(\ell+1)} - \ell\right). \tag{11.50}$$

Therefore, from above equation we could be tempted to write the exact energy eigenvalues of the θ-dependent part for the symmetric Rosen-Morse potential (11.45) as

$$E_\ell = -(\ell - n)^2. \tag{11.51}$$

In the present case $E_\ell = -(m^2 + V_2)$, this yields

$$\ell = n_\ell + (m^2 + V_2)^{1/2}. \tag{11.52}$$

Second, we consider the radial equation (11.39) where the effective potential looks like

$$V_{\text{eff.}}(r) = \frac{V_1}{r} + \frac{\ell(\ell+1)}{r^2}. \tag{11.53}$$

By a similar procedure, we have

$$r_A + r_B = \frac{V_1}{E_{n_r}}, \qquad r_A r_B = -\frac{\ell(\ell+1)}{E_{n_r}} \tag{11.54}$$

with

$$E_0 = -\frac{V_1^2}{4(\ell+1)^2}. \tag{11.55}$$

Then, the two integrals in the quantization rule are evaluated exactly

$$\int_{r_A}^{r_B} k_n(r)dr = \pi\left(V_1\sqrt{\frac{1}{-4E_{n_r}}} - \sqrt{\ell(\ell+1)}\right), \tag{11.56}$$

$$\int_{r_{0A}}^{r_{0B}} k_0(r)dr = \pi(\ell + 1 - \sqrt{\ell(\ell+1)}). \tag{11.57}$$

The proper quantization rule in Eq. (11.25) takes the following form

$$\pi\left(V_1\sqrt{\frac{1}{-4E_{n_r}}} - \sqrt{\ell(\ell+1)}\right) = n_r\pi + \pi\left[\ell + 1 - \sqrt{\ell(\ell+1)}\right]. \tag{11.58}$$

Therefore, from above equation one can immediately obtain the well-known energy eigenvalues for the Coulomb potential E_n as

$$E_n = -\frac{V_1^2}{4n^2}, \quad n = n_r + \ell + 1, \tag{11.59}$$

where n is the principal quantum number. This result essentially agrees well with that of Ref. [385], in which the corresponding energy spectra has been calculated by the path integral method. One can also verify that the present result is consistent with that given in other works [383, 386].

6.3 Energy Spectra for the Manning-Rosen Effective Potential

Finally, we shall apply the proper quantization rule to generate the exact solutions of the Manning-Rosen effective potential [387–390] in D dimensions

$$V_{\text{eff.}}(r) = \frac{\hbar^2}{2M}\left[\frac{\beta(\beta-1)\alpha^2 e^{-2\alpha r}}{(1-e^{-\alpha r})^2} - \frac{A\alpha^2 e^{-\alpha r}}{1-e^{-\alpha r}} + \frac{\alpha^2 e^{-\alpha r}\ell'(\ell'+1)}{(1-e^{-\alpha r})^2}\right], \tag{11.60}$$

where the modified ℓ' is given by [380]

$$\ell' = \ell + \frac{D-3}{2}. \tag{11.61}$$

Introduce a new variable y

$$y = \frac{1}{e^{\alpha r}-1}, \qquad \frac{d}{dr}y(r) = -\alpha y(1+y). \tag{11.62}$$

Denote the coefficients in the effective potential by parameters V_2 and V_1, respectively

$$\begin{aligned}
&V_{\text{eff.}}(r) = V_2 y^2 + V_1 y,\\
&V_1 = \frac{\alpha^2\gamma^2\hbar^2}{2M}, \qquad V_2 = \frac{\alpha^2\sigma^2\hbar^2}{2M},\\
&\sigma^2 = \beta(\beta-1) + \ell'(\ell'+1),\\
&\gamma^2 = \ell'(\ell'+1) - A.
\end{aligned} \tag{11.63}$$

The integral of the momentum $k_n(r)$ in Eq. (11.25) gives

$$\begin{aligned}
\int_{r_A}^{r_B} k_n(r)dr &= \int_{y_A}^{y_B} \frac{\sqrt{2MV_2}}{-\hbar\alpha y(1+y)}\sqrt{(y-y_A)(y_B-y)}dy\\
&= -\pi\sigma\left(\sqrt{\frac{-V_1}{V_2} - \frac{E_{n\ell}}{V_2} + 1} - \sqrt{\frac{-E_{n\ell}}{V_2}} - 1\right).
\end{aligned} \tag{11.64}$$

After some algebraic calculation and simplification, one may arrive at an expression for the Maslov index

$$\mu = \frac{4}{\pi}\int_{r_1}^{r_2} k_0(r)dr = 4(\sigma + m), \quad m = \frac{1}{2} + \sqrt{\frac{1}{4} + \sigma^2}. \tag{11.65}$$

By successively using the relations in Eqs. (11.64) and (11.65), the proper quantization rule in Eq. (11.25) can be expressed as

$$-\pi\sigma\left(\sqrt{\frac{-V_1}{V_2}-\frac{E_{n\ell}}{V_2}+1}-\sqrt{\frac{-E_{n\ell}}{V_2}}-1\right)=\left(n+\frac{\mu}{4}\right)\pi. \tag{11.66}$$

Hence, one can immediately obtain the exact energy eigenvalues [388, 389] for the D-dimensional Manning-Rosen potential as

$$E_{n\ell}=-\frac{\hbar^2}{2M}\left[\frac{(\beta(\beta-1)+A)\alpha}{2(n+m)}-\frac{(n+m)\alpha}{2}\right]^2. \tag{11.67}$$

7 The Langer Modification and Maslov Index in D Dimensions

In the last years some attempts [341–343, 374, 391, 392] have been made to avoid the Langer modification. For example, the Langer modification can be avoided by introducing the concept of non-integral, energy-dependent Maslov index μ. Since there is no any correction in the application of the present proper quantization rule, we will investigate the general relationships between the Langer modification, the Maslov index and the proper quantization rule as follows.

As shown in previous sections, the proper quantization rule for three dimensional Schrödinger equation with a spherically symmetric potential [348, 349] takes the same form as that in one dimension simply by replacing the variable x with r. On the other hand, it is shown in previous study [380] that this methodology also holds for D dimensional Schrödinger equation with a spherically symmetric potential, where the effective potential can be expressed as

$$V_{\text{eff.}}(r)=V(r)+\frac{\hbar^2\ell'(\ell'+1)}{2Mr^2}, \tag{11.68}$$

where ℓ' is given in Eq. (11.61).

Thus, we have the explicit formulation of proper quantization rule in D dimensions

$$\int_{r_A}^{r_B}k_n(r)dr-\int_{r_{0A}}^{r_{0B}}k_0(r)dr=n\pi, \tag{11.69}$$

with the momentum between two turning points

$$k_n(r)=\sqrt{2M[E_n-V_{\text{eff.}}(r)]}/\hbar, \quad E_n\geq V_{\text{eff.}}(r). \tag{11.70}$$

Equation (11.69) may be rewritten in the form

$$\int_{r_A}^{r_B}k_n(r)dr=\left(n+\frac{\mu}{4}\right)\pi, \quad \mu=\frac{4}{\pi}\int_{r_{0A}}^{r_{0B}}k_0(r)dr. \tag{11.71}$$

In the following, we will investigate the hydrogen atom, harmonic oscillator and Manning-Rosen potential respectively to show the procedure to obtain the general form of the Langer modification and Maslov index with the aid of Eq. (11.71).

First, the effective potential for the hydrogen atom in D dimensions is

$$V_{\text{eff.}}(r) = \frac{\hbar^2 \ell'(\ell'+1)}{2Mr^2} - \frac{\hbar^2}{2Mr}. \tag{11.72}$$

The integral of the momentum $k_n(r)$ in the quantization rule is calculated to be

$$\int_{r_1}^{r_2} k_n(r)dr = \left[\frac{1}{\sqrt{-2E_n}} - \sqrt{\ell'(\ell'+1)}\right]\pi. \tag{11.73}$$

The Maslov Index for hydrogen atom in D dimensions may be obtained

$$\mu = \frac{4}{\pi}\int_{r_1}^{r_2} k_0(r)dr = 4\left(\ell - \sqrt{\ell'(\ell'+1)} + \frac{D-1}{2}\right). \tag{11.74}$$

By using the quantization rule Eq. (11.71), one may obtain the energy expression [37, 380]

$$E_n = -\frac{e^4}{2}\frac{1}{(n+\frac{D-3}{2})^2}. \tag{11.75}$$

The Langer modification in D dimensions may be derived as

$$\ell'(\ell'+1) \to \ell' + 1/2, \tag{11.76}$$

which is consistent with that once obtained by Watson's method [393].

Next, the proper quantization rule is applied to study the harmonic oscillator in arbitrary dimensions. As is well known, the effective potential for the D-dimensional harmonic oscillator is

$$V_{\text{eff.}}(r) = \frac{1}{2}M\omega^2 r^2 + \frac{\hbar^2 \ell'(\ell'+1)}{2Mr^2}. \tag{11.77}$$

The integral of the momentum $k_n(r)$ in the quantization rule Eq. (11.71) is calculated to be

$$\int_{r_A}^{r_B} k_n(r)dr = \frac{\pi}{2}\left[\frac{E_n}{\hbar\omega} - \sqrt{\ell'(\ell'+1)}\right]. \tag{11.78}$$

One may also evaluate the Maslov index for harmonic oscillator in D dimensions

$$\mu = \frac{4}{\pi}\int_{r_1}^{r_2} k_0(r)dr = D + 2\big(\ell - \sqrt{\ell'(\ell'+1)}\big). \tag{11.79}$$

Therefore, the proper quantization rule gives the familiar result for the energy levels of the D-dimensional harmonic oscillator

$$E_n = \hbar\omega(n + D/2), \quad n = 2n_r + \ell, \tag{11.80}$$

where n represents the principal quantum number and n_r the radial quantum number, which is equal to n in the proper quantization rule.

In the case of $D = 3$, one may obtain the Maslov index from Eq. (11.79)

$$\mu = 3 + 2[\ell - \sqrt{\ell(\ell+1)}]. \tag{11.81}$$

This result is consistent with that $\mu = 1 + 2[\ell + 1/2 - \sqrt{\ell(\ell+1)}] + 1$ derived by the phase loss method [341–343] and that $\eta = 1/2 + [\ell + 1/2 - \sqrt{\ell(\ell+1)}]/2$ via periodic orbit theory [370, 371] by noting that $\eta = \mu/4$. Obviously, those previous results [341–343, 370, 371] are the special cases of the general result in Eq. (11.79) obtained by the proper quantization rule.

8 Calculations of Logarithmic Derivatives of Wavefunction

As shown above, we have illustrated how to obtain the energy levels of some solvable quantum potentials. For completeness, we are going to show how we calculate the logarithmic derivative of wavefunction so that we might get the eigenfunctions.

For this purpose, we can get it by solving the non-linear Riccati equation. It is known that the logarithmic derivative of wavefunction $\phi(x)$ has $(N = n + 1)$ zeros and n nodes, thus we might express it into an algebraic fraction. As mentioned above, we only take the logarithmic derivative of the ground state $\phi_0(x)$ to calculate the quantum correction since it is independent of the nodes of the wavefunction. In fact, one is able to take the logarithmic derivative of arbitrary state of wavefunction to calculate the corresponding quantum correction Q.

As an illustration, we present one-dimensional harmonic oscillator $V(x) = M\omega^2x^2/2$. In the case of the ground state, take $\phi_0(x) = -\alpha^2 x$. Substituting it into the non-linear Riccati equation (11.5) allows us to obtain $E_0 = \hbar\omega/2$. For the first excited state, we can write down the non-linear Riccati equation (11.5) as

$$\frac{d}{dx}\phi_n(x) = -\frac{2M}{\hbar^2}E_n + \alpha^4 x^2 - \phi_n^2(x), \quad \alpha = \sqrt{\frac{M\omega}{\hbar}}. \tag{11.82}$$

Define

$$\phi_1(x) = \frac{c_2x^2 + c_0}{x}, \quad c_2 < 0. \tag{11.83}$$

Substituting this into Eq. (11.82) leads to the following expression

$$c_2 - c_0x^{-2} = -\frac{2ME_1}{\hbar^2} + \left(\frac{M\omega}{\hbar}\right)^2 x^2 - c_2^2x^2 - 2c_2c_0 - c_0^2x^{-2}, \tag{11.84}$$

from which we have

$$c_0 = 1, \qquad c_2 = -\frac{M\omega}{\hbar} = -\alpha^2, \qquad E_1 = -\frac{3\hbar^2c_2}{2M} = \frac{3}{2}\hbar\omega. \tag{11.85}$$

In a similar way, for the second excited state we define

$$\phi_2(x) = \frac{c_3x^3 + c_1x}{x^2 + c_2}, \quad c_3 < 0. \tag{11.86}$$

After substituting it into the Riccati equation (11.82), we obtain

$$\begin{aligned}&(3c_3x^2 + c_1)(x^2 + c_2) - 2x(c_3x^3 + c_1x)\\&= \left[-\frac{2ME_2}{\hbar^2} + \left(\frac{M\omega}{\hbar}\right)^2 x^2\right](c_2 + x^2)^2 - c_3^2x^6 - 2c_3c_1x^4 - c_1^2x^2.\end{aligned} \tag{11.87}$$

This allows us to obtain those coefficients as follows:

$$c_3 = -\alpha^2, \qquad c_2 = -\frac{1}{2\alpha^2}, \qquad c_1 = \frac{5}{2}, \qquad E_2 = \frac{5}{2}\hbar\omega. \tag{11.88}$$

Now, we attempt to study the third excited state. Define

$$\phi_3(x) = \frac{c_4x^4 + c_2x^2 + c_0}{x^3 + c_1x}, \qquad c_4 < 0. \tag{11.89}$$

Substitution of this into the non-linear Riccati equation (11.82) yields

$$\begin{aligned}&(c_0 + c_2x^2 + c_4x^4)^2 + c_1c_2x^2 - (c_2 - 3c_1c_4)x^4 + c_4x^6 - c_0(c_1 + 3x^2)\\ &= \left[-\frac{2ME_3}{\hbar^2} + \left(\frac{M\omega}{\hbar}\right)^2 x^2\right]x^2(c_1 + x^2)^2,\end{aligned} \tag{11.90}$$

from which we obtain

$$c_0 = c_1 = -\frac{3}{2\alpha^2}, \qquad c_2 = \frac{9}{2}, \qquad c_4 = -\alpha^2, \qquad E_3 = \frac{7}{2}\hbar\omega. \tag{11.91}$$

Finally, we discuss a more complicated case, i.e., the fourth excited state. We define

$$\phi_4(x) = \frac{c_5x^5 + c_3x^3 + c_1x}{x^4 + c_2x^2 + c_0}, \qquad c_5 < 0. \tag{11.92}$$

Substituting this into Riccati equation (11.82) allows us to obtain

$$\begin{aligned}&c_0c_1 + (c_1^2 - c_1c_2 + 3c_0c_3)x^2 + [c_2c_3 + c_1(-3 + 2c_3) + 5c_0c_5]x^4\\ &\quad + (-c_3 + c_3^2 + 2c_1c_5 + 3c_2c_5)x^6 + (1 + 2c_3)c_5x^8 + c_5^2x^{10}\\ &= \left[-\frac{2ME_4}{\hbar^2} + \left(\frac{M\omega}{\hbar}\right)^2 x^2\right](c_0 + c_2x^2 + x^4)^2,\end{aligned} \tag{11.93}$$

from which we have

$$\begin{aligned}&c_0 = \frac{3}{4}\alpha^{-4}, \qquad c_1 = -\frac{27}{4}\alpha^{-2}, \qquad c_2 = -3\alpha^{-2},\\ &c_3 = 7, \qquad c_5 = -\alpha^2, \qquad E_4 = \frac{9}{2}\hbar\omega.\end{aligned} \tag{11.94}$$

Likewise, we are able to calculate those for the higher excited states. In general, we may define the logarithmic derivatives of the wavefunction with the form

$$\begin{aligned}\phi_{2n+1}(x) &= \frac{\sum_{m=0}^{n+1} c_{2m}x^{2m}}{x^{2n+1} + \sum_{m=0}^{n-1} c_{2m+1}x^{2m+1}},\\ \phi_{2n}(x) &= \frac{\sum_{m=0}^{n} c_{2m+1}x^{2m+1}}{x^{2n} + \sum_{m=0}^{n-1} c_{2m}x^{2m}}.\end{aligned} \tag{11.95}$$

More precisely, they can also be unified to

$$\phi_n(x) = -\alpha^2x + 2n\alpha\frac{H_{n-1}(\alpha x)}{H_n(\alpha x)}, \tag{11.96}$$

from which we may obtain the wavefunction of harmonic oscillator by integrating

$$\psi_n(x) = N_n e^{-\frac{\alpha^2}{2}x^2} H_n(\alpha x), \tag{11.97}$$

where N_n is the normalization factor, $H_n(\alpha x)$ denote the nth Hermitian polynomial. The energy levels are given by $E_n = (n + 1/2)\hbar\omega$.

As illustrated above, it indicates that the quantum correction Q is independent of the nodes of the wavefunction. Once the logarithmic derivative of the wavefunction $\phi_n(x)$ is obtained, it is very easy to get the corresponding wavefunction $\psi_n(x)$ by integrating with respect to variable x. Certainly, the wavefunctions of other quantum systems could also treated in a similar way.

9 Conclusions

Since the exact quantization rule approach was proposed by Ma-Xu in 2004, this new formalism has been applied to determine eigenvalues of most known analytically solvable potentials and find the relativistic and non-relativistic solutions for a wide class of physical problems. The advantage of this method is to estimate the energy eigenvalues with the logarithmic derivative $\phi_0(x)$ of wavefunction from the ground state only, without ever having to solve the Schrödinger equation by the standard method. As an illustration, we have solved the asymmetric trigonometric Rosen-Morse potential by this exact quantization rule method and found that the integral calculations become rather tedious and complicated.

To overcome this difficulty, on the basis of this exact quantization rule method, the proper quantization rule proposed by Qiang-Dong Eqs. (11.24) and (11.25), more symmetric than the original one, greatly simplifies the calculations of the complicated integrals in the previous studies. One needs to calculate only one of two integrals. With the proper quantization rule, the ground state energy is sufficient to determine the energy levels of the quantum system. Thus, due to the fact that whenever the number of the nodes of $\phi(x)$ or the number of the nodes of the wavefunction $\psi(x)$ increases by one, the momentum integral $\int_{x_A}^{x_B} k(x)dx$ will increase by π.

The proper quantization rule method is shown to be convenient to investigate one dimensional or spherically symmetric D dimensional solvable potentials. It is verified to be also efficient for the noncentral but separable potentials. After separating the variables, the radial and the angular pieces of the Schrödinger equation can both be treated within the same framework. It is shown that the energy levels of the modified Rosen-Morse potential, the Coulombic ring-shaped noncentral Hartmann system and the Manning-Rosen effective potential are derived only from the ground state energy. The procedure is general and the extensions of the present method to other noncentral separable potentials is straightforward.

Besides semiempirical hypothesis, this Chapter presents an alternative analytic means for the treatment of the quantization rule, Langer modification and Maslov index. If the correct energies are expected when one deals with radial quantum systems in WKB frame, the Langer modification is necessary. If the Langer prescription is to be avoided, the ansatz of the non-integer Maslov indices may be introduced as suggested in some proposals. Within the new quantization rule scheme, the non-integer value of Maslov indices may be directly derived in a natural way and one may get energy eigenvalues, without any correction, as accurate as those obtained from the conventional approaches.

For completeness, we have also shown how to calculate the logarithmic derivatives of the wavefunction in terms of the non-linear Riccati equation. Once they are available, it is not difficult to obtain the corresponding wavefunction by integrating them with respect to the argument.

We would like to mention that the semiclassical EBK quantization rule and the Maslov index are related to the advance of the perihelion of the classical orbit, which is used to derive the quantum defect parameterization and Ritz expansion [394, 395]. It will be interesting to consider in future how to establish a conceptual relationship between the present results and both the advance of the classical perihelion and the quantum defect. It is also important to give further investigation of the properties of Maslov index for all exactly solvable potentials.

It should also be noted that the well-known quantum Hamilton-Jacobi method [372, 373, 396, 397], which also gives energy levels without need for solving for wavefunction and also gives the wavefunctions for all exactly solvable models, and the present method are not identical. Further investigations of the present method to other quantum solvable models are necessary in order to explain in more detail the properties of energy eigenvalues and wavefunctions of those systems. A number of investigations of this fundamental quantum problem can be expected in the near future.

Before ending this Chapter, we want to give some useful remarks. First, related to present study we have noticed that any l-state solutions of the Woods-Saxon potential in arbitrary dimensions within the new improved quantization rule have been studied recently [398]. Second, Yin *et al.* have shown why SWKB approximation is exact for all shape invariant potentials [399] by analytical transfer matrix theory. It should be noted that this theory is closely related to the exact and proper quantization rules. Third, Grandatia *et al.* have also shown that the exact quantization rule results from the exactness of the modified JWKB quantization condition[1] proved by Barclay [402] and proposed a very direct alternative way to calculate the appropri-

[1]It should be noted that the validity of the Ma-Xu formula follows from a Barclay's result. He found that for these potentials the JWKB series can be resumed beyond the lowest-order giving an energy-independent correction which can be absorbed into the Maslov index and written in a closed analytical expression. Moreover, he showed equally that this result is directly correlated to the exactness of the lowest-order SJWKB quantization condition [374, 400]. The starting point is the definition of two classes of potentials, each characterized by a specific change of variable which brings the potential into a quadratic form. It is shown that this two classes coincide with the Barclay-Maxwell classes [401], which are based upon a functional characterization of superpotentials and which cover the whole set of translationally shape invariant potentials.

ate correction for the whole class of translationally shape invariant potentials [403]. The SJWKB quantization rule is written as

$$\oint_{E_n} k_n(x)dx = 2\int_{x_a}^{x_b} k_n(x)dx = 2(n\pi + \gamma), \tag{11.98}$$

where γ is an energy-independent correction characteristic of the studied potentials. In Ma-Xu's quantization rule, the constant γ is nothing but the quantum correction Q_0. More than ten years later Bhaduri *et al.* [404] proposed another interesting derivation of this result which replies on periodic orbit theory.

Chapter 12
Schrödinger Equation with Position-Dependent Mass

1 Introduction

Generally speaking, the effective mass is taken as a constant in the traditional wave equations. Recently, the study of the non-relativistic equation with the position-dependent effective mass has attracted a lot of attention to many authors [405–439]. This is because such systems have been found to have wide applications in various fields such as the electronic properties of the semiconductors [412], ^{3}He clusters [413], quantum wells, wires and dots [405, 406, 414], quantum liquids [416], the graded alloys, semiconductor heterostructures [417] and others. Recently, the algebraic method has also been used to study these systems [436–438].

This Chapter is organized as follows. In Sect. 2 we employ a point canonical transformation to study the D-dimensional position-dependent effective mass Schrödinger equation. Two typical examples such as the harmonic oscillator and Coulomb potential are carried out in Sect. 3. Some concluding remarks are given in Sect. 4.

2 Formalism

According to recent contributions [70, 440, 441], it is shown that the position-dependent effective mass Schrödinger equation with physical potentials is given by

$$\nabla_D\left(\frac{1}{m}\nabla_D\psi(\mathbf{r})\right)+2[E-V(r)]\psi(\mathbf{r})=0, \tag{12.1}$$

where $m=m_0 m(r)=m(r)$. For D-dimensional spherical symmetry, we take the wavefunction $\psi(\mathbf{r})$ as follows:

$$\psi(\mathbf{r})=r^{-(D-1)/2}R(r)Y^{l}_{l_{D-2},\ldots,l_1}(\hat{\mathbf{x}}). \tag{12.2}$$

S.-H. Dong, *Wave Equations in Higher Dimensions*,
DOI 10.1007/978-94-007-1917-0_12, © Springer Science+Business Media B.V. 2011

Based on the following formula

$$\nabla_D \frac{1}{m} \nabla_D \psi(\mathbf{r}) = \left(\nabla_D \frac{1}{m}\right) \cdot [\nabla_D \psi(\mathbf{r})] + \frac{1}{m} \nabla_D^2 \psi(\mathbf{r}), \tag{12.3}$$

substituting (12.2) into (12.1) allows us to obtain the radial position-dependent mass Schrödinger equation with a spherically symmetric potential

$$\left\{\frac{d^2}{dr^2} + \frac{m'}{m}\left(\frac{D-1}{2r} - \frac{d}{dr}\right) - \frac{\eta^2 - 1/4}{r^2} + 2m[E - V(r)]\right\} R(r) = 0, \tag{12.4}$$

where $m' = dm(r)/dr$ and $\eta = |l - 1 + D/2|$.

On the other hand, it is well known that the radial Schrödinger equation with constant mass ($m = m_0 = 1$) is given by

$$\left\{\frac{d^2}{d\varrho^2} - \frac{\eta_1^2 - 1/4}{\varrho^2} + 2[\varepsilon - U(\varrho)]\right\} \psi(\varrho) = 0, \quad \eta_1 = |\Lambda - 1 + D/2|, \tag{12.5}$$

where Λ denotes the angular momentum.

By performing the following transformations on above Eq. (12.5)

$$\varrho = q(r), \qquad \psi(\varrho) = f(r) R(r), \tag{12.6}$$

we have

$$\begin{aligned} &\left\{\frac{d^2}{dr^2} + \left(2\frac{f'}{f} - \frac{q''}{q'}\right)\frac{d}{dr} + \left(\frac{f''}{f} - \frac{q''}{q'}\frac{f'}{f}\right)\right. \\ &\left. \quad - (\eta_1^2 - 1/4)(q'/q)^2 + 2(q')^2[\varepsilon - U(q(r))]\right\} R(r) = 0. \end{aligned} \tag{12.7}$$

In comparison Eq. (12.4) with Eq. (12.7) we find that

$$f(r) = \sqrt{\frac{q'}{m}} \tag{12.8}$$

$$\begin{aligned} V(r) - E + \frac{\eta^2 - 1/4}{2m r^2} = {} & \frac{(q')^2}{m}[U(q(r)) - \varepsilon] + \frac{(D-1)m'}{4m^2 r} \\ & + \frac{\eta_1^2 - 1/4}{2m}(q'/q)^2 + \frac{1}{4m}[g(m) - g(q')] \end{aligned} \tag{12.9}$$

with

$$g(x) = \frac{x''}{x} - \frac{3}{2}\left(\frac{x'}{x}\right)^2. \tag{12.10}$$

In the calculation, we have used the following results

$$\begin{aligned} \frac{f'}{f} &= \frac{1}{2}\left\{\frac{q''}{q'} - \frac{m'}{m}\right\}, \\ \frac{f''}{f} &= \frac{1}{4}\left\{\frac{3m'^2}{m^2} - \frac{2m''q' + 2m'q''}{mq'} - \frac{(q'')^2 - 2q'q'''}{q'^2}\right\}. \end{aligned} \tag{12.11}$$

3 Applications to Harmonic Oscillator and Coulomb Potential

In this section, we are ready to illustrate the harmonic oscillator. The solutions are given by

$$\begin{aligned} U(\varrho) &= \frac{1}{2}\omega^2\varrho^2, \\ \varepsilon_n &= (2n + \Lambda + D/2)\omega, \\ \psi(\varrho) &= C_n(\omega\varrho)^{\Lambda+(D-1)/2}e^{-\omega\varrho^2/2}\,{}_1F_1(-n; \Lambda + D/2; \omega\varrho^2), \end{aligned} \tag{12.12}$$

where Λ is related to η_1 given in Eq. (12.5).

Take $m(r) = \alpha r^\tau$ and $q(r) = r^\nu$, where α, τ and ν are non-zero real parameters. For simplicity, we only consider the case $\nu = 1 + \tau/2$ with $\tau \neq 2$. Thus, $(q')^2/m$ becomes a constant. Substituting them into Eq. (12.9) allows us to obtain

$$\begin{aligned} V(r) &= \frac{\nu^2}{\alpha}U(q(r)) = \frac{\alpha}{2}c^2 r^{\tau+2}, \\ E_n &= \frac{\nu^2}{\alpha}\varepsilon_n = \frac{\tau+2}{2}c(2n + \Lambda(l) + D/2), \\ R(r) &= c_n(\beta r)^{(1+\tau/2)[\Lambda(l)+(D-1)/2]+\tau/4}e^{-\beta r^{(\tau+2)}/2} \\ &\quad \times {}_1F_1[-n; \Lambda(l) + D/2; \beta r^{\tau+2}], \end{aligned} \tag{12.13}$$

where c is a real potential parameter, $\omega = \beta = 2\alpha c/(2 + \tau)$ and $\Lambda(l)$ satisfies a constraint

$$16(\eta^2 - 1/4) = 4(2+\tau)^2(\eta_1^2 - 1/4) - 3\tau^2 + (8D - 12)\tau. \tag{12.14}$$

When $\tau = 0$, m is independent of position r and then $\Lambda = l$. Thus, it is straight to see that Eqs. (12.13) agree with Eqs. (12.12) completely.

Finally, we briefly carry out the Coulomb potential. The solutions are given by

$$\begin{aligned} U(\varrho) &= -\frac{\xi}{\varrho}, \\ \varepsilon_n &= -\frac{\xi^2}{2[n + (D-3)/2]^2}, \\ \psi(\varrho) &= B_n(\beta\varrho)^{\Lambda+(D-1)/2}e^{-\beta\varrho} \\ &\quad \times {}_1F_1[-(n - \Lambda - 1); 2\Lambda + D - 1; 2\beta\varrho], \end{aligned} \tag{12.15}$$

where

$$\beta = \frac{\xi}{n + (D-3)/2}, \quad n = 1, 2, 3, \ldots. \tag{12.16}$$

With the same process as that of the harmonic oscillator, we have

$$\begin{aligned} V(r) &= -\frac{d}{r^{1+\tau/2}}, \\ E_n &= -\frac{2\alpha d^2}{(2+\tau)^2}\frac{1}{[n_1+(D-3)/2]^2}, \\ R(r) &= b_n r^{(1+\tau/2)[\Lambda(l)+(D-1)/2]+\tau/4} e^{-\gamma r^{1+\tau/2}} \\ &\quad \times {}_1F_1[-(n_1-\Lambda(l)-1); 2\Lambda(l)+D-1; 2\gamma r^{1+\tau/2}], \end{aligned} \tag{12.17}$$

where

$$\begin{aligned} n_1 &= n_r + \Lambda(l) + 1, \\ \gamma &= \frac{4\alpha d}{(2+\tau)^2}\frac{1}{n+(D-3)/2}, \\ d &= \frac{(2+\tau)^2\xi}{4\alpha}, \end{aligned} \tag{12.18}$$

where the $\Lambda(l)$ satisfies the same constraint as Eq. (12.14).

Before ending this section, we give a useful remark on the recent work [442]. The authors Ballesteros *et al.* claimed that they have found a new exactly solvable quantum model in N dimensions with the Hamiltonian

$$H = -\frac{\hbar^2}{2(1+\lambda r^2)}\nabla^2 + \frac{\omega^2 r^2}{2(1+\lambda r^2)}. \tag{12.19}$$

They addressed that the spectrum of this model is shown to be hydrogen-like (should be harmonic oscillator-like), and their eigenvalues and eigenfunctions are explicitly obtained by deforming appropriately the symmetry properties of the N-dimensional harmonic oscillator. We pinpoint that such an understanding is incorrect since the kinetic energy term was not defined as Eq. (12.3). This means that the operator ∇ does not commute with the position-dependent mass $m(r)$. Therefore, this system does not permit exact solutions at all. Furthermore, the choice of the position-dependent mass $m(r) = (1+\lambda r^2)$ has no physical meaning since the mass $m(r) \to \infty$ when $r \to \infty$. In particular, the mass $m(r)$ was taken as $1/(1+\lambda r^2) = 1/m(r)$ for the harmonic oscillator term.

4 Conclusions

In this Chapter we have employed a point canonical transformation to study the D-dimensional position-dependent effective mass Schrödinger equation. By mapping this wave equation into a well-known solvable D-dimensional Schrödinger equation with a constant mass, the exact bound state solutions have been derived for a given spatial dependent mass distribution. As illustrations, we have carried out two typical examples such as the harmonic oscillator and Coulomb potential. Before ending this Chapter, we give two useful remarks on this topic. The advantage of this method is easy to obtain the corresponding solutions for given position-dependent mass $m(r)$

and variable $q(r)$. However, some constraints on the parameters will be appearing unavoidably. On the other hand, note that Ikhdair and Sever have worked over the exactly solvable effective mass D-dimensional Schrödinger equation for pseudo-harmonic and modified Kratzer problems [443]. However, it should be pointed out that these problems are essentially same as the harmonic oscillator and Coulomb potential except for a slight modification.

Part IV
Applications in Relativistic Quantum Mechanics

Part IV
Applications in Relativistic Quantum Mechanics

Chapter 13
Dirac Equation with the Coulomb Potential

1 Introduction

The exact solutions of quantum system with a $1/r$ type potential are of importance in quantum mechanics [1, 2, 444]. Due to the recent interest of the higher-dimensional field theory, many problems related to the Schrödinger equation and Klein-Gordon equation in $(D+1)$ dimensions have been discussed. To fill in the gap between them, we have carried out the Dirac equation with this potential in $(D+1)$ dimensions [91].

The purposes of this Chapter are two-fold. First, we exhibit the exact solutions of the hydrogen atom by the confluent hypergeometric equation approach. The second is to investigate the variations of energy difference $\Delta E(n,l,D)$ and the energy levels $E(n,l,D)$ on the dimension D [87] and also to study the variations of energy levels $E(n,l,\xi)$ on the potential strength $\xi = Z\alpha$ for a given D.

This Chapter is organized as follows. The exact solutions of the radial equations will be displayed via the confluent hypergeometric equation approach in Sect. 2. The variations of energy difference $\Delta E(n,l,D)$ and energy levels $E(n,l,D)$ on the dimension D as well as the variations of energy levels $E(n,l,\xi)$ on the potential strength $\xi = Z\alpha$ shall be elucidated in Sect. 3. The Dirac equation with a Coulomb potential plus a scalar potential will be discussed in Sect. 4. Some concluding remarks are given in Sect. 5.

2 Exact Solutions of Hydrogen-like Atoms

As demonstrated in Chap. 4, the D-dimensional radial Dirac equations are written as

$$\begin{aligned}
&\frac{d}{dr}G_{KE}(r)+\frac{K}{r}G_{KE}(r)=[E-V(r)-M]F_{KE}(r),\\
&-\frac{d}{dr}F_{KE}(r)+\frac{K}{r}F_{KE}(r)=[E-V(r)+M]G_{KE}(r),
\end{aligned} \tag{13.1}$$

S.-H. Dong, *Wave Equations in Higher Dimensions*,
DOI 10.1007/978-94-007-1917-0_13,

with

$$K = \pm(2l + D - 1)/2. \tag{13.2}$$

Let us study the solutions of radial equations (13.1) by the confluent hypergeometric equation approach. This is different from power series expansion method used in Chap. 4.

Consider the Coulomb-like potential, i.e.,

$$V(r) = -\frac{\xi}{r}, \quad \xi = Z\alpha. \tag{13.3}$$

Introduce a new variable ρ for bound states $|E| < M$,

$$\rho = 2r\sqrt{M^2 - E^2}. \tag{13.4}$$

Substitution of this, together with Eq. (13.3), into Eq. (13.1) leads to

$$\begin{aligned} \frac{d}{d\rho}G_{KE}(\rho) + \frac{K}{\rho}G_{KE}(\rho) &= \left(-\frac{1}{2}\sqrt{\frac{M-E}{M+E}} + \frac{\xi}{\rho}\right)F_{KE}(\rho), \\ \frac{d}{d\rho}F_{KE}(\rho) - \frac{K}{\rho}F_{KE}(\rho) &= \left(-\frac{1}{2}\sqrt{\frac{M+E}{M-E}} - \frac{\xi}{\rho}\right)G_{KE}(\rho). \end{aligned} \tag{13.5}$$

Define the wavefunction $\Phi_\pm(\rho)$ of the forms

$$\begin{aligned} G_{KE}(\rho) &= \sqrt{M-E}[\Phi_+(\rho) + \Phi_-(\rho)], \\ F_{KE}(\rho) &= \sqrt{M+E}[\Phi_+(\rho) - \Phi_-(\rho)]. \end{aligned} \tag{13.6}$$

Substitutions of them into Eq. (13.5) allow us to write down

$$\begin{aligned} &\left\{\frac{d}{d\rho}\Phi_+(\rho) + \frac{d}{d\rho}\Phi_-(\rho)\right\} + \frac{K}{\rho}[\Phi_+(\rho) + \Phi_-(\rho)] \\ &\quad = \left\{-\frac{1}{2} + \frac{\xi}{\rho}\sqrt{\frac{M+E}{M-E}}\right\}[\Phi_+(\rho) - \Phi_-(\rho)], \\ &\left\{\frac{d}{d\rho}\Phi_+(\rho) - \frac{d}{d\rho}\Phi_-(\rho)\right\} - \frac{K}{\rho}[\Phi_+(\rho) - \Phi_-(\rho)] \\ &\quad = \left\{-\frac{1}{2} - \frac{\xi}{\rho}\sqrt{\frac{M-E}{M+E}}\right\}[\Phi_+(\rho) + \Phi_-(\rho)]. \end{aligned} \tag{13.7}$$

Their addition and subtraction yield

$$\begin{aligned} &\frac{d}{d\rho}\Phi_\pm(\rho) \mp \left(\frac{\xi E}{\rho\sqrt{M^2 - E^2}} - \frac{1}{2}\right)\Phi_\pm(\rho) \\ &\quad = -\left(\frac{K}{\rho} \pm \frac{\xi M}{\rho\sqrt{M^2 - E^2}}\right)\Phi_\mp(\rho). \end{aligned} \tag{13.8}$$

Define the following notations

$$\tau = \frac{\xi E}{\sqrt{M^2 - E^2}}, \qquad \tau' = \frac{\xi M}{\sqrt{M^2 - E^2}}. \tag{13.9}$$

Equation (13.8) is simplified to

$$\frac{d}{d\rho}\Phi_{\pm}(\rho)\mp\left(\frac{\tau}{\rho}-\frac{1}{2}\right)\Phi_{\pm}(\rho)=-\frac{K\pm\tau'}{\rho}\Phi_{\mp}(\rho), \tag{13.10}$$

from which we can obtain an important second-order differential equation

$$\left\{\frac{d^2}{d\rho^2}+\frac{1}{\rho}\frac{d}{d\rho}+\left(-\frac{1}{4}+\frac{\tau\pm 1/2}{\rho}-\frac{\eta^2}{\rho^2}\right)\right\}\Phi_{\pm}(\rho)=0, \tag{13.11}$$

where

$$\eta^2=K^2-\xi^2. \tag{13.12}$$

For a weak potential, we have

$$\eta=\sqrt{K^2-\xi^2}>0. \tag{13.13}$$

It indicates that Eq. (13.11) is a special case of the Tricomi equation [20]

$$\frac{d^2y}{dx^2}+\left(a+\frac{b}{x}\right)\frac{dy}{dx}+\left(\alpha+\frac{\beta}{x}+\frac{\xi}{x^2}\right)y=0. \tag{13.14}$$

From the behaviors of the wavefunction at the origin and at infinity, we define

$$\Phi_{\pm}(\rho)=\rho^{\eta}e^{-\rho/2}R_{\pm}(\rho). \tag{13.15}$$

Substitution of this into (13.11) leads to

$$\begin{aligned}&\frac{d^2}{d\rho^2}R_{\pm}(\rho)+\left(-1+\frac{1+2\eta}{\rho}\right)\frac{d}{d\rho}R_{\pm}(\rho)\\&\quad+\frac{\tau-\eta-1/2\pm 1/2}{\rho}R_{\pm}(\rho)=0,\end{aligned} \tag{13.16}$$

whose solutions are the confluent hypergeometric functions

$$\begin{aligned}R_{+}(\rho)&=a_{0\,1}F_1(\eta-\tau;2\eta+1;\rho),\\R_{-}(\rho)&=b_{0\,1}F_1(1+\eta-\tau;2\eta+1;\rho).\end{aligned} \tag{13.17}$$

It is shown from Eqs. (13.6), (13.15) and (13.17) that $G_{KE}(\rho)$ and $F_{KE}(\rho)$ can be directly obtained by the combinations of the confluent hypergeometric functions.

We now study the relation between the coefficients a_0 and b_0. Before proceeding, we recall the following recurrence relations between the confluent hypergeometric functions [192]

$$\begin{aligned}&\gamma\frac{d}{dz}{}_1F_1(\alpha;\gamma;z)=\alpha\,{}_1F_1(\alpha+1;\gamma+1;z)\\&\qquad=\frac{\alpha\gamma}{z}[\gamma\,{}_1F_1(\alpha+1;\gamma;z)-\gamma\,{}_1F_1(\alpha;\gamma;z)],\\&\alpha\,{}_1F_1(\alpha+1;\gamma+1;z)=(\alpha-\gamma)\,{}_1F_1(\alpha;\gamma+1;z)+\gamma\,{}_1F_1(\alpha;\gamma;z),\\&\alpha\,{}_1F_1(\alpha+1;\gamma;z)=(z+2\alpha-\gamma)\,{}_1F_1(\alpha;\gamma;z)\\&\qquad\qquad+(\gamma-\alpha)\,{}_1F_1(\alpha-1;\gamma;z).\end{aligned} \tag{13.18}$$

It is shown from Eqs. (13.11), (13.17) and (13.18) that

$$\left(\frac{\eta-\tau}{\rho}a_0+\frac{\tau'+K}{\rho}b_0\right){}_1F_1(1+\eta-\tau;2\eta+1;\rho)=0. \tag{13.19}$$

Since both a_0 and b_0 cannot be vanishing, we obtain the following relation between them

$$b_0=\frac{\tau-\eta}{\tau'+K}a_0. \tag{13.20}$$

From Eq. (13.6) we thus have

$$\begin{aligned}G_{KE}(\rho)&=N_{KE}\sqrt{M-E}\rho^{\eta}e^{-\rho/2}\\&\quad\times[(\tau'+K){}_1F_1(\eta-\tau;2\eta+1;\rho)\\&\quad+(\tau-\eta){}_1F_1(1+\eta-\tau;2\eta+1;\rho)],\\F_{KE}(\rho)&=N_{KE}\sqrt{M+E}\rho^{\eta}e^{-\rho/2}\\&\quad\times[(\tau'+K)\,{}_1F_1(\eta-\tau;2\eta+1;\rho)\\&\quad-(\tau-\eta)\,{}_1F_1(1+\eta-\tau;2\eta+1;\rho)],\end{aligned} \tag{13.21}$$

where the normalization factor

$$N_{KE}=a_0(\tau'+K)^{-1}\left(2\sqrt{M^2-E^2}\right)^{-1/2} \tag{13.22}$$

is to be determined.

We now study the eigenvalues of this quantum system. The quantum condition is obtained from the finiteness of the solutions at infinity

$$\tau-\eta=n'=0,1,2,\ldots. \tag{13.23}$$

When $n'=0$, $\eta=\tau$, and

$$K^2=\tau^2+\xi^2=(\tau')^2. \tag{13.24}$$

Therefore, K has to be positive to avoid a trivial solution.

Introduce the principal quantum number n

$$\begin{aligned}n&=|K|-(D-3)/2+n'\\&=|K|-(D-3)/2+\tau-\eta\\&=l+1+n'=1,2,\ldots.\end{aligned} \tag{13.25}$$

The n may be equal to 1 only for $K=(D-1)/2$ and is equal to other positive integers for both signs of K. The energy E can be obtained from Eqs. (13.9), (13.11) and (13.25)

$$E(n,l,D)=M\frac{\xi}{|\xi|}\left\{1+\frac{\xi^2}{(\sqrt{K^2-\xi^2}+n-l-1)^2}\right\}^{-1/2}. \tag{13.26}$$

For a large D, we have

$$E(n,D)\simeq M\frac{\xi}{|\xi|}[1-2\xi^2D^{-2}+4\xi^2(2n-3)D^{-3}-\cdots], \tag{13.27}$$

which implies that the energy is independent of l for a large D.

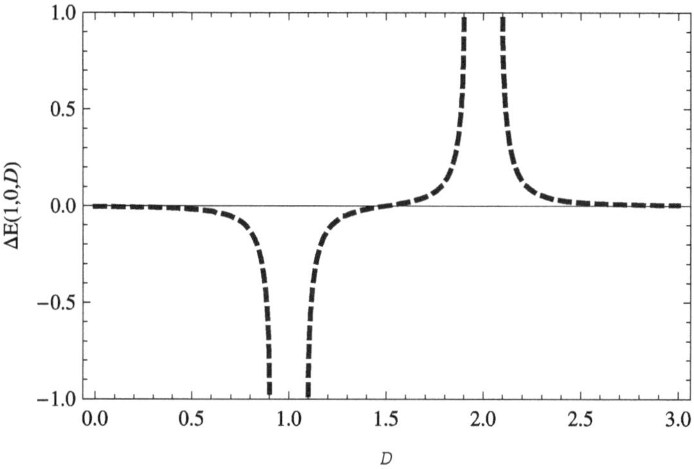

Fig. 13.1 The energy difference $\Delta E(1,0,D)$ decreases with the dimension $D \in (0, 0.9]$, while increases with the dimension $D \in [1.1, 1.9]$ and then decreases again with the dimension $D \geq 2.1$. The $\Delta E(1,0,D)$ is symmetric with respect to the point $(1.5, 0)$, so are those $\Delta E(n,0,D)$. The parameter $\xi = 0.05$ is taken here and also in Figs. 13.2–13.6

For a small ξ, we have

$$E(n,l,D) \simeq M\left\{1 - \frac{\xi^2}{2[n+(D-3)/2]^2} - \frac{\xi^4}{2[n+(D-3)/2]^4}\left(\frac{2n+D-3}{2l+D-1} - \frac{3}{4}\right)\right\}, \tag{13.28}$$

where the first term on the right hand side is the rest energy M $(c^2 = 1)$, the second one is from the solutions of the Schrödinger equation, and the third one is the fine structure energy, which removes the degeneracy between the states with the same n. It is found from Eq. (13.28) that the energy levels $E(n,l,D)$ are almost independent of the quantum number l for a small potential parameter ξ as shown in Fig. 13.1.

We now determine the normalization factor N_{KE} from the normalization condition

$$\int \Psi_{KE}^{\dagger}\Psi_{KE}dV = 1. \tag{13.29}$$

Since $n' = \tau - \eta$ is a non-negative integer, we may express the confluent hypergeometric functions by the associated Laguerre polynomials (5.21). Through a direct calculation, we obtain the normalization factor

$$N_{KE} = \frac{(M^2 - E^2)^{1/4}}{\Gamma(2\eta+1)}\left[\frac{\Gamma(\tau+\eta+1)}{2M\tau'(K+\tau')(\tau-\eta)!}\right]^{1/2}. \tag{13.30}$$

3 Analysis of the Eigenvalues

We now analyze the properties of eigenvalues. Recently, Nieto has made use of the concept of the continuous dimension D to study the bound states of quantum system with a special potential [87]. With this spirit, we attempt to discuss what happened to a continuous dimension D as shown in Ref. [100].

It is shown from Eq. (13.25) that $n = 1, 2, 3, \ldots$ and $l = 0, 1, \ldots, n-1$. If the principal quantum number n and the angular momentum quantum number l are fixed, the energy difference $\Delta E(n, l, D)$ between eigenvalues for dimensions D and $D-1$ can be calculated by

$$\begin{aligned}\Delta E(n,l,D) &= E(n,l,D) - E(n,l,D-1)\\ &= M\frac{\xi}{|\xi|}\left\{\frac{1}{\sqrt{1+\frac{\xi^2}{(\sigma+\sqrt{K^2-\xi^2})^2}}} - \frac{1}{\sqrt{1+\frac{\xi^2}{(\sigma+\sqrt{K_1^2-\xi^2})^2}}}\right\},\end{aligned} \tag{13.31}$$

where

$$\sigma = n - l - 1, \qquad K_1 = (2l + D - 2)/2. \tag{13.32}$$

It is found from Eq. (13.31) that the relation between $\Delta E(n, l, D)$ and dimension D is more complicated than that of the Schrödinger equation case [87]. The problem arises from the factor $\sqrt{K^2 - \xi^2}$ (or $\sqrt{K_1^2 - \xi^2}$). Therefore, for a weak potential it is shown from Eqs. (13.2), (13.13) and (13.31) that

$$D = 2(1/2 - l + K)\begin{cases} \geq 1 + 2|\xi|, & \text{when } l = 0 \text{ and } K > |\xi|,\\ \leq 1 - 2|\xi|, & \text{when } l = 0 \text{ and } K < -|\xi|,\end{cases} \tag{13.33}$$

and

$$D = 2(1 - l + K)\begin{cases} \geq 2 + 2|\xi|, & \text{when } l = 0 \text{ and } K_1 > |\xi|,\\ \leq 2 - 2|\xi|, & \text{when } l = 0 \text{ and } K_1 < -|\xi|.\end{cases} \tag{13.34}$$

We present the variations of energy difference $\Delta E(n, l, D)$ on the dimension D in Figs. 13.1, 13.2, 13.3. We take the parameters $M = 1$ and $\xi = 0.05$ for definiteness. It is shown from Fig. 13.1 that there exist two singular points around $D \sim 1$ and $D \sim 2$ for $\Delta E(1, 0, D)$. As the dimension D increases, the energy difference $\Delta E(1, 0, D)$ first decreases with the dimension D in the region $(0, 0.9]$ and then increases with the dimension D in the region $[1.1, 1.9]$, and finally decreases again with the dimension $D \geq 2.1$. In particular, it is found that the energy difference $\Delta E(1, 0, D)$ is symmetric with respect to the point $(1.5, 0)$. The singular points can be easily explained from Eqs. (13.31), (13.33) and (13.34). That is, notice that there exist singular points for $K^2 = \xi^2$ and $K_1^2 = \xi^2$. Thus, it is shown from Eqs. (13.33) and (13.34) that $D = 1 \pm 2\xi$ and $D = 2 \pm 2\xi$ for $l = 0$, respectively. It is shown in Fig. 13.2, where $n = 2$ and $l = 1, 0$, that there are no bound states for $D < 0.1$. This can also be explained by Eqs. (13.33) and (13.34). As the dimension D increases, the energy difference $\Delta E(2, 1, D)$ monotonically decreases with dimension $D \geq 0.1$. Finally, we show the $\Delta E(3, l, D)$ as a function of dimension D in Fig. 13.3, where

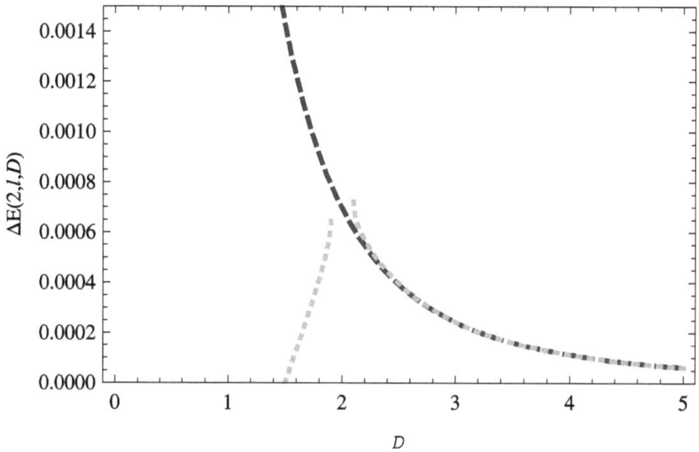

Fig. 13.2 The plot of energy differences $\Delta E(2, l, D)$ $(l = 0, 1)$ as a function of dimension D. Note that $\Delta E(2, 1, D)$ decreases with the increasing dimension $D \geq 0.1$ and the variation of the $\Delta E(2, 0, D)$ on the dimension D is very similar to $\Delta E(1, 0, D)$

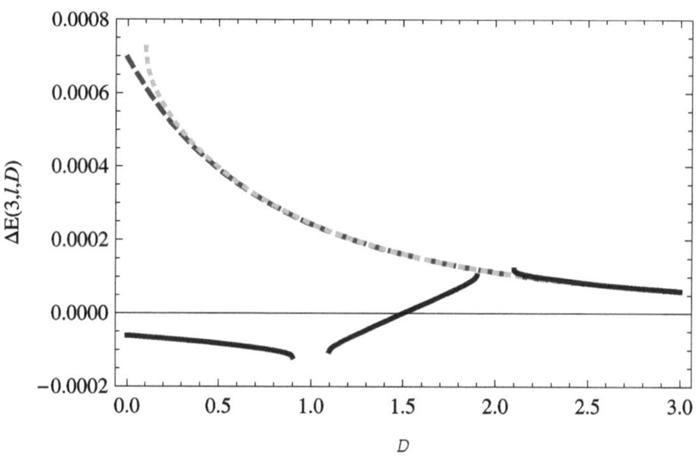

Fig. 13.3 The plot of energy difference $\Delta E(3, l, D)$ as a function of dimension D. The *red dashed*, *green dotted* and *blue solid lines* correspond to $l = 2, 1, 0$, respectively

the red dashed, green dotted and blue solid lines correspond to $l = 2, 1, 0$, respectively. Note that the energy difference $\Delta E(3, 2, D)$ decreases monotonically with the dimension D. The variations of the $\Delta E(n, 2, D)$, $\Delta E(n, 1, D)$ and $\Delta E(n, 0, D)$ are very similar to the $\Delta E(3, 2, D)$, $\Delta E(2, 1, D)$ and $\Delta E(1, 0, D)$, respectively.

Second, we display the variations of energy $E(n, l, D)$ on the dimension D in Figs. 13.4, 13.5, 13.6. It is shown from Fig. 13.4 that the energy $E(1, 0, D)$ decreases with the increasing dimension $D \in (0, 0.9]$, but increases with the dimension $D \geq 1.1$. Likewise, this singular point can be explained by Eqs. (13.26) or

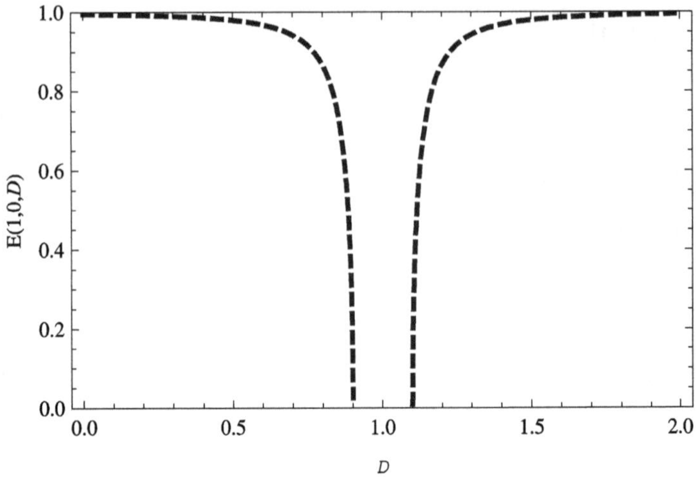

Fig. 13.4 The energy $E(1, 0, D)$ decreases with the increasing dimension $D \in (0, 0.9]$, but increases with the increasing dimension $D \geq 1.1$. Note that it is symmetric with respect to axis $D = 1$

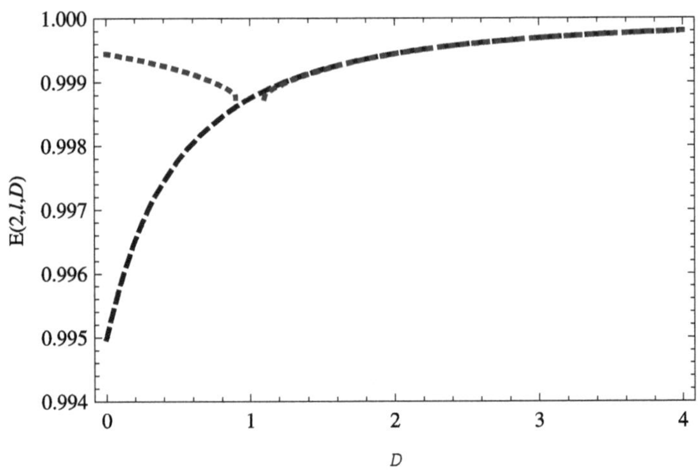

Fig. 13.5 The variation of energy $E(2, 0, D)$ (*red dotted line*) on the dimension D is very similar to $E(1, 0, D)$. The energy $E(2, 1, D)$ (*blue dashed*) increases with the increasing dimension D. Specially, for $D > 1$ the energy $E(2, 1, D)$ almost overlaps $E(2, 0, D)$

(13.33), i.e., $K^2 \geq \xi^2$. This means that there are no solutions in $D \in (1-2\xi, 1+2\xi)$. Figure 13.5 shows the properties of $E(2, l, D)$, in which the red dotted and blue dashed lines correspond to $l = 0, 1$, respectively. The variation of the $E(2, 0, D)$ on D is similar to energy $E(1, 0, D)$. But energy $E(2, 1, D)$ increases monotonically as the dimension D increases. It is found that, when $D > 1$ the energies $E(2, l, D)$ ($l = 0, 1$) are almost overlapped. This kind of phenomenon occurs in

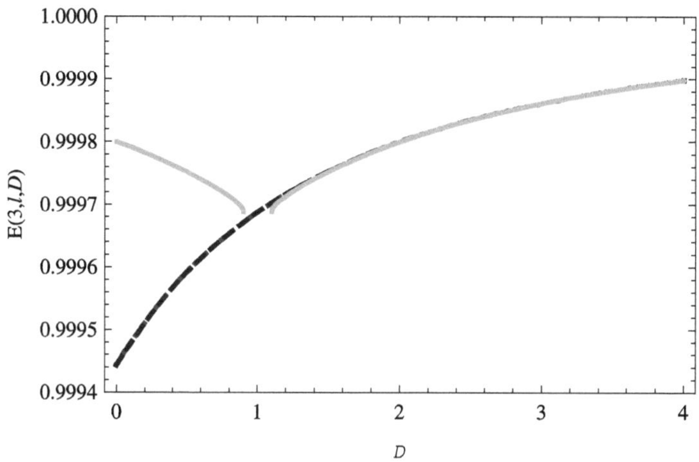

Fig. 13.6 The variations of energy levels $E(3, l, D)$ $(l = 2, 1, 0)$ on dimension D are similar to $E(2, l, D)$ $(l = 1, 0)$. The energy $E(3, 2, D)$ almost overlaps energy $E(3, 1, D)$. This can be explained well by Eq. (13.28). When $D > 1$, the energies $E(3, l, D)$ $(l = 2, 1, 0)$ are almost overlapped. The *red dotted*, *blue dashed* and *green solid lines* correspond to quantum numbers $l = 2, 1, 0$, respectively

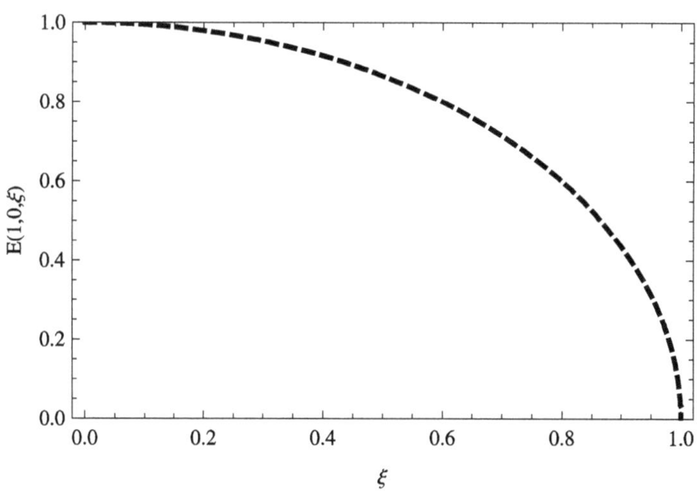

Fig. 13.7 The energy $E(1, 0, \xi)$ decreases with the increasing potential parameter $\xi \leq 1$

$n = 3$ $(l = 0, 1, 2)$ as shown in Fig. 13.6. This can be explained well through the series expansion for $1/D$ as given in Eq. (13.27).

Third, we exhibit the variations of energy $E(n, l, \xi)$ on the potential strength ξ in Figs. 13.7, 13.8, 13.9. It is found from Fig. 13.7 that the energy $E(1, 0, \xi)$ decreases with $\xi \leq 1$. This can be explained by Eq. (13.26). That is to say, $\xi \leq l + 1$. Figures 13.8 and 13.9 show the variations of the $E(2, l, \xi)$ and $E(3, l, \xi)$ on the

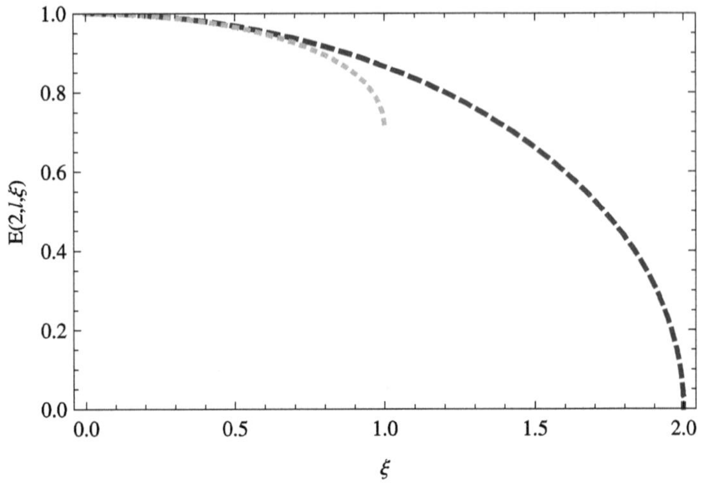

Fig. 13.8 The energies $E(2, l, \xi)$ $(l = 0, 1)$ decrease with the increasing potential parameter ξ. Notice that $\xi \leq 1$ for $l = 0$ (*green dotted line*), while $\xi \leq 2$ for $l = 1$ (*red dashed line*)

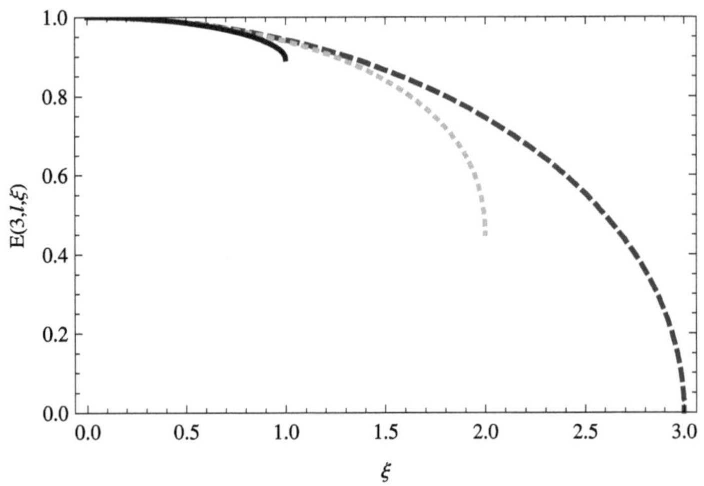

Fig. 13.9 The energies $E(3, l, \xi)$ $(l = 0, 1, 2)$ decrease with the increasing potential parameter ξ. Note that $\xi \leq 1$ for $l = 0$ (*blue solid line*), while $\xi \leq 2$ for $l = 1$ (*green dotted line*) and $\xi \leq 3$ for $l = 2$ (*red dashed line*). Generally, the energies $E(n, l, \xi)$ decrease with the potential parameter $\xi \leq l + 1$ for a given l

potential parameter ξ. The energies $E(n, l, \xi)$ decrease with potential parameter $\xi \leq l + 1$. In Fig. 13.8, the green dotted and red dashed lines correspond to $l = 0, 1$, respectively. In Fig. 13.9, the blue solid, green dotted and red dashed lines correspond to $l = 0, 1, 2$, respectively. Generally, the energies $E(n, l, \xi)$ decrease with the potential parameter $\xi \leq l + 1$ for a given l.

Fourth, it is worth pointing out the interest of the singularity at $|K| = \xi$, i.e., $l + (D-1)/2 = \xi = Z\alpha$. This is a famous phenomenon to make the energy formula broken down for potential strength with $Z\alpha \geq 1$ in three dimensions. When the Dirac equation with the $1/r$ potential is generalized to the $(D+1)$-dimensional case, however, the singularity survives and moves to big values for $Z\alpha$, e.g., $Z\alpha = 3/2$ for $D = 4$. Such a kind of phenomenon has been addressed in Ref. [445].

Finally, we want to briefly address the one-dimensional hydrogen atom as discussed by Moss [189]. The interest arises from its relevance to the behavior of hydrogen-like atoms in the strong magnetic fields and to hydrogenic impurities confined in the quantum-well wire structures [188]. It is found that when $D = 1$ and $Z = 1$, the energy spectrum (13.26) does not exist at all since the factor $\sqrt{K^2 - \xi^2}$ becomes imaginary. This is because from the definition $|K| = l + (D-1)/2$, K is equal to zero regardless of the value of potential parameter ξ if $D = 1$ and $l = 0$. This conclusion agrees well with that of Ref. [189].

4 Generalization to the Dirac Equation with a Coulomb Potential Plus a Scalar Potential

4.1 Introduction

The Dirac equation with a Coulomb potential has been discussed above. Recently, the bound states and the S-matrix in the quantum scattering theory of the Dirac equation with a Coulomb plus a scalar potential have been investigated in $3+1$ dimensions [446–448]. Apart from this, the bound states of the Dirac equation with the Coulomb plus scalar potential have been carried out both in two dimensions and in D dimensions [105, 449]. In this section we are going to study the Dirac equation with a Coulomb potential plus a scalar potential.

4.2 Exact Solutions

We now consider the Dirac equation with a mixed potential including a Coulomb potential and a scalar potential. The Coulomb potential is taken as

$$V_c = -\frac{v}{r}. \tag{13.35}$$

Also, the scalar potential is chosen as

$$V_s = -\frac{s}{r}, \tag{13.36}$$

which is added to the mass term of the Dirac equation. The v and s denote the electrostatic and scalar coupling constants, respectively.

After some algebraic manipulations, we observe that the radial components $F_{KE}(r)$ and $G_{KE}(r)$ satisfy the following first-order differential equations [105]

$$\begin{aligned} \frac{d}{dr}G_{KE}(r) + \frac{K}{r}G_{KE}(r) &= \left(E - M + \frac{v+s}{r}\right)F_{KE}(r), \\ -\frac{d}{dr}F_{KE}(r) + \frac{K}{r}F_{KE}(r) &= \left(E + M + \frac{v-s}{r}\right)G_{KE}(r). \end{aligned} \tag{13.37}$$

Because this equation keeps invariant by interchanging $F_{KE} \leftrightarrow G_{-K-E}$ and $V_c(r) \leftrightarrow -V_c(r)$, we only discuss the attractive Coulomb potential case $V_c(r)$ ($v > 0$). The solutions for the repulsive Coulomb potential $V_c(r)$ ($v < 0$) can be obtained from the former by interchanging $F_{KE} \leftrightarrow G_{-K-E}$. The scalar potential V_s, however, has to be discussed by two different cases $s > 0$ and $s < 0$.

Likewise, we introduce $\rho = 2r\sqrt{M^2 - E^2}$ for the bound states $|E| < M$. Thus, Eq. (13.37) becomes

$$\begin{aligned} \frac{d}{dr}G_{KE}(\rho) + \frac{K}{\rho}G_{KE}(\rho) &= \left(-\frac{1}{2}\sqrt{\frac{M-E}{M+E}} + \frac{v+s}{\rho}\right)F_{KE}(\rho), \\ \frac{d}{dr}F_{KE}(\rho) - \frac{K}{\rho}F_{KE}(\rho) &= \left(-\frac{1}{2}\sqrt{\frac{M+E}{M-E}} - \frac{v-s}{\rho}\right)G_{KE}(\rho). \end{aligned} \tag{13.38}$$

Similarly, define the wavefunction $\Psi_{\pm}(\rho)$ with the forms

$$\begin{aligned} G_{KE}(\rho) &= \sqrt{M-E}\,[\Psi_+(\rho) + \Psi_-(\rho)], \\ F_{KE}(\rho) &= \sqrt{M+E}\,[\Psi_+(\rho) - \Psi_-(\rho)]. \end{aligned} \tag{13.39}$$

Substitution of them into Eq. (13.38) leads to

$$\begin{aligned} &\frac{d}{d\rho}\Psi_+(\rho) \pm \frac{d}{d\rho}\Psi_-(\rho) \pm \frac{K}{\rho}[\Psi_+(\rho) \pm \Psi_-(\rho)] \\ &= \left[-\frac{1}{2} \pm \frac{v \pm s}{\rho}\sqrt{\frac{M \pm E}{M \mp E}}\right][\Psi_+(\rho) \mp \Psi_-(\rho)]. \end{aligned} \tag{13.40}$$

Their addition and subtraction allow us to obtain

$$\begin{aligned} &\frac{d}{d\rho}\Psi_{\pm}(\rho) \mp \left(\frac{vE + sM}{\rho\sqrt{M^2 - E^2}} - \frac{1}{2}\right)\Psi_{\pm}(\rho) \\ &= -\left(\frac{K}{\rho} \pm \frac{vM + sE}{\rho\sqrt{M^2 - E^2}}\right)\Psi_{\mp}(\rho). \end{aligned} \tag{13.41}$$

Take the following conventions

$$\tau = \frac{vE + sM}{\sqrt{M^2 - E^2}}, \qquad \tau' = \frac{vM + sE}{\sqrt{M^2 - E^2}}. \tag{13.42}$$

Then, Eq. (13.41) is simplified to

$$\frac{d}{d\rho}\Psi_{\pm}(\rho) \mp \left(\frac{\tau}{\rho} - \frac{1}{2}\right)\Psi_{\pm}(\rho) = \mp\frac{\tau' \pm K}{\rho}\Psi_{\mp}(\rho). \tag{13.43}$$

Based on this relation, we are able to obtain the following important second-order differential equations[1]

$$\left\{\frac{d^2}{d\rho^2}+\frac{1}{\rho}\frac{d}{d\rho}+\left(-\frac{1}{4}+\frac{\tau\pm 1/2}{\rho}-\frac{\lambda^2}{\rho^2}\right)\right\}\Psi_\pm(\rho)=0, \tag{13.44}$$

where

$$\lambda^2=K^2-\upsilon^2+s^2. \tag{13.45}$$

For a weak Coulomb potential, we take

$$\lambda=\sqrt{K^2-\upsilon^2+s^2}>0. \tag{13.46}$$

From the behaviors of the wavefunction at the origin and at infinity, we define

$$\Psi_\pm(\rho)=\rho^\lambda e^{-\rho/2}R_\pm(\rho). \tag{13.47}$$

Substitution of this into (13.44) yields

$$\begin{aligned}&\frac{d^2}{d\rho^2}R_\pm(\rho)+\left(-1+\frac{1+2\lambda}{\rho}\right)\frac{d}{d\rho}R_\pm(\rho)\\&+\frac{\tau-\lambda-1/2\pm 1/2}{\rho}R_\pm(\rho)=0,\end{aligned} \tag{13.48}$$

whose solutions are the confluent hypergeometric functions

$$\begin{aligned}R_+(\rho)&=a_0\,{}_1F_1(\lambda-\tau;2\lambda+1;\rho),\\R_-(\rho)&=b_0\,{}_1F_1(1+\lambda-\tau;2\lambda+1;\rho).\end{aligned} \tag{13.49}$$

Let us study the relation between a_0 and b_0. With the same technique as above, we have

$$\left(\frac{\lambda-\tau}{\rho}a_0+\frac{\tau'+K}{\rho}b_0\right){}_1F_1(1+\lambda-\tau;2\lambda+1;\rho)=0. \tag{13.50}$$

Since both a_0 and b_0 are not equal to zero, we obtain

$$b_0=\frac{\tau-\lambda}{\tau'+K}a_0. \tag{13.51}$$

From Eq. (13.39) one has

$$\begin{aligned}G_{KE}(\rho)&=N_{KE}\sqrt{M-E}\rho^\lambda e^{-\rho/2}\\&\quad\times[(\tau'+K)\,{}_1F_1(\lambda-\tau;2\lambda+1;\rho)\\&\quad+(\tau-\lambda)\,{}_1F_1(1+\lambda-\tau;2\lambda+1;\rho)],\\F_{KE}(\rho)&=N_{KE}\sqrt{M+E}\rho^\lambda e^{-\rho/2}\\&\quad\times[(\tau'+K)\,{}_1F_1(\lambda-\tau;2\lambda+1;\rho)\\&\quad-(\tau-\lambda)\,{}_1F_1(1+\lambda-\tau;2\lambda+1;\rho)],\end{aligned} \tag{13.52}$$

[1] Notice that Eq. (13.44) is a special case of the Tricomi equation as given in Eq. (13.14).

where $N_{KE} = a_0(\tau' + K)^{-1}(2\sqrt{M^2 - E^2})^{-1/2}$.

We now study the eigenvalues of this quantum system. The quantum condition is given by

$$\tau - \lambda = n' = 0, 1, 2, \ldots. \tag{13.53}$$

When $n' = 0$, $\lambda = \tau$, and

$$K^2 = \tau^2 + v^2 - s^2 = (\tau')^2. \tag{13.54}$$

Therefore, K has to be positive to avoid a trivial solution.

Introduce a principal quantum number

$$\begin{aligned} n &= |K| - (D-3)/2 + n' \\ &= |K| - (D-3)/2 + \tau - \lambda \\ &= l + 1 + n' = 1, 2, \ldots. \end{aligned} \tag{13.55}$$

Based on Eqs. (13.42) and (13.53), we have

$$\frac{Ev + Ms}{\sqrt{M^2 - E^2}} = n - |K| + \frac{D-3}{2} + \lambda = n' + \lambda = n - l - 1 + \lambda \equiv \kappa. \tag{13.56}$$

The energy E can be solved from Eq. (13.56)

$$E(n, l, D) = M\left\{-\frac{vs}{v^2 + \kappa^2} \pm \left[\left(\frac{vs}{v^2 + \kappa^2}\right)^2 - \frac{s^2 - \kappa^2}{v^2 + \kappa^2}\right]^{1/2}\right\}. \tag{13.57}$$

We now consider a few special cases. First, if $v = 0$, then $\lambda = \sqrt{K^2 + s^2}$. Thus, the energy becomes

$$E(n, l, D) = \pm M\left\{1 - \frac{s^2}{(n - l - 1 + \sqrt{K^2 + s^2})^2}\right\}^{1/2}. \tag{13.58}$$

It implies that there are two branches of symmetric solutions for the positive and negative energies.

For a large D, we have

$$E(n, D) \simeq \pm M[1 - 2s^2 D^{-2} + 4s^2(2n - 3)D^{-3} - \cdots], \tag{13.59}$$

which implies that the energy is independent of l for a large D.

For a small s, we have

$$\begin{aligned} E(n, l, D) \simeq \pm M\bigg\{1 &- \frac{s^2}{2[n + (D-3)/2]^2} \\ &+ \frac{s^4}{2[n + (D-3)/2]^4}\left(\frac{2n + D - 3}{2l + D - 1} - \frac{1}{4}\right)\bigg\}, \end{aligned} \tag{13.60}$$

where the first term on the right hand side is the rest energy M ($c^2 = 1$), the second one is from the solutions of the Schrödinger equation, and the third one is the fine structure energy, which removes the degeneracy between the states with the same n. On the other hand, it is shown from λ given in (13.46) that the scalar potential parameter s can be taken arbitrarily for the vector potential parameter $v = 0$.

Second, if $s = 0$, then $\lambda = \sqrt{K^2 - v^2}$. It is found from Eq. (13.46) that E has the same sign as v when $K^2 > v^2$. For the attractive Coulomb potential ($v > 0$) we have the positive energy $E(n, l, D)$

$$E(n,l,D) = M\left\{1 + \frac{v^2}{(n-l-1+\sqrt{K^2-v^2})^2}\right\}^{-1/2}. \tag{13.61}$$

This coincides with the conclusion from the Sturm-Liouville theorem for a weak attractive potential [184, 247, 248]. For a large D we have the similar result as Eq. (13.59) (s is replaced by v).

For a small v, we have

$$\begin{aligned} E(n,l,D) \simeq M\frac{v}{|v|}\bigg\{1 &- \frac{v^2}{2[n+(D-3)/2]^2} \\ &- \frac{v^4}{2[n+(D-3)/2]^4}\left(\frac{2n+D-3}{2l+D-1} - \frac{3}{4}\right)\bigg\}. \end{aligned} \tag{13.62}$$

Similarly, the physical meanings of three terms are similar to those of Eq. (13.60) except for the different expansion coefficients. This case has been studied in our recent work [104, 450], to which the reader can refer for more information.

Third, if $v = s$, from Eq. (13.56) both v and s have to be positive, $\lambda = |K|$, and the positive κ is given by

$$\kappa = n - l - 1 + |K| = \begin{cases} n + \frac{D-3}{2}, & \text{when } K \geq 0, \\ n - 2l - \frac{1+D}{2}, & \text{when } K < 0. \end{cases} \tag{13.63}$$

The energy is written as

$$E(n,l,D) = \begin{cases} M\{1 - \frac{2v^2}{v^2+(n+\frac{D-3}{2})^2}\}, & \text{for } K \geq 0, \\ M\{1 - \frac{2v^2}{v^2+(n-2l-\frac{1+D}{2})^2}\}, & \text{for } K < 0, \end{cases} \tag{13.64}$$

which implies that the energy is independent of the angular momentum quantum number l for $K \geq 0$, but for $K < 0$ it depends upon l. If we choose the negative sign in the result, we have $E = -M$, which is a singular solution of Eq. (13.56).

We now determine the normalization factor N_{KE} from the normalization condition

$$\int \Psi_{KE}^{\dagger} \Psi_{KE} dV = 1. \tag{13.65}$$

Similarly, through a direct calculation we get

$$N_{KE} = \frac{(M^2 - E^2)^{1/4}}{\Gamma(2\lambda + 1)}\left[\frac{\Gamma(\tau + \lambda + 1)}{2M\tau'(K + \tau')(\tau - \lambda)!}\right]^{1/2}. \tag{13.66}$$

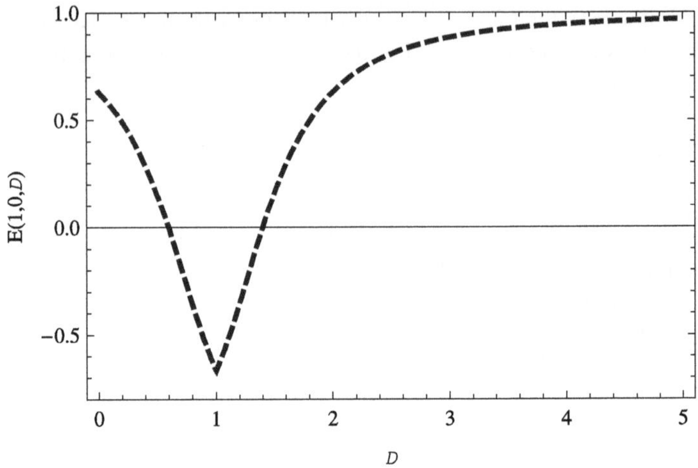

Fig. 13.10 The energy $E(1, 0, D)$ decreases with the increasing dimension $D \in (0, 1)$, but increases with the increasing dimension $D \geq 1$. The parameters $v = 0.2$ and $s = 0.3$ are taken here and also in Figs. 13.11 and 13.12

4.3 Analysis of the Energy Level

We now analyze the variation of the energy $E(n, l, D)$ on the dimension D. It is shown from Eqs. (13.46) and (13.56) that the energy $E(n, l, D)$ given in Eq. (13.57) is closely related with the following two conditions,

$$s^2 + K^2 > v^2, \qquad Ev + Ms > 0, \tag{13.67}$$

from which we can determine whether the energy is positive or negative. Recall that we only consider the case with positive v. For the case of negative v, the energy $E(n, l, D)$ changes its sign. For simplicity, we take sign "+" for the second term in Eq. (13.57). In fact, the energy levels are also valid if we take sign "−" for the second term, while the variations shall change with the different choice of the signs.

First, we consider the case $s > 0$. We discuss the energy $E(n, l, D)$ in two cases $s > v$ and $s < v$. For the first case $s > v$, due to Eq. (13.67) the bound states with $0 < E < M$ are physically acceptable. We take $v = 0.2$ and $s = 0.3$ for the weak attractive Coulomb potential. It is shown in Figs. 13.10, 13.11, 13.12 that the energies $E(n, 0, D)$ first decrease for $D \in (0, 1)$ and then increase with the increasing dimension $D \geq 1$, while the energies $E(n, l, D)$ $(l \neq 0)$ are almost independent of the quantum number l. Also, the energies $E(n, l, D)$ are almost overlapped for a large D. It is shown that the energies $E(n, 0, D)$ are symmetric with respect to $D = 1$ for $D \in (0, 2)$. For the second case $s < v$, we take $v = 0.3$ and $s = 0.2$. It is found that there does not permit bound state for $l = 0$ and $D \in (0.55, 1.45)$. This can be explained well from Eq. (13.46), i.e., $((D-1)/2)^2 + s^2 \geq v^2$ for $l = 0$. Consequently, there exist the bound states for $D \geq 1 + 2\sqrt{v^2 - s^2}$ and $D \leq 1 - 2\sqrt{v^2 - s^2}$. The corresponding variations of energies $E(n, l, D)$ on the dimension D are illustrated in Figs. 13.13, 13.14, 13.15, 13.16. It is shown in Fig. 13.13 that the

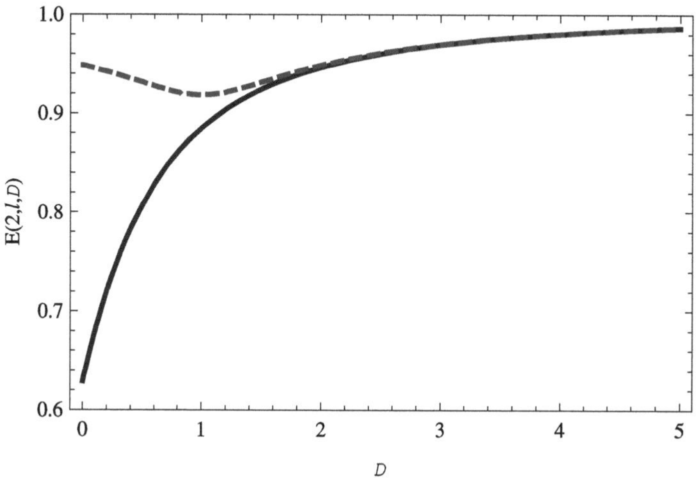

Fig. 13.11 The variation of energy $E(2, 0, D)$ (*red dashed line*) on the dimension D is similar to $E(1, 0, D)$. The energy $E(2, 1, D)$ (*blue solid line*) increases with the dimension D. For $D > 2$ the energy $E(2, 1, D)$ almost overlaps $E(2, 0, D)$

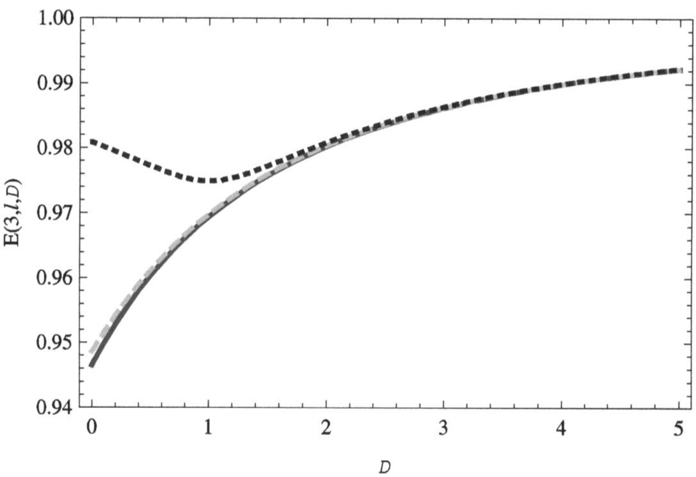

Fig. 13.12 The variations of energy levels $E(3, l, D)$ $(l = 2, 1, 0)$ on dimension D are similar to $E(2, l, D)$ $(l = 1, 0)$. The energy $E(3, 2, D)$ almost overlaps energy $E(3, 1, D)$. For $D > 2$, the energies $E(3, l, D)$ $(l = 2, 1, 0)$ are almost overlapped. The *blue dotted*, *green dashed* and *red solid lines* correspond to $l = 0, 1, 2$, respectively

energy $E(1, 0, D)$ decreases with the increasing dimension $D \in (0, 0.55]$, but increases with the increasing dimension $D \geq 1.45$. There does not exist bound state for $D \in (0.55, 1.45)$. It is shown in Figs. 13.15 and 13.16 that the variations of energies $E(n, 0, D)$ $(n = 2, 3)$ are very similar to $E(1, 0, D)$. The energies $E(n, 1, D)$ $(n = 2, 3)$ increase with the dimension D. Note that the energies $E(n, l, D)$ are al-

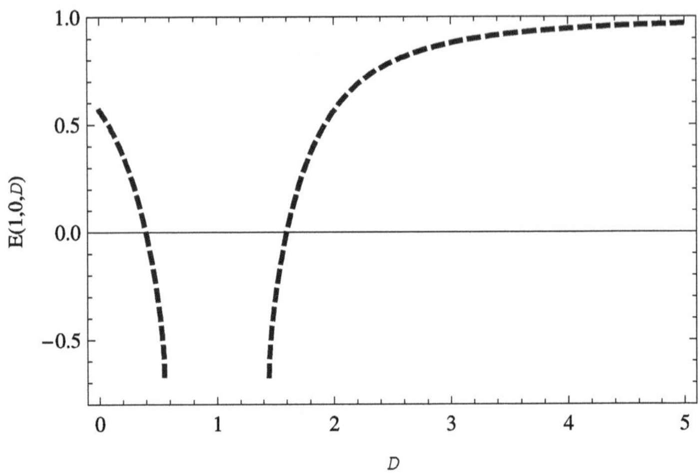

Fig. 13.13 The energy $E(1, 0, D)$ decreases with the increasing dimension $D \in (0, 0.55]$, but increases with the increasing dimension $D \geq 1.45$. There are no bound states for $D \in (0.55, 1.45)$. The parameters $v = 0.3$ and $s = 0.2$ are taken here and also in Figs. 13.15 and 13.16

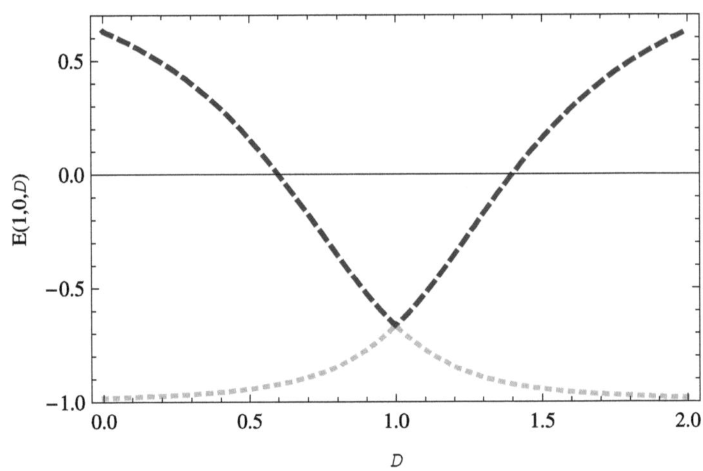

Fig. 13.14 The plot of energy spectra $E(1, 0, D)$ as a function of dimension D. Two signs "±" are considered. Note that their variations are absolutely different from each other

most independent of the quantum number l for a large D. It should be addressed that the variation range of dimension D without permitting bound state depends on the values of the potential parameters v and s. On the contrary, if we choose the sign "−" for the second term of Eq. (13.57), the variation will be opposite to that of the sign "+" as illustrated in Fig. 13.14.

Second, we consider the case $v > 0$ and $s < 0$. If $-s = |s| \geq v$, there is no bound state according to Eq. (13.67). However, if $-s = |s| < v$, the variations of ener-

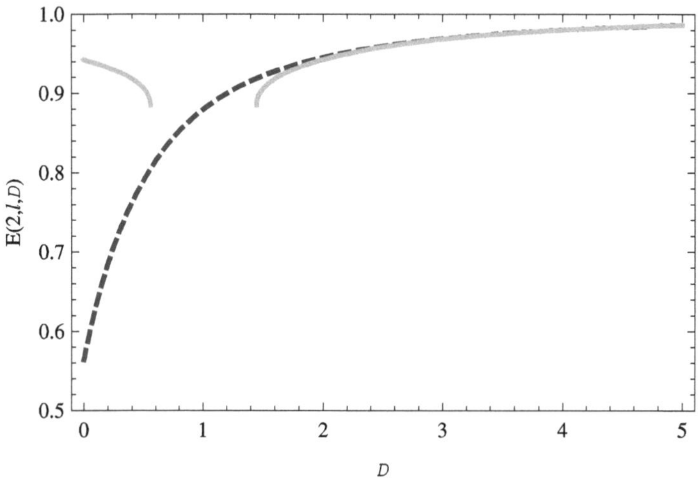

Fig. 13.15 The variation of energy $E(2, 0, D)$ (*green solid line*) on the dimension D is similar to $E(1, 0, D)$. The energy $E(2, 1, D)$ (*red dashed line*) increases with dimension D. The energy $E(2, 1, D)$ almost overlaps $E(2, 0, D)$ for $D \geq 2$

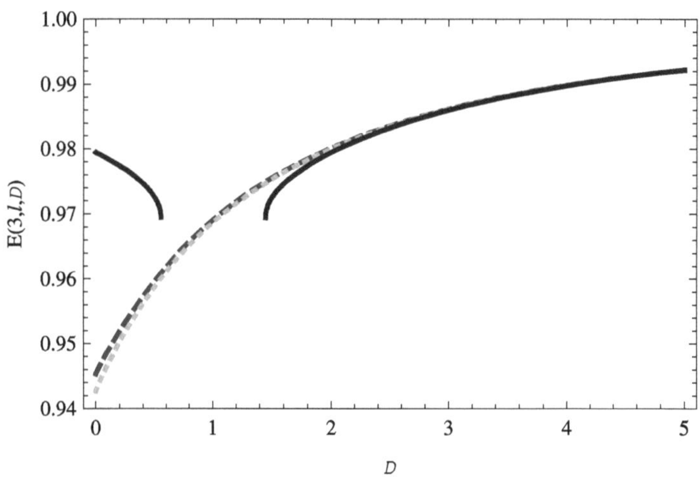

Fig. 13.16 The variations of energy levels $E(3, l, D)$ on the dimension D are similar to $E(2, l, D)$ $(l = 1, 0)$. For $D \geq 2$ the energies $E(3, l, D)$ $(l = 2, 1, 0)$ are almost overlapped. The *red dashed*, *green dotted* and *blue solid lines* correspond to the $l = 2, 1, 0$, respectively

gies $E(n, l, D)$ are similar to those of the case $0 < s < v$ as studied above (see Figs. 13.13, 13.14, 13.15, 13.16). We do not show them for simplicity.

We now consider some special cases. First, for special case $s = 0$ and $v \neq 0$, i.e., the Dirac equation with the Coulomb potential has been investigated in our recent work [104, 450]. Second, for special case $v = 0$, but $s \neq 0$, it is shown from Eq. (13.67) that there are bound states for $s > 0$, say $s = 0.2$. The variations of the

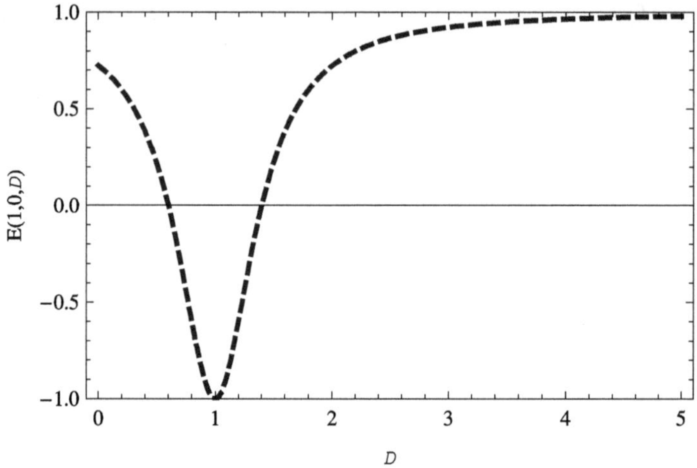

Fig. 13.17 The energy $E(1, 0, D)$ decreases with the increasing dimension $D \in (0, 1)$, but increases with the increasing dimension $D \geq 1$. The parameters $v = s = 0.2$ are taken here and also in Figs. 13.18 and 13.19

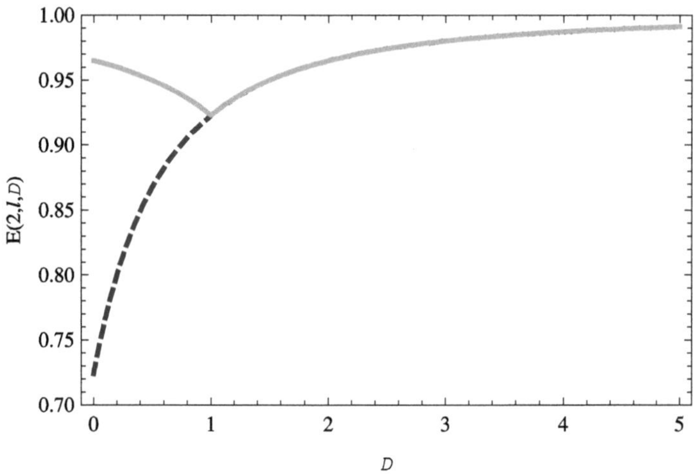

Fig. 13.18 The variation of energy $E(2, 0, D)$ (*green solid line*) on the dimension D is similar to $E(1, 0, D)$. The energy $E(2, 1, D)$ (*red dashed*) increases with the dimension D. The energy $E(2, 1, D)$ completely overlaps $E(2, 0, D)$ for $D \geq 1$

energies on the dimension D are similar to those of the case $s = 0.3$ and $v = 0.2$ given in Figs. 13.10, 13.11, 13.12. Likewise, we do not show them here. Third, for special case $v = s$, there are bound states for positive $s = v > 0$. The corresponding variations of the energies $E(n, l, D)$ on the dimension D are displayed in Figs. 13.17, 13.18, 13.19. Basically, the variations of the energies on the dimension D are similar to those of the case $v = 0.2$ and $s = 0.3$ given in Figs. 13.10,

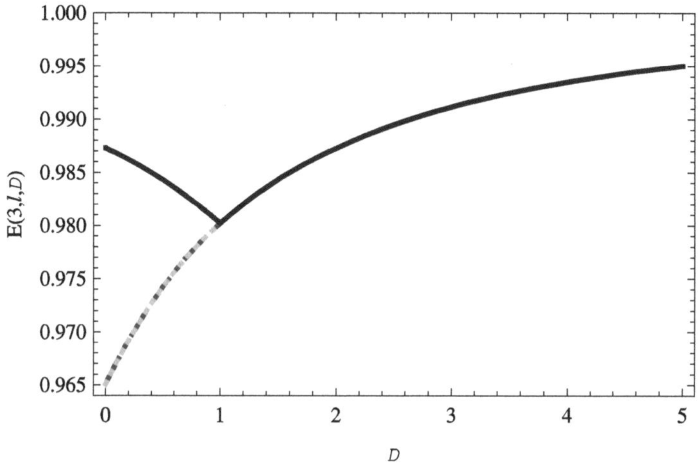

Fig. 13.19 The variation of energy $E(3, 0, D)$ on the dimension D is similar to $E(n, 0, D)$ $(n = 2, 1)$. The energy $E(3, 2, D)$ overlaps $E(3, 1, D)$ completely. For $D \geq 1$ the energies $E(3, l, D)$ $(l = 2, 1, 0)$ are completely overlapped. The *red dashed*, *green dotted* and *blue solid lines* correspond to $l = 2, 1, 0$, respectively

13.11, 13.12. However, the energies $E(n, l, D)$ $(l \neq 0)$ are completely independent of quantum number l. This can be explained well from Eq. (13.64) for $K > 0$. For $D \geq 2$ the energies $E(3, l, D)$ $(l = 2, 1, 0)$ are overlapped. However, the energy $E(n, 0, D)$ is closely related to the quantum number l as shown in Figs. 13.17, 13.18, 13.19. Also, this can be explained well by Eq. (13.64) for $K < 0$. It should be pointed out that there exists a singular solution $E = -M = -1$ for $D = 1$ and $l = 0$.

We now briefly study the variations of energies on the potential strengths v and s for a given $D = 3$ even though such a study which is not less important in physics. The constraints on the potential parameters v and s are closely related to the λ defined in Eq. (13.46). That is to say, $v^2 \leq K^2 + s^2 = (l + 1)^2 + s^2$ but $s^2 \geq v^2 - K^2 = v^2 - (l + 1)^2$. Generally speaking, there is no constraint on the potential strength s for a small v (weak potential strength). However, there is a constraint on the potential strength v, i.e., $|v| \leq \sqrt{K^2 + s^2} = \sqrt{(l + 1)^2 + s^2}$. These features are shown in Figs. 13.20 and 13.21. The detailed study for the constraint on the potential strength v can be found in Ref. [450].

5 Concluding Remarks

In this Chapter we have studied the $(D + 1)$-dimensional Dirac equation with the $1/r$ type potential through the confluent hypergeometric equation approach. This is different from the method used in our previous work [91]. The eigenfunctions are analytically obtained. The eigenvalues and the fine structure energy have also

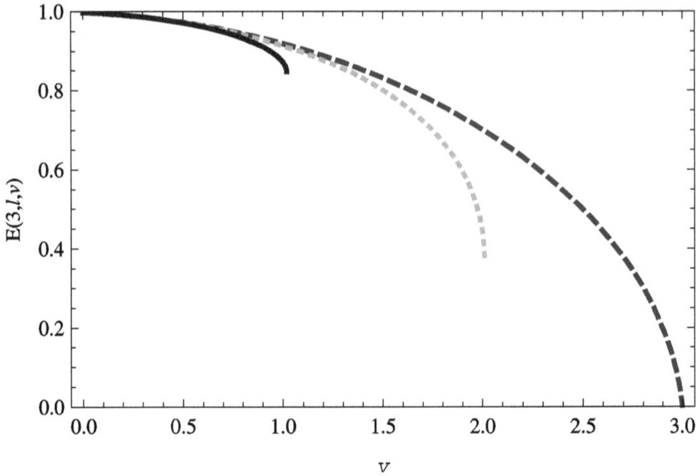

Fig. 13.20 The plot of energy levels $E(3,l,v)$ $(l=0,1,2)$ as a function of parameter v. They decrease with the parameter v. The $v \leq \sqrt{(l+1)^2+s^2}$ for $D=3$. The *red dashed*, *green dotted* and *blue solid lines* correspond to $l=2,1,0$, respectively. The parameters $D=3$ and $s=0.2$ are taken

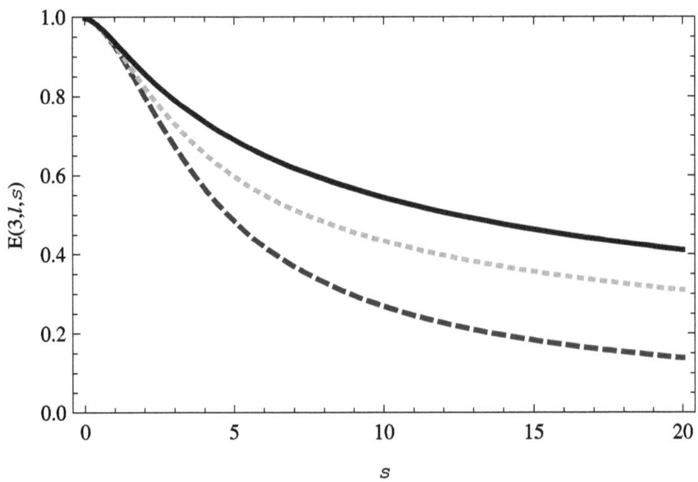

Fig. 13.21 The plot of energy levels $E(3,l,s)$ $(l=2,1,0)$ as a function of parameter s. The energy decreases with s. The *red dashed*, *green dotted* and *blue solid lines* correspond to $l=2,1,0$, respectively. The parameters $D=3$ and $v=0.2$ are chosen

been studied. The present work has paid more attention to the variations of energy differences $\Delta E(n,l,D)$ and the energy levels $E(n,l,D)$ on the continuous dimension D. The variations of energy differences $\Delta E(n,l,D)$ on the dimension D have been analyzed. Generally speaking, there exist three different kinds of variations. First, note that the energy difference $\Delta E(n,0,D)$ first decreases, then increases, and finally decreases as the dimension D increases. That is, the $\Delta E(1, 0, D)$ is symmetric with respect to the point $(1.5, 0)$. Second, we have found that $\Delta E(n, 1, D)$ decreases with the D, but there are no bound states when $D < 0.1$. Third, notice that $\Delta E(n, l, D)$ decreases monotonically with the dimension D.

For the energy $E(n, l, D)$, we have following properties. For the energy levels $E(n, 0, D)$, there exists a singular point at $D = 1$. That is, the $E(n, 0, D)$ is symmetric with respect to axis $D = 1$ for $D \in (0, 2)$. The energies $E(n, 0, D)$ first decrease with the dimension D and then increase with it. The energies $E(n, l, D)$ $(l > 0)$ are almost independent of quantum number l for a large D.

The variations of energy levels $E(n, l, \xi)$ on potential strength ξ have also been analyzed. For a given dimension $D = 3$, the energy $E(n, l, \xi)$ decreases with the potential parameter $\xi \leq l + 1$.

As a generalization, we have studied the Dirac equation with a Coulomb potential plus a scalar potential. The eigenvalues and some special cases have been carried out. We have elucidated the variations of energies $E(n, l, D)$ on the dimension D and found following typical properties. First, the energies $E(n, 0, D)$ first decrease with the dimension D and then increase with it. The energy levels $E(n, l, D)$ $(l \neq 0)$ increase with the dimension D. Second, the energies $E(n, l, D)$ are almost independent of quantum number l and the $E(n, l, D)$ $(l \neq 0)$ are almost overlapped for a large D. Third, the energies $E(n, 0, D)$ are symmetric with the respect to $D = 1$ for $D \in (0, 2)$. This is different from the case without the scalar potential. The variations of energies on potential parameters v and s are also studied for $D = 3$. We have found that the constraints on the potential parameters v and s are closely related to the parameter λ, i.e., $v^2 \leq K^2 + s^2 = (l+1)^2 + s^2$ but $s^2 \geq v^2 - K^2 = v^2 - (l+1)^2$. Generally speaking, there is no constraint on the potential strength s for a small v. However, there is a constraint on the potential strength v, i.e., $|v| \leq \sqrt{K^2 + s^2} = \sqrt{(l+1)^2 + s^2}$. We have found that the energy levels $E(n, l, v)$ and $E(n, l, s)$ decrease with the parameters v and s. In particular, it is interesting to observe that the $E(n, l, v)$ decreases with parameter $v \leq l + 1$ for a given l.

Chapter 14
Klein-Gordon Equation with the Coulomb Potential

1 Introduction

The exact solutions of non-relativistic and relativistic equations with the Coulomb potential have become an important subject in quantum mechanics. During the past several decades, the Klein-Gordon equation with the Coulomb potential has been studied in three dimensions such as the operator analysis [185], in an intense laser field [186], in two dimensions [187] and in one dimension [188–191]. With the interest of the higher-dimensional field theory, the Schrödinger equation and the Dirac equation with a Coulomb potential have been studied in $(D+1)$ dimensions. The Klein-Gordon equation with a Coulomb potential in $(D+1)$ dimensions has been discussed by the different approaches such as the large-N expansion approximate method [93] and the associated Laguerre equation approach [64], which paid more attention to the hydrogen atom case than the relativistic pi-mesonic atom one. The purposes of this Chapter are two-fold. The first one is to re-study this problem following the confluent hypergeometric equation approach. The another one, which is the main purpose of this Chapter, is to analyze the variations of the eigenvalues on the dimension D [64, 87].

This Chapter is organized as follows. Section 2 is devoted to the derivation of the eigenfunctions and eigenvalues. The properties of the eigenvalues are analyzed in Sect. 3. The Klein-Gordon equation with the Coulomb potential plus a scalar potential will be considered in Sect. 4. The comparison theorem is studied in Sect. 5. Some conclusions are given in Sect. 6.

2 Eigenfunctions and Eigenvalues

For simplicity the atomic units $\hbar = c = 1$ are employed if not explicitly stated otherwise. Considering the motion of a particle in a spherically symmetric potential $V(r)$ in D dimensions, the time-independent Klein-Gordon equation is taken as

$$(-\nabla^2 + M^2)\Psi(\mathbf{r}) = [E - V(r)]^2\Psi(\mathbf{r}), \tag{14.1}$$

where M and E denote the mass and the energy of the particle, respectively.

S.-H. Dong, *Wave Equations in Higher Dimensions*,
DOI 10.1007/978-94-007-1917-0_14, © Springer Science+Business Media B.V. 2011

As illustrated in previous Chapters, take the wavefunction of the form

$$\Psi(\mathbf{r}) = r^{-(D-1)/2} R_l(r) Y^l_{l_{D-2}\ldots l_1}(\hat{\mathbf{x}}). \tag{14.2}$$

Substitution of this into Eq. (14.1) yields

$$\left\{\frac{d^2}{dr^2} - \frac{\kappa^2 - 1/4}{r^2}\right\} R_l(r) = -\{[E - V(r)]^2 - M^2\} R_l(r), \tag{14.3}$$

where $\kappa \equiv l - 1 + D/2$.

For the present work, the symmetric potential $V(r)$ is taken as the Coulomb-type one

$$V(r) = -\frac{\xi}{r}, \quad \xi = Z\alpha. \tag{14.4}$$

We will discuss the weak potential, say $|\xi| < 1/2$. The radial equation (14.3) thus becomes

$$\frac{d^2 R_l(r)}{dr^2} + \frac{\xi^2 - \kappa^2 + 1/4}{r^2} R_l(r) + \frac{2E\xi}{r} R_l(r) + (E^2 - M^2) R_l(r) = 0. \tag{14.5}$$

It is convenient to take a new variable ρ for the bound states:

$$\rho = 2r\sqrt{M^2 - E^2}, \quad |E| < M. \tag{14.6}$$

As a result, Eq. (14.5) is changed to

$$\frac{d^2 R_l(\rho)}{d\rho^2} + \frac{1}{\rho^2}\left(\xi^2 - \kappa^2 + \frac{1}{4}\right) R_l(\rho) + \frac{\tau}{\rho} R_l(\rho) - \frac{1}{4} R_l(\rho) = 0, \tag{14.7}$$

with

$$\tau = \frac{E\xi}{\sqrt{M^2 - E^2}}. \tag{14.8}$$

From the behaviors of the radial function at the origin and at infinity, we define

$$R_l(\rho) = \rho^{\lambda + 1/2} e^{-\rho/2} \phi(\rho), \quad \lambda = \sqrt{\kappa^2 - \xi^2} > 0, \tag{14.9}$$

where we assume $\kappa^2 > \xi^2$. $\phi(\rho)$ satisfies the following confluent hypergeometric equation

$$\rho \frac{d^2\phi(\rho)}{d\rho^2} + (2\lambda + 1 - \rho)\frac{d\phi(\rho)}{d\rho} + (\tau - \lambda - 1/2)\phi(\rho) = 0. \tag{14.10}$$

Finally, the radial function can be expressed as

$$R_l(\rho) = N_l \rho^{\lambda + 1/2} e^{-\rho/2} {}_1F_1(\lambda - \tau + 1/2; 2\lambda + 1; \rho), \tag{14.11}$$

where N_l is the normalization factor to be determined.

We now discuss the eigenvalues. From the consideration of the finiteness of the solutions at infinity, the general quantum condition is obtained from Eq. (14.11)

$$\tau - \lambda - 1/2 = n' = 0, 1, 2, \ldots. \tag{14.12}$$

Introduce a principal quantum number

$$n = n' + \kappa - D/2 + 2 = n' + l + 1. \quad (14.13)$$

Based on Eqs. (14.8) and (14.12), we obtain

$$\frac{E\xi}{\sqrt{M^2 - E^2}} = \tau = n - l - 1/2 + \sqrt{\kappa^2 - \xi^2} > 0. \quad (14.14)$$

Therefore, we can obtain E with the same sign as ξ

$$E(n, l, D) = M\frac{\xi}{|\xi|}\left\{1 + \frac{\xi^2}{(n - l - 1/2 + \sqrt{\kappa^2 - \xi^2})^2}\right\}^{-1/2}, \quad (14.15)$$

which essentially coincides with that of Ref. [87] except that the factor $\xi/|\xi|$ was not considered there.

For a large D, we have

$$E(n, D) = M\frac{\xi}{|\xi|}[1 - 2\xi^2 D^{-2} + 4\xi^2(2n - 3)D^{-3} - \cdots]. \quad (14.16)$$

For a small ξ, we have

$$\begin{aligned} E(n, l, D) = M\frac{\xi}{|\xi|}\bigg\{1 &- \frac{\xi^2}{2[n + (D-3)/2]^2} \\ &+ \frac{\xi^4(D + 6l - 4n)}{4(2l - 2 + D)[n + (D-3)/2]^4}\bigg\}, \end{aligned} \quad (14.17)$$

where the first term on the right hand side is the rest energy M ($c = 1$), the second one is from the solutions of the Schrödinger equation with this potential, and the third one is the fine structure energy, which removes the degeneracy between the states with the same n.

We now calculate the normalization factor N_l from the normalization condition

$$\int_0^\infty R_l(\rho)^2 dr = 1. \quad (14.18)$$

Since $n' = \tau - \lambda - 1/2$ is a non-negative integer so that we can express the confluent hypergeometric functions ${}_1F_1(-n'; \beta + 1; \rho)$ by the associated Laguerre polynomial $L_{n'}^{\beta}(\rho)$. Based on Eq. (5.21) and through a direct calculation, we obtain

$$\begin{aligned} N_l &= \frac{(M^2 - E^2)^{1/4}}{\Gamma(2\lambda + 1)}\left[\frac{2\Gamma(n' + 2\lambda + 1)}{n'!(2n' + 2\lambda + 1)}\right]^{1/2}, \\ \lambda &= [(l - 1 + D/2)^2 - \xi^2]^{1/2}. \end{aligned} \quad (14.19)$$

3 Analysis of the Eigenvalues

We now analyze the properties of eigenvalues. It is shown from Eq. (14.14) that $n = 1, 2, 3, \ldots$ and $l = 0, 1, \ldots, n - 1$. If the principal quantum number n and the angular momentum quantum number l are fixed, the energy difference $\Delta E(n, l, D)$ between eigenvalues for dimensions D and $D-1$ is written as

$$\begin{aligned}\Delta E(n,l,D) &= E(n,l,D) - E(n,l,D-1)\\ &= M\frac{\xi}{|\xi|}\left\{\frac{1}{\sqrt{1+\frac{\xi^2}{(\sigma+\sqrt{\kappa^2-\xi^2})^2}}} - \frac{1}{\sqrt{1+\frac{\xi^2}{(\sigma+\sqrt{\kappa_1^2-\xi^2})^2}}}\right\},\end{aligned} \tag{14.20}$$

where

$$\sigma = n-l-1/2, \qquad \kappa_1 = (2l+D-3)/2. \tag{14.21}$$

It is found from Eq. (14.20) that the relation between $\Delta E(n,l,D)$ and the dimension D is more complicated than that of the Schrödinger equation for the hydrogen atom. The problem arises from the factor $\sqrt{\kappa^2-\xi^2}$ (or $\sqrt{\kappa_1^2-\xi^2}$). For a bound state, it is found from Eq. (14.2) that λ has to be a real number. Therefore, for a weak potential, $|\xi| < 1/2$, we obtain from Eqs. (14.3) and (14.9)

$$D = 2(1-l+\kappa)\begin{cases}\geq 2+2|\xi|, & \text{when } l=0 \text{ and } \kappa > |\xi|,\\ \leq 2-2|\xi|, & \text{when } l=0 \text{ and } \kappa < -|\xi|,\\ \geq 2|\xi|, & \text{when } l=1,\\ \geq 0, & \text{when } l \geq 2.\end{cases} \tag{14.22}$$

Now, we first show the variation of the energy difference $\Delta E(n,l,D)$ on the dimension D in Figs. 14.1, 14.2, 14.3. We take the parameters $M=1$ and $\xi = 11/137$ ($Z=11$) for definiteness. Recall that the energy and ξ has the same sign as shown in Eq. (14.15). These three figures display three typical variations of $\Delta E(n,l,D)$ on D.

It is shown from Fig. 14.1 that there exist two singular points at $D \sim 2$ and $D \sim 3$ for $E(1,0,D)$. As the dimension D increases, the energy difference $\Delta E(1,0,D)$ first decreases with the dimension D in the region $(0, 1.8]$ and then increases with the dimension D in the region $[2.2, 2.8]$, and finally decreases again with the dimension $D \geq 3.2$. Recall that the variant range of dimension D is related with the parameter ξ as shown in Eq. (14.22). The variation of $\Delta E(2,0,D)$ on the dimension D is very similar to $\Delta E(1,0,D)$. We do not show $\Delta E(2,0,D)$ for simplicity. In Fig. 14.2 we find that there exists one singular point near $D=1$ for $\Delta E(2,1,D)$. As the dimension D increases, the energy difference $\Delta E(2,1,D)$ first increases with the dimension D and then decreases with it. We show the $\Delta E(3,l,D)$ ($l=2,1,0$) in Fig. 14.3. Notice that the energy difference $\Delta E(3,2,D)$ decreases monotonically as the dimension D increases. In particular, it is found that $\Delta E(n,0,D)$ is symmetrical with respect to point $(2.5, 0)$.

Second, we analyze the variations of the energy $E(n,l,D)$ on the dimension D in Figs. 14.4, 14.5, 14.6. First of all, we emphasize that there is no bound state for s wave ($l=0$) in two-dimensional space $D=2$, because from Eqs. (14.3) and (14.2) we find that $\kappa=0$ and λ is imaginary. Notice that if the two-dimensional space is considered as a special case of three-dimensional space where all physical quantities

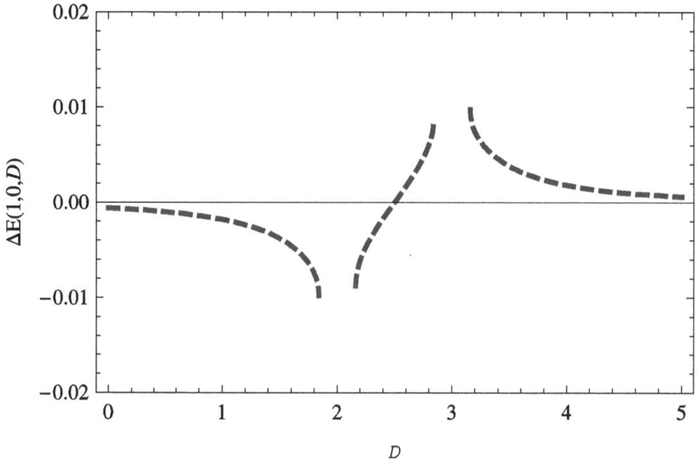

Fig. 14.1 For $D \in (0, 1.8)$, the energy difference $\Delta E(1, 0, D)$ decreases with the increasing dimension D, while increases with the increasing dimension $D \in [1.8, 2.8)$ and then decreases again with the increasing dimension $D \geq 3.2$. There exist two singular points around $D = 2$ and $D = 3$. Note that $\Delta E(1, 0, D)$ is symmetrical with respect to point (2.5, 0). The parameter $\xi = 11/137$ is taken here and also in Figs. 14.2–14.6

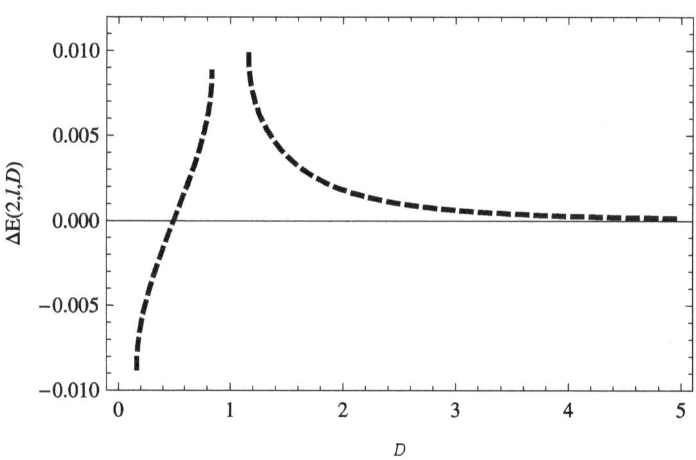

Fig. 14.2 Note that the energy difference $\Delta E(2, 1, D)$ increases with the increasing dimension $D \in (0, 0.8)$, and then decreases with the increasing dimension $D \geq 1.2$

do not depend upon the coordinate in one direction, the Coulomb potential becomes logarithmic instead of r^{-1}.

It is shown from Fig. 14.4 that the energy $E(1, 0, D)$ decreases as D increases in the region (0, 1.8), but increases with it in the region $D > 2$. Figure 14.5 shows the properties of $E(2, l, D)$ ($l = 1$ or 0). The variation of the $E(2, 0, D)$ on D is similar to energy $E(1, 0, D)$. However, the energy $E(2, 1, D)$ increases monotonically as

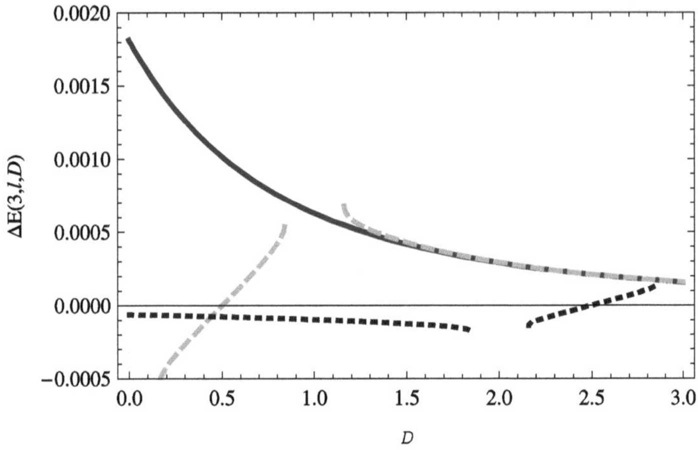

Fig. 14.3 The variation of energy difference $\Delta E(3, l, D)$ on the dimension D. It is noted that $\Delta E(3, 2, D)$ (*red solid line*) decreases monotonically with the increasing dimension $D > 0$. The $\Delta E(3, 1, D)$ (*green dashed line*) and $\Delta E(3, 0, D)$ (*blue dotted line*) are very similar to $\Delta E(2, 1, D)$ and $\Delta E(1, 0, D)$, respectively

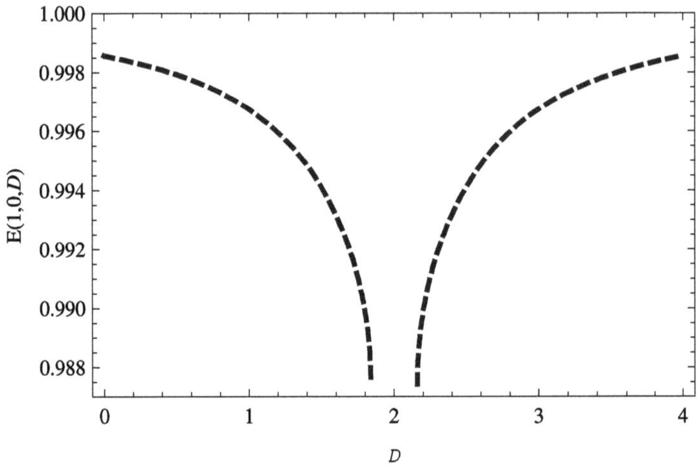

Fig. 14.4 Note that the energy $E(1, 0, D)$ decreases with the increasing dimension $D \in (0, 1.8)$, but increases with the increasing dimension $D \geq 2.2$. It should be stressed that there exists a singular point around $D \sim 2$. Note that the energy $E(1, 0, D)$ is symmetric with respect to axis $D = 2$ for $D \in (0, 4)$, so are those energy levels $E(n, 0, D)$

D increases. It is found that the $E(2, 1, D)$ is almost overlapped with the $E(2, 0, D)$ when $D > 2$. This kind of phenomenon occurs to $n = 3$ as shown in Fig. 14.6. That is to say, the energy $E(3, 2, D)$ almost overlaps with the energy $E(3, 1, D)$. This can be explained well through the series expansion for $1/D$ as given in Eq. (14.16).

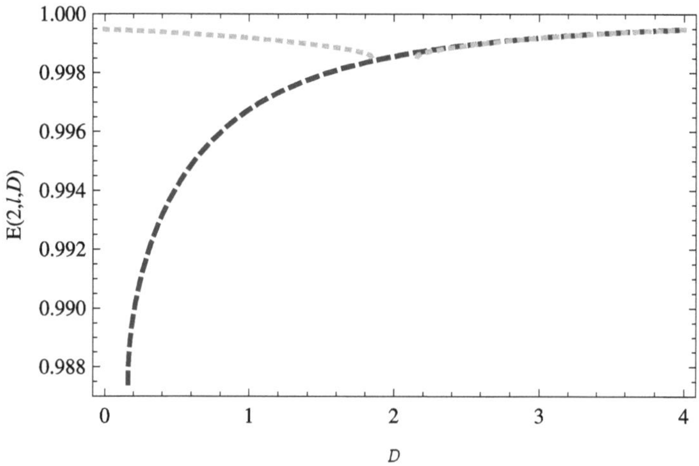

Fig. 14.5 The variation of energy $E(2, 0, D)$ (*green dotted line*) on the dimension D is very similar to $E(1, 0, D)$. The energy $E(2, 1, D)$ (*red dashed line*) increases with the increasing dimension D. Specially, it is found for $D > 2$ that the energy $E(2, 1, D)$ almost overlaps $E(2, 0, D)$

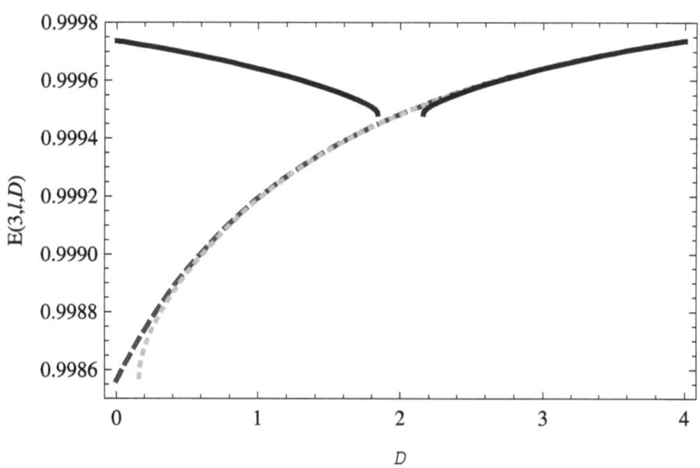

Fig. 14.6 The variation of $E(3, l, D)$ $(l = 2, 1, 0)$ on the dimension D is similar to $E(2, l, D)$ $(l = 1, 0)$. Note that the energy $E(3, 2, D)$ (*red dashed line*) almost overlaps that of the energy $E(3, 1, D)$ (*green dotted line*). When $D > 2$ the energies $E(3, l, D)$ $(l = 2, 1, 0)$ are almost overlapped

Before ending this part, we turn to one-dimensional hydrogen atom case due to its importance in physics. Actually, this problem has been studied by many authors [188–191]. When $D = 1$ and $Z = 1$ ($\xi_1 = 1/137$), the energy spectrum (14.15) reduces to the one-dimensional hydrogen atom,

$$E = M\frac{\xi_1}{|\xi_1|}\left\{1 + \frac{\xi_1^2}{(n - \frac{1}{2} + \frac{1}{2}\sqrt{1 - 4\xi_1^2})^2}\right\}^{-1/2}, \tag{14.23}$$

which coincides with the result given by Spector and Lee [188], in which Eq. (3) has a typo on S. The correct expression should be $S = 1/2 \pm 1/2(1 - 4Z^2\alpha^2)^{1/2}$. Here we only take the positive sign for S if considering the behavior of the radial function at the origin. The quantum number n' used here is n in [188]. The energy spectrum of the ground state ($n = 1$) can be obtained as

$$E = M\frac{\xi_1}{|\xi_1|}\left\{1 + \frac{4\xi_1^2}{(1 + \sqrt{1 - 4\xi_1^2})^2}\right\}^{-1/2}. \tag{14.24}$$

4 Generalization to the Klein-Gordon Equation with a Coulomb Potential Plus a Scalar Potential

4.1 Introduction

As mentioned above, the Klein-Gordon equation with the Coulomb potential has been carried out widely in three dimensions. Motivated by our recent work on the Dirac equation with a Coulomb plus a scalar potential in two dimensions [449] and in D dimensions [105], we are going to study the exact solutions of the Klein-Gordon equation with a Coulomb plus a scalar potential in D dimensions. On the other hand, we attempt to analyze the variations of the eigenvalues $E(n, l, D)$ on the continuous dimension D.

4.2 Eigenfunctions and Eigenvalues

Considering the motion of a particle in a spherically symmetric Coulomb potential $V_c(r)$ and the scalar potential $V_s(r)$ in D dimensions, we take the time-independent Klein-Gordon equation as [451]

$$\{-\nabla_D^2 + [M + V_s(r)]^2\}\Psi(\mathbf{r}) = [E - V_c(r)]^2\Psi(\mathbf{r}), \tag{14.25}$$

where M and E denote the mass and the energy of the particle, respectively. The Coulomb potential and the scalar potential are taken as

$$V_c = -\frac{v}{r}, \qquad V_s = -\frac{s}{r}. \tag{14.26}$$

It is evident that Eq. (14.25) keeps invariant by interchanging $V_c(r) \to -V_c(r)$ and $E \to -E$. For simplicity we only discuss the attractive Coulomb potential case $V_c(r)$ $(v > 0)$. The solutions for the repulsive Coulomb potential case $V_c(r)$ $(v < 0)$ can be obtained from the former by changing the sign of the energy E. The scalar potential V_s, however, will be discussed in two different cases $s > 0$ and $s < 0$.

Following Refs. [20, 112, 242], take the wavefunction as

$$\Psi(\mathbf{r}) = r^{-(D-1)/2} R_l(r) Y^l_{l_{D-2}\dots l_1}(\hat{\mathbf{x}}). \tag{14.27}$$

Substitution of this into Eq. (14.25) allows us to obtain the radial Klein-Gordon equation in D dimensions

$$\frac{d^2 R_l(r)}{dr^2} + \left\{ E^2 - M^2 + \frac{\upsilon^2 - s^2 - \kappa^2 + \frac{1}{4}}{r^2} + \frac{2(E\upsilon + Ms)}{r} \right\} R_l(r) = 0. \tag{14.28}$$

Take a new variable ρ for the bound states:

$$\rho = 2r\sqrt{M^2 - E^2}, \quad |E| < M. \tag{14.29}$$

Equation (14.28) becomes

$$\frac{d^2 R_l(\rho)}{d\rho^2} + \frac{1}{\rho^2}\left(\upsilon^2 - s^2 - \kappa^2 + \frac{1}{4}\right) R_l(\rho) + \frac{\tau}{\rho} R_l(\rho) - \frac{1}{4} R_l(\rho) = 0, \tag{14.30}$$

with

$$\tau = \frac{E\upsilon + Ms}{\sqrt{M^2 - E^2}}. \tag{14.31}$$

From the behaviors of the radial function at the origin and at infinity, we define

$$R_l(\rho) = \rho^{\lambda + \frac{1}{2}} e^{-\rho/2} \phi(\rho), \quad \lambda = \sqrt{\kappa^2 + s^2 - \upsilon^2} > 0. \tag{14.32}$$

We only consider the weak Coulomb potential case, i.e., $\kappa^2 + s^2 > \upsilon^2$. Otherwise the solution becomes oscillatory. $\phi(\rho)$ satisfies the following confluent hypergeometric equation

$$\rho \frac{d^2 \phi(\rho)}{d\rho^2} + (2\lambda + 1 - \rho) \frac{d\phi(\rho)}{d\rho} + \left(\tau - \lambda - \frac{1}{2}\right) \phi(\rho) = 0, \tag{14.33}$$

from which we finally obtain

$$R_l(\rho) = N_l \rho^{\lambda + \frac{1}{2}} e^{-\rho/2} \, {}_1F_1\left(\lambda - \tau + \frac{1}{2}; 2\lambda + 1; \rho\right), \tag{14.34}$$

where N_l is the normalization factor to be determined.

We now discuss the eigenvalues. As before, the general quantum condition is obtained from Eq. (14.34)

$$\tau - \lambda - \frac{1}{2} = n' = 0, 1, 2, \dots. \tag{14.35}$$

Introduce a principal quantum number

$$n = n' + \kappa - \frac{D}{2} + 2 = n' + l + 1. \tag{14.36}$$

It is shown from Eqs. (14.31), (14.35) and (14.36) that

$$\frac{E\upsilon + Ms}{\sqrt{M^2 - E^2}} = \tau = n - l - \frac{1}{2} + \sqrt{\kappa^2 + s^2 - \upsilon^2} \equiv \beta > 0. \tag{14.37}$$

The energy E can be solved from Eq. (14.37)

$$E(n,l,D) = M\left\{-\frac{vs}{v^2+\beta^2} \pm \left[\left(\frac{vs}{v^2+\beta^2}\right)^2 - \frac{s^2-\beta^2}{v^2+\beta^2}\right]^{\frac{1}{2}}\right\}. \tag{14.38}$$

We now consider some special cases. First, if $v = 0$, s has to be positive due to Eq. (14.37) so $\lambda = \sqrt{\kappa^2 + s^2}$. Thus, we have a pair of solutions $\pm E(n,l,D)$ for the energy

$$E(n,l,D) = \pm M\left\{1 - \frac{s^2}{(n-l-\frac{1}{2}+\sqrt{\kappa^2+s^2})^2}\right\}^{\frac{1}{2}}. \tag{14.39}$$

For a large D, we have

$$E(n,l,D) \simeq M[1 - 2s^2D^{-2} + 4s^2(2n-3)D^{-3} - \cdots], \tag{14.40}$$

which implies that the energy is independent of the l for $v = 0$.

For a small s, we have

$$E(n,l,D) \simeq M\left\{1 - \frac{s^2}{2[n+(D-3)/2]^2} + \frac{s^4}{2[n+(D-3)/2]^4}\left(\frac{2n+D-3}{2l+D-2} - \frac{1}{4}\right)\right\}. \tag{14.41}$$

Second, if $s = 0$, then $\lambda = \sqrt{\kappa^2 - v^2}$. Due to Eq. (14.37), the energy E has the same sign as v when $\kappa^2 > v^2$. For the attractive Coulomb potential ($v > 0$), we have the positive energy

$$E(n,l,D) = M\left\{1 + \frac{v^2}{(n-l-\frac{1}{2}+\sqrt{\kappa^2-v^2})^2}\right\}^{-\frac{1}{2}}. \tag{14.42}$$

It coincides with the conclusion from the Sturm-Liouville theorem for a weak attractive potential [184]. For a large D, we have the similar result to Eq. (14.40) (s is replaced by v).

For a small v, we have

$$E(n,l,D) \simeq M\frac{v}{|v|}\left\{1 - \frac{v^2}{2[n+(D-3)/2]^2} - \frac{v^4}{2[n+(D-3)/2]^4}\left(\frac{2n+D-3}{2l+D-2} - \frac{3}{4}\right)\right\}. \tag{14.43}$$

Third, if $v = s$, from Eq. (14.37) both v and s have to be positive, $\lambda = |\kappa|$, and the positive β is given by

$$\beta = n - l - \frac{1}{2} + |\kappa| = \begin{cases} n + \frac{D-3}{2}, & \text{when } \kappa \geq 0, \\ n - 2l + \frac{1-D}{2}, & \text{when } \kappa < 0. \end{cases} \tag{14.44}$$

When $\kappa \geq 0$, we obtain from Eq. (14.37)

$$E(n,l,D) = M\left\{1 - \frac{2v^2}{v^2 + (n+\frac{D-3}{2})^2}\right\}. \tag{14.45}$$

The energy is independent of the l. However, when $\kappa < 0$, the energy depends upon l.

We now calculate the normalization factor N_l from the normalization condition

$$\int_0^\infty R_l(\rho)^2 dr = 1 \tag{14.46}$$

as

$$\begin{aligned} N_l &= \frac{(M^2 - E^2)^{1/4}}{\beta(2\lambda + 1)} \left[\frac{2\beta(n' + 2\lambda + 1)}{n'!(2n' + 2\lambda + 1)} \right]^{\frac{1}{2}}, \\ \lambda &= \left\{ \left(l - 1 + \frac{D}{2} \right)^2 + s^2 - v^2 \right\}^{\frac{1}{2}}. \end{aligned} \tag{14.47}$$

4.3 Analysis of the Energy Levels

We now analyze the variation of energy $E(n, l, D)$ on the continuous dimension D. It is shown from Eqs. (14.32) and (14.37) that the energy $E(n, l, D)$ given in Eq. (14.39) is closely related with the following two conditions,

$$s^2 + \kappa^2 > v^2, \qquad Ev + Ms > 0, \tag{14.48}$$

from which we can determine whether the energy is positive or negative. Recall that we only consider the case with positive v. For the negative v case, the energy $E(n, l, D)$ changes its sign. For simplicity we take $M = 1$.

First, we consider the case $s > 0$. We discuss the energy $E(n, l, D)$ in two cases with $s > v$ and $s < v$. When $s > v$, due to Eq. (14.48) the bound states with $0 < E < M$ is an acceptable solution. We take $v = 0.2$ and $s = 0.3$ for the weak attractive Coulomb potential. On the other hand, the sign "+" is taken for the second term of Eq. (14.38). If we take the sign "−", the variations are absolutely different from each other. It is found from Figs. 14.7, 14.8, 14.9 that the energy $E(n, 0, D)$ first decreases for $D \in (0, 2)$ and then increases with the dimension $D \geq 2$, while the energy $E(n, l, D)$ $(l \neq 0)$ is almost independent of the angular momentum quantum number l. Also, the energy $E(n, l, D)$ are almost overlapped for a large D. It is shown that the energy $E(n, 0, D)$ is symmetric with respect to $D = 2$ for $D \in (0, 4)$. When $s < v$, we take $v = 0.3$ and $s = 0.2$. Note that there is no bound state for $l = 0$ and near $D \sim 2$ since $\lambda < 0$. The corresponding variations of energy $E(n, l, D)$ on the dimensions D are illustrated in Figs. 14.10, 14.11, 14.12. It is shown from Fig. 14.10 that the energy $E(1, 0, D)$ decreases with the increasing dimension $D \in (0, 1.4]$, but increases with the increasing dimension $D \geq 2.4$. There is no bound state for $D \in [1.6, 2.4]$. It is shown from Figs. 14.11 and 14.12 that variation of the energy $E(n, 0, D)$ $(n = 2, 3)$ on the dimension D is very similar to $E(1, 0, D)$. For the same reason $(\lambda < 0)$, there is no bound state when $l = 1$ and $D \in (0, 0.4]$. The energy $E(n, 1, D)$ $(n = 2, 3)$ increases with the dimension $D \geq 0.4$. It is found that the energy $E(n, l, D)$ is almost independent of the quantum number l for a large D. It should be pointed out that the variant range of the dimension D in which

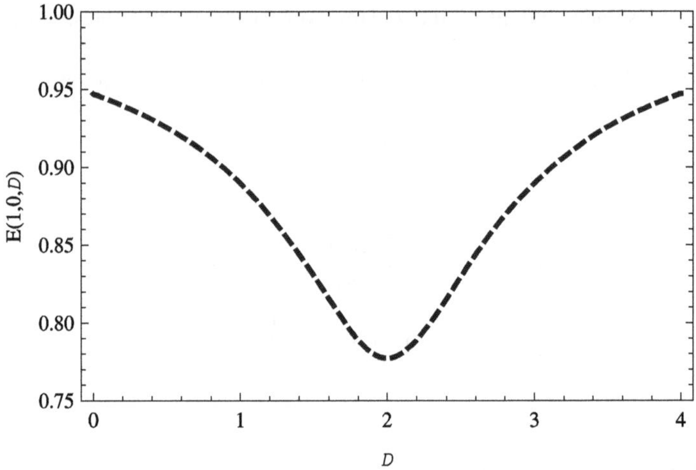

Fig. 14.7 The energy $E(1, 0, D)$ decreases with the dimension $D \in (0, 2)$, but increases with the increasing dimension $D \geq 2$. The parameters $v = 0.2$ and $s = 0.3$ are taken here and also in Figs. 14.8 and 14.9. Note that the energy $E(1, 0, D)$ is symmetric with respect to axis $D = 2$

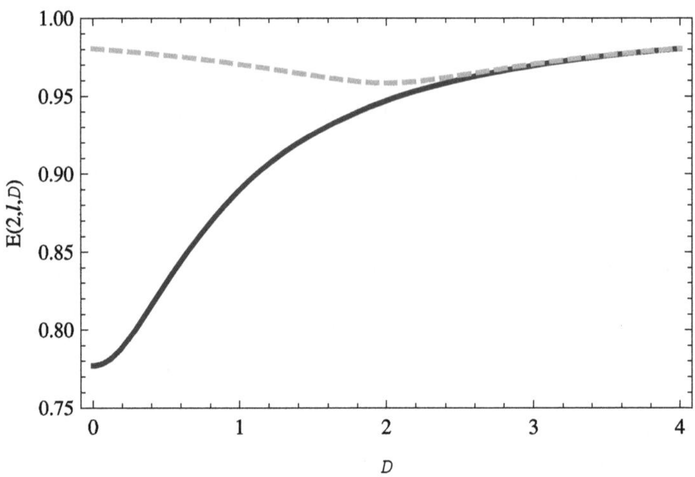

Fig. 14.8 The variation of energy $E(2, 0, D)$ (*green dashed line*) on the dimension D is very similar to $E(1, 0, D)$. The energy $E(2, 1, D)$ (*red solid line*) increases with the increasing dimension D. For $D > 2$ the energy $E(2, 1, D)$ almost overlaps $E(2, 0, D)$

the bound state is not admissible is closely related with the values of the potential parameters v and s.

Second, we consider the case $v > 0$ and $s < 0$. If $-s = |s| \geq v$, there is no bound state due to Eq. (14.37). However, if $-s = |s| < v$, the variation of the energy $E(n, l, D)$ on the dimension D is similar to that of the case $0 < s < v$ as shown above (see Figs. 14.10, 14.11, 14.12).

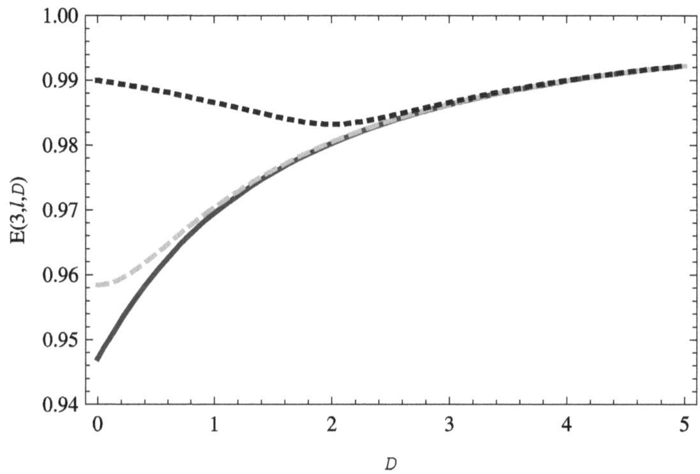

Fig. 14.9 The variations of the $E(3, l, D)$ ($l = 2, 1, 0$) on the dimension D are similar to $E(2, l, D)$ ($l = 1, 0$). The energy $E(3, 2, D)$ (*red solid line*) almost overlaps the energy $E(3, 1, D)$ (*green dashed line*) for a large D. For $D > 2$, the energies $E(3, l, D)$ ($l = 2, 1, 0$) are almost overlapped. The *red solid*, *green dashed* and *blue dotted lines* correspond to $l = 2, 1, 0$, respectively

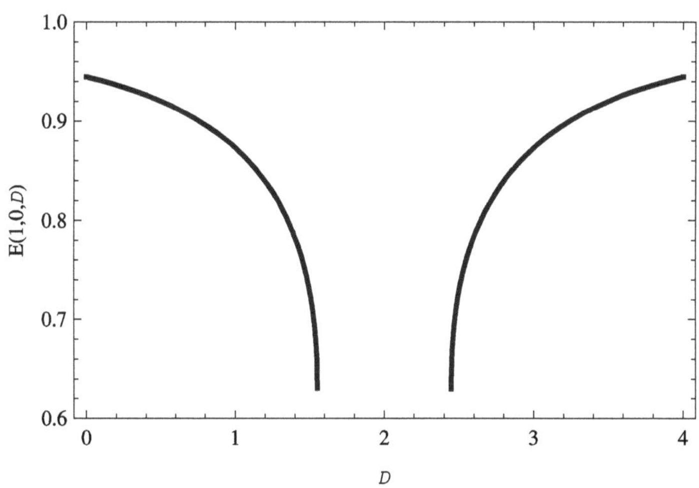

Fig. 14.10 The energy $E(1, 0, D)$ decreases with the increasing dimension $D \in (0, 1.6]$, but increases with the increasing dimension $D \geq 2.4$. There are no bound states for $D \in [1.6, 2.4]$. The parameters $\upsilon = 0.3$ and $s = 0.2$ are taken here and also in Figs. 14.11 and 14.12

We now consider some special cases below. First, for the special case $s = 0$ and $\upsilon \neq 0$, namely, the Klein-Gordon equation with the Coulomb potential has been carried out in our recent work [100]. Second, for the special case $\upsilon = 0$ but $s \neq 0$, it is shown from Eq. (14.48) that there are bound states for $s > 0$, say $s = 0.2$. The variations of energy levels on the dimension D shown in Figs. 14.13, 14.14,

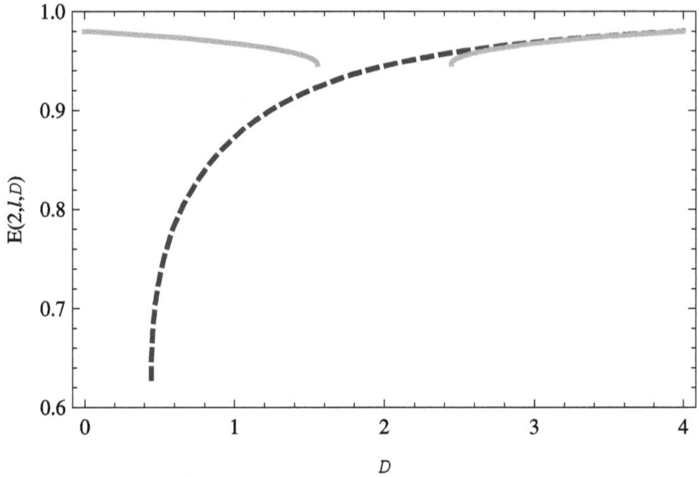

Fig. 14.11 The variation of the energy $E(2, 0, D)$ (*green solid line*) on the dimension D is very similar to $E(1, 0, D)$. The energy $E(2, 1, D)$ (*red dashed line*) increases with the dimension $D \geq 0.4$. The energy $E(2, 1, D)$ almost overlaps $E(2, 0, D)$ for $D \geq 2.6$

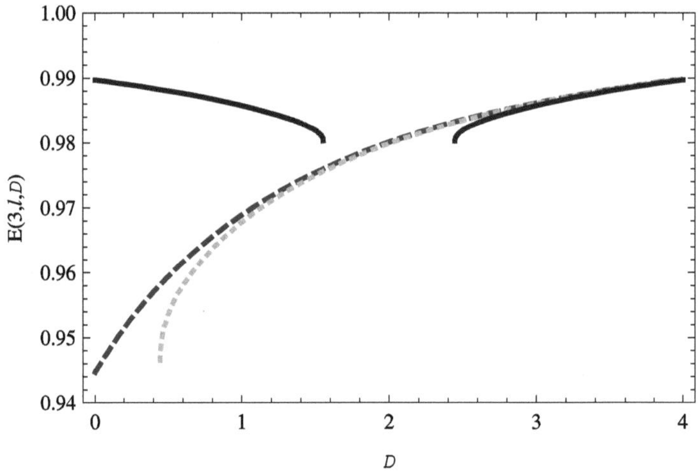

Fig. 14.12 The variation of the $E(3, l, D)$ $(l = 2, 1, 0)$ on the dimension D is similar to $E(2, l, D)$ $(l = 1, 0)$. For $D \geq 2.6$ the energies $E(3, l, D)$ $(l = 2, 1, 0)$ are almost overlapped. The *red dashed*, *green dotted* and *blue solid lines* correspond to $l = 2, 1, 0$, respectively

14.15 are similar to those of the case $s = 0.3$ and $v = 0.2$ in Figs. 14.7, 14.8, 14.9. Third, for the special case $v = s$, there are bound states for the positive $s = v > 0$. The corresponding variations of energy $E(n, l, D)$ on the dimension D are shown in Figs. 14.16, 14.17, 14.18. Basically, their variations are similar to those of case $v = 0.2$ and $s = 0.3$ as shown in Figs. 14.7, 14.8, 14.9. However, the energies $E(n, l, D)$ $(l \neq 0)$ are completely independent of the quantum

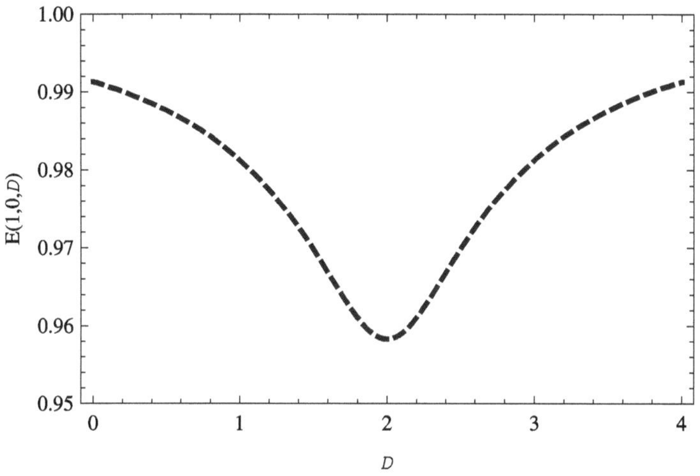

Fig. 14.13 The energy $E(1, 0, D)$ decreases with the increasing dimension $D \in (0, 2)$, but increases with the increasing dimension $D \geq 2$. The parameters $v = 0$ and $s = 0.2$ are taken and also in Fig. 14.14. Note that its variation is very similar to that illustrated in Fig. 14.7 except for a different amplitude

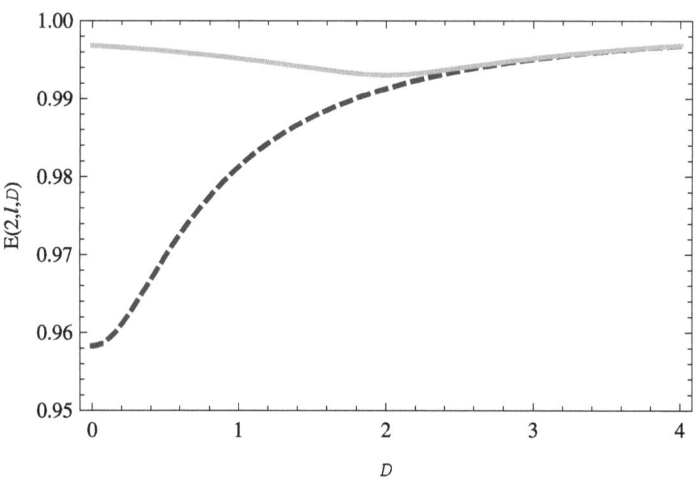

Fig. 14.14 The variation of energy $E(2, 0, D)$ (*green solid line*) on the dimension D is very similar to $E(1, 0, D)$. The energy $E(2, 1, D)$ (*red dashed line*) increases with dimension D. The energy $E(2, 1, D)$ almost overlaps $E(2, 0, D)$ for $D > 2.6$

number l since $\kappa > 0$ for $l \neq 0$. This can be explained well from Eq. (14.45). For $D > 2.6$ the energies $E(3, l, D)$ $(l = 2, 1, 0)$ are almost overlapped. However, the energy $E(n, 0, D)$ is closely related with the angular momentum quantum number l as shown by Figs. 14.16, 14.17, 14.18.

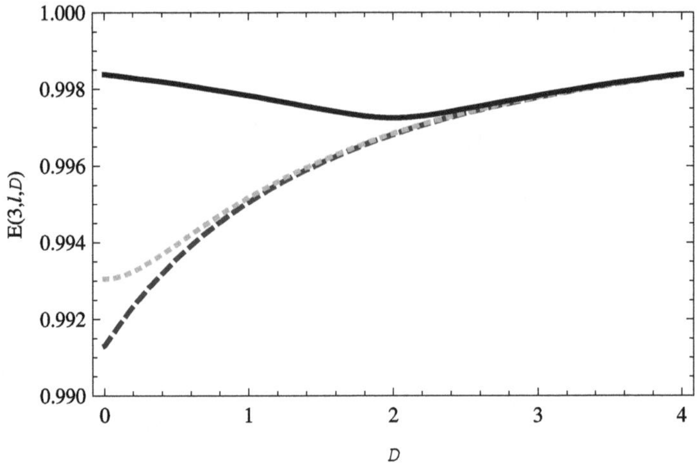

Fig. 14.15 The variations of the $E(3, l, D)$ $(l = 2, 1, 0)$ on the dimension D are similar to $E(2, l, D)$ $(l = 1, 0)$. The energy $E(3, 2, D)$ (*red dashed line*) almost overlaps energy $E(3, 1, D)$ (*green dotted line*). For $D > 2.6$, the energies $E(3, l, D)$ $(l = 2, 1, 0)$ are almost overlapped

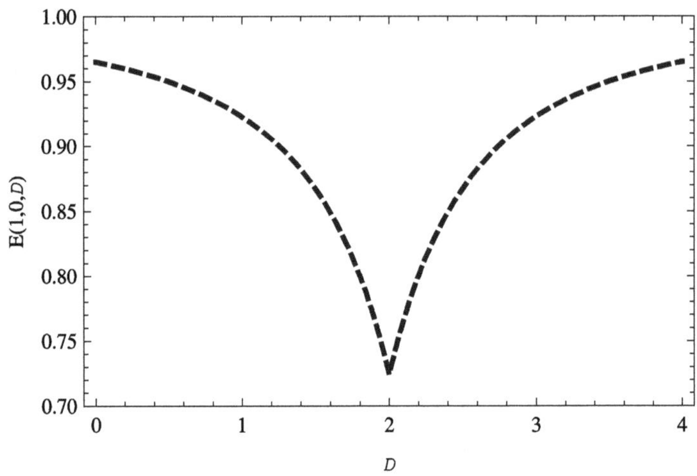

Fig. 14.16 The energy $E(1, 0, D)$ decreases with the increasing dimension $D \in (0, 2)$, but increases with the increasing dimension $D \geq 2$. Note that it is symmetric with respect to the axis $D = 2$. The parameters $v = s = 0.2$ are chosen here and also in Figs. 14.17 and 14.18

We now briefly study the variations of energy levels on the potential strengths v and s for $D = 3$. The constraints on the potential parameters v and s are closely related to the λ given in Eq. (14.32). That is, $v^2 \leq \kappa^2 + s^2 = (l + 1)^2 + s^2$ but $s^2 \geq v^2 - \kappa^2 = v^2 - (l + 1/2)^2$. In general, there is no constraint on the potential strength s for the small v (weak potential strength). However, there is a constraint on the potential strength v, i.e., $|v| \leq \sqrt{\kappa^2 + s^2} = \sqrt{(l + 1/2)^2 + s^2}$. These features

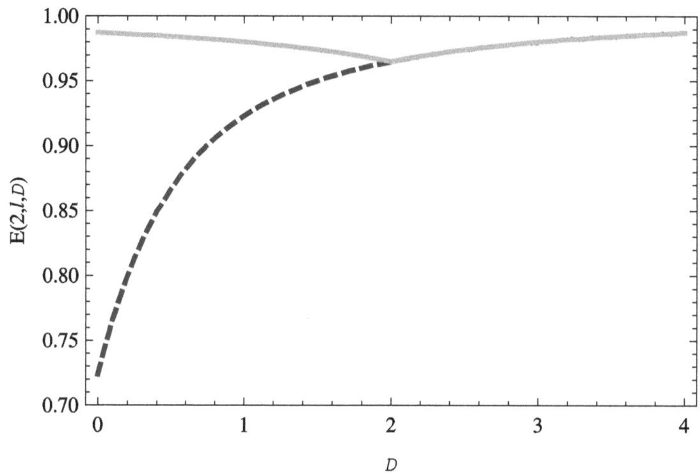

Fig. 14.17 The variation of the energy $E(2, 0, D)$ (*green solid line*) on the dimension D is very similar to $E(1, 0, D)$. The energy $E(2, 1, D)$ increases with the dimension D. The energy $E(2, 1, D)$ (*red dashed line*) completely overlaps $E(2, 0, D)$ for $D \geq 2$

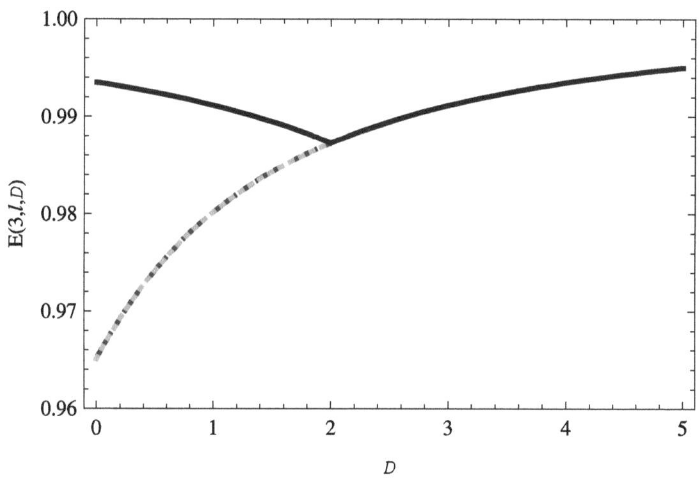

Fig. 14.18 The variation of the $E(3, 0, D)$ (*blue solid line*) on the dimension D is similar to $E(n, 0, D)$ $(n = 2, 1)$. The energy $E(3, 2, D)$ (*red dashed line*) overlaps the $E(3, 1, D)$ (*green dotted line*) completely. For $D \geq 2$ the energies $E(3, l, D)$ $(l = 2, 1, 0)$ are completely overlapped

are displayed in Figs. 14.19 and 14.20. It is interesting to notice that the energy $E(n, l, v)$ decreases with the parameter $v \leq l + 1$ for a given l.

Let us turn to one-dimensional case. When $D = 1$, the energy $E(n, l, D)$ reduces to a similar expression to Eq. (14.39), but with a different β

$$\beta = n - \frac{1}{2} + \sqrt{1/4 + s^2 - v^2}. \tag{14.49}$$

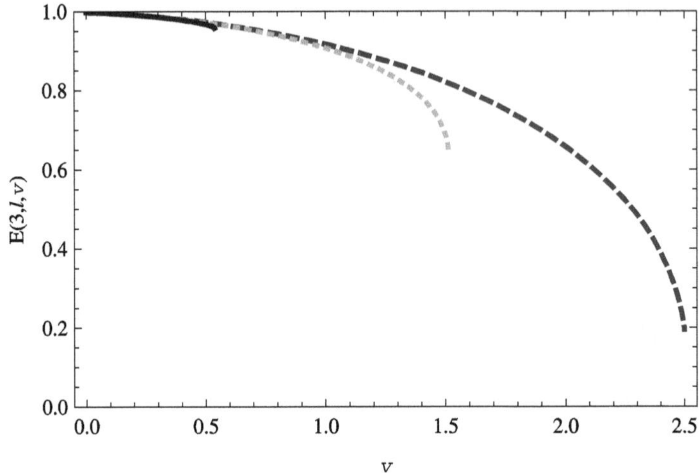

Fig. 14.19 The variation of the energy levels $E(3, l, v)$ ($l = 0, 1, 2$) on the parameter v. The energy decreases with the parameter v. The *red dashed*, *green dotted* and *blue solid lines* correspond to $l = 2, 1, 0$, respectively. The parameters $D = 3$ and $s = 0.2$ are taken

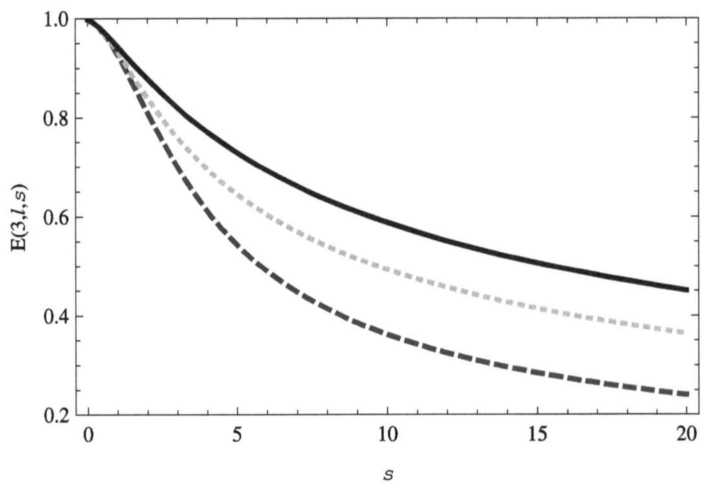

Fig. 14.20 The variation of the energy levels $E(3, l, s)$ on the parameter s. The energy decreases with the v. The *red dashed*, *green dotted* and *blue solid lines* correspond to the $l = 2, 1, 0$, respectively. The parameters $D = 3$ and $v = 0.2$ are taken

The energy $E(n, l, D)$ is determined by two conditions

$$\frac{1}{4} + s^2 > v^2, \qquad Ev + Ms > 0. \tag{14.50}$$

The discussion is similar to that of the general case as studied above. We do not mention them for simplicity.

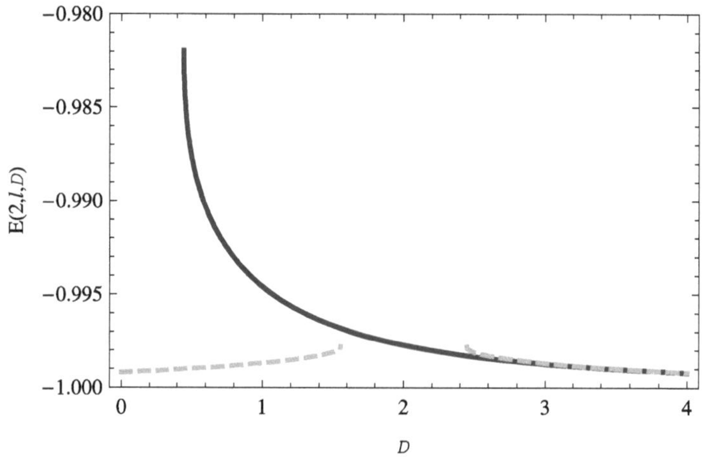

Fig. 14.21 The plot of energy levels $E(2, l, D)$ as a function of dimension D for energy (14.38) with sign "−". The *red solid* and *green dashed lines* correspond to $l = 1, 0$, respectively. The $v = 0.3$ and $s = 0.2$ are taken

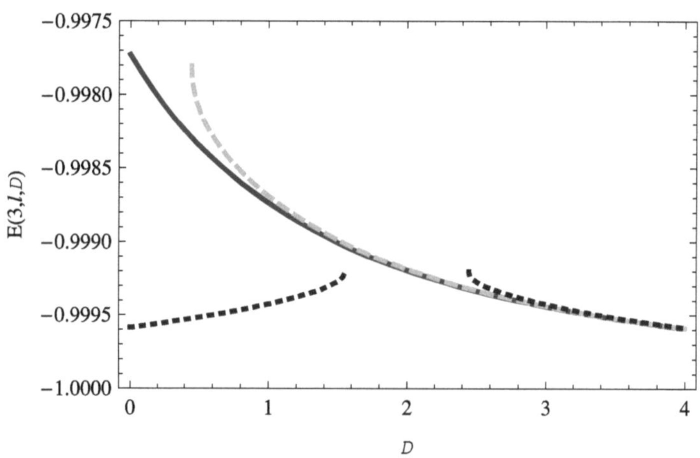

Fig. 14.22 The plot of energy levels $E(3, l, D)$ as a function of dimension D. The *red solid*, *green dashed* and *blue dotted lines* correspond to $l = 2, 1, 0$, respectively. Same parameters are taken as those in Fig. 14.21

Before ending this section, let us mention the case of Eq. (14.38) with sign "−". To show this, we are going to illustrate it in Figs. 14.21, 14.22, 14.23, 14.24 by choosing suitable parameters. It is found that its variation is absolutely different from the case with "+" sign. Their absolute values of energy spectrum are same but with different signs.

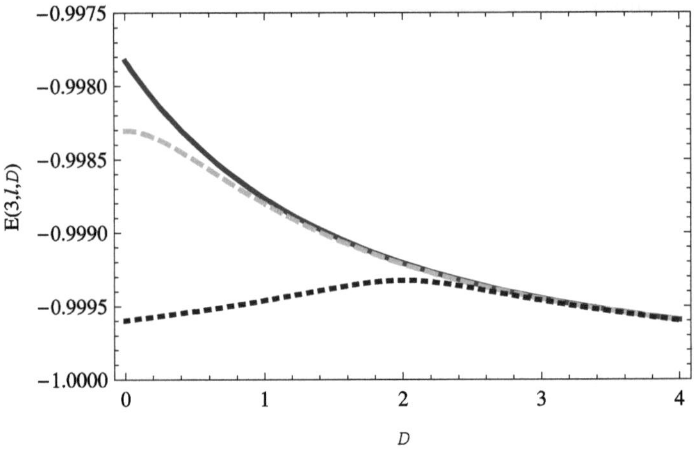

Fig. 14.23 The plot of energy levels $E(3, l, D)$ as a function of dimension D. The *red solid*, *green dashed* and *blue dotted lines* correspond to $l = 2, 1, 0$, respectively. The parameters $\upsilon = 0.2$ and $s = 0.3$ are taken

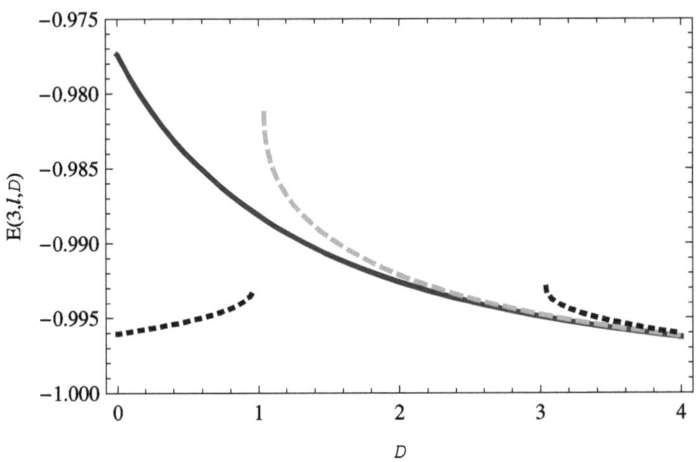

Fig. 14.24 The plot of energy levels $E(3, l, D)$ as a function of dimension D. The *red solid*, *green dashed* and *blue dotted lines* correspond to $l = 2, 1, 0$, respectively. Here the parameters $\upsilon = 0.6$ and $s = 0.3$ are chosen

5 Comparison Theorem

Finally, we present the comparison theorem for Klein-Gordon equation in arbitrary dimensions D. In non-relativistic quantum mechanics, this theorem is an immediate consequence of the variational characterization of the bound state spectrum as shown in Chap. 9. In the Klein-Gordon equation case, recently Hall and Aliyu have dealt with this problem [103]. It is found that if ψ_1 and ψ_2 are node-free ground states corresponding to positive energies $E_1 \geq 0$ and $E_2 \geq 0$ and

$V_1(r) \le V_2(r) \le 0$, then one has $E_1 \le E_2$. On the other hand, if $V(r,a)$ depends on a parameter $a \in (a_1, a_2)$, and $E(a)$ is any positive eigenvalue, then one has $\partial V/\partial a \ge 0 \Rightarrow E'(a) \ge 0$ and $\partial V/\partial a \le 0 \Rightarrow E'(a) \le 0$. This can be easily proved by using Eq. (14.3). For ordered central potentials $V_1(r) \le V_2(r)$, suppose that the respective ground states $\{\psi_1(r), \psi_2(r)\}$ in Eq. (14.3) are node-free and write the corresponding energy levels as $\{E_1, E_2\}$. The two eigenfunctions are written by

$$-\psi_1''(r) + \frac{C}{r^2}\psi_1(r) = \{[E_1 - V_1(r)]^2 - M^2\}\psi_1(r), \tag{14.51}$$

$$-\psi_2''(r) + \frac{C}{r^2}\psi_2(r) = \{[E_2 - V_2(r)]^2 - M^2\}\psi_2(r), \tag{14.52}$$

with $C = (2l + D - 1)(2l + D - 3)/4$.

With these two equations we calculate the difference between Eq. (14.51) multiplied by $\psi_2(r)$ and Eq. (14.52) multiplied by $\psi_1(r)$ and integrating it over the argument $r \in [0, \infty)$ to give the following equation

$$(E_2 - E_1)\int_0^\infty W(r)\psi_1(r)\psi_2(r)dr$$
$$= \int_0^\infty [V_2(r) - V_1(r)]W(r)\psi_1(r)\psi_2(r)dr, \tag{14.53}$$

with

$$W(r) = E_1 + E_2 - V_1(r) - V_2(r). \tag{14.54}$$

Thus, for positive energy levels E_1, E_2 belonging to node-free ground states $\psi_1(r)$ and $\psi_2(r)$, we have $E_1 \le E_2$ if $V_1(r) \le V_2(r) \le 0$.

6 Concluding Remarks

In this Chapter we have studied the D-dimensional Klein-Gordon equation with a Coulomb potential. The exact solutions have been analytically obtained. It is worth pointing out that when the Coulomb potential is not very strong, an attractive potential leads to the bound states with positive energies, and a repulsive potential leads to those with negative energies.

The variation of the energy difference $\Delta E(n, l, D)$ on the dimension D has been analyzed in detail. In general, there are three kinds of variations. First, the energy difference $\Delta E(n, 0, D)$ first decreases, then increases, and finally decreases as the dimension D increases. In other words, we have found that $\Delta E(n, l, D)$ is symmetric with respect to the point $(2.5, 0)$. Second, it is seen that $\Delta E(n, 1, D)$ first increases and then decreases as D increases. There exists a singular point around $D = 1$. Third, we have noticed that $\Delta E(n, 2, D)$ decreases monotonically as D increases. As far as the energy $E(n, l, D)$ is concerned, the following properties are displayed. First, when $l = 0$ there is no bound state around $D = 2$, i.e., $E(n, 0, D)$

first decreases and then increases with $D > 2.4$. More precisely, $E(n, l, D)$ is symmetric with respect to axis $D = 2$. The energy $E(n, l, D)$ with $l > 0$ is almost independent of the quantum number l for a large D.

As a generalization, we have also studied the D-dimensional Klein-Gordon equation with a Coulomb plus a scalar potential. The eigenfunctions have been gotten analytically. We have investigated the variation of the energy $E(n, l, D)$ on the dimension D in some detail. We have observed following typical properties. First, the positive energy $E(n, 0, D)$ first decreases with the dimension D and then increases with it. We have found that $E(n, 0, D)$ is symmetric with respect to axis $D = 2$. When $\upsilon < s$, there is no singular point at $D = 2$, while there exists one singular point when $\upsilon > s$. Second, the positive $E(n, l, D)$ $(l \neq 0)$ increases with the dimension D. Third, the variations of the negative $E(n, 0, D)$ and $E(n, l, D)$ $(l \neq 0)$ on the dimension D are completely opposite to those of the positive cases. Fourth, the energy $E(n, l, D)$ is almost independent of the angular momentum quantum number l for a large D and the $E(n, l, D)$ $(l \neq 0)$ are almost overlapped. Also, the variations of the energy levels $E(n, l, \upsilon)$ and $E(n, l, s)$ on the parameters υ and s have been studied. It is found that the energy levels decrease with those parameters. Interestingly, we have found that the energy $E(n, l, \upsilon)$ decreases with the parameter $\upsilon \leq l + 1$ for a given l.

Finally, the comparison theorem has been established by the traditional method. It is found that for positive energy levels E_1, E_2 belonging to node-free ground states $\psi_1(r)$ and $\psi_2(r)$ we have $E_1 \leq E_2$ if $V_1(r) \leq V_2(r) \leq 0$.

Chapter 15
The Levinson Theorem for Dirac Equation

1 Introduction

The Levinson theorem [109] is an important theorem in the quantum scattering theory, which sets up the relation between the number of bound states and the phase shift at zero momentum. It has been generalized and applied to different fields in modern physics [184, 247–296]. With the interest of higher-dimensional field theory, the Levinson theorem for the Schrödinger equation in arbitrary D dimensions was studied [112].

As mentioned in Part I, we have generalized the Dirac equation with a spherically symmetric potential to arbitrary $(D+1)$ dimensions and constructed the eigenfunctions of the total angular momentum. We have also derived the radial equations [91]. It is found that the total (or orbital, spinor) angular momentum in D-dimensional space can be described by an irreducible representation of the SO(D) group denoted by the highest weight, instead of only one parameter j (or l, s) in three-dimensional space.

In this Chapter we shall uniformly study the Levinson theorem for the Dirac equation in $(D+1)$ dimensions by the Sturm-Liouville theorem. In Sect. 2 we study the generalized Sturm-Liouville theorem. The number of bound states will be calculated in Sect. 3. In Sect. 4 the Levinson theorem is established by proving the number of bound states to be equal to the sum of the phase shifts of the scattering states at $E = \pm M$ with the given angular momentum. The critical cases are also analyzed there. Some discussions are given in Sect. 5. Due to the connection between the Levinson theorem and the Friedel theorem, we shall present in Sect. 6 the Friedel theorem in arbitrary dimensions D. We shall outline the comparison theorem for radial Dirac equations in arbitrary dimensions D in Sect. 7. Finally, we conclude this Chapter in Sect. 8.

S.-H. Dong, *Wave Equations in Higher Dimensions*,
DOI 10.1007/978-94-007-1917-0_15,

2 Generalized Sturm-Liouville Theorem

As shown in Chap. 4, we have obtained the radial equations as follows:

$$\begin{aligned}&\frac{d}{dr}G(r)+\frac{K}{r}G(r)=[E-V(r)-M]F(r),\\&-\frac{d}{dr}F(r)+\frac{K}{r}F(r)=[E-V(r)+M]G(r),\end{aligned}\tag{15.1}$$

with $K=\mp[j+(D-2)/2]$. For example, in three-dimensional case we take $K=\kappa=-(l+1)$ for spin-up $j=l+1/2$ while $\kappa=l$ for spin-down $j=l-1/2$. When $D=4$, the SO(4) group is homomorphism to SU(2) × SU(2), and the representations j_1 and j_2 belong to two different SU(2) groups, respectively. When $D=2$, the SO(2) group is an Abelian group, and $K=\pm j=\pm 1/2,\pm 3/2,\ldots$. However, Eq. (15.1) still holds for these cases.

The spherically symmetric potential $V(r)$ has to satisfy the boundary condition at the origin for the nice behavior of wavefunction

$$\int_0^1 r|V(r)|dr<\infty.\tag{15.2}$$

For simplicity, we firstly discuss the case where the potential $V(r)$ is a cutoff one at a sufficiently large radius r_0:

$$V(r)=0,\quad \text{when } r\geq r_0.\tag{15.3}$$

The general case where the potential $V(r)$ has a tail at infinity will be discussed in Sect. 5.

Introduce a parameter λ for the potential $V(r)$:

$$V(r,\lambda)=\lambda V(r),\qquad V(r,1)=V(r).\tag{15.4}$$

As λ increases from zero to one, the potential $V(r,\lambda)$ changes from zero to the given potential $V(r)$. If λ changes its sign, the potential $V(r,\lambda)$ changes sign, too.

Although the spherical spinor functions and eigenvalues K are different for the $D=2N+1$ case and the $D=2N$ case, the forms of the radial equations are uniform:

$$\begin{aligned}&\frac{d}{dr}G_{KE}(r,\lambda)+\frac{K}{r}G_{KE}(r,\lambda)=[E-V(r,\lambda)-M]F_{KE}(r,\lambda),\\&-\frac{d}{dr}F_{KE}(r,\lambda)+\frac{K}{r}F_{KE}(r,\lambda)=[E-V(r,\lambda)+M]G_{KE}(r,\lambda),\\&K=\pm 1/2,\pm 1,\pm 3/2,\ldots.\end{aligned}\tag{15.5}$$

It is easy to see that the solutions with a negative K can be obtained from those with a positive K by interchanging $F_{KE}(r,\lambda)\longleftrightarrow G_{-K-E}(r,-\lambda)$, so that in the following we only discuss the solutions with a positive K. The main results for the case with a negative K will be indicated in the text.

The physically acceptable solutions are finite, continuous, vanishing at the origin, and square integrable:

$$F_{KE}(r,\lambda) = G_{KE}(r,\lambda) = 0, \quad \text{when } r = 0, \tag{15.6}$$

$$\int_0^\infty dr\{|F_{KE}(r,\lambda)|^2 + |G_{KE}(r,\lambda)|^2\} < \infty. \tag{15.7}$$

The solutions for $|E| > M$ describe the scattering states, while those for $|E| \le M$ describe the bound states. We will solve Eq. (15.5) in two regions, $0 \le r < r_0$ and $r_0 < r < \infty$, and then match two solutions at r_0 by the matching condition:

$$A_K(E,\lambda) \equiv \left.\frac{F_{KE}(r,\lambda)}{G_{KE}(r,\lambda)}\right|_{r=r_0^-} = \left.\frac{F_{KE}(r,\lambda)}{G_{KE}(r,\lambda)}\right|_{r=r_0^+}. \tag{15.8}$$

When r_0 is the zero point of $G_{KE}(r,\lambda)$, the matching condition can be replaced by its inverse $G_{KE}(r,\lambda)/F_{KE}(r,\lambda)$ instead. The merit of using this matching condition is that we need not care the normalization factor in the solutions.

The key point to prove the Levinson theorem is that $F_{KE}(r,\lambda)/G_{KE}(r,\lambda)$ is monotonic with respect to the energy E. From Eq. (15.5) we have

$$\begin{aligned}&\frac{d}{dr}\{F_{kE_1}(r,\lambda)G_{KE}(r,\lambda) - G_{KE_1}(r,\lambda)F_{KE}(r,\lambda)\}\\ &\quad = -(E_1 - E)\{F_{KE_1}(r,\lambda)F_{KE}(r,\lambda) + G_{KE_1}(r,\lambda)G_{KE}(r,\lambda)\}.\end{aligned} \tag{15.9}$$

From the boundary condition that both solutions vanish at the origin, we integrate Eq. (15.9) in the region $0 \le r \le r_0$ and obtain

$$\begin{aligned}&\{F_{kE_1}(r,\lambda)G_{KE}(r,\lambda) - G_{KE_1}(r,\lambda)F_{KE}(r,\lambda)\}|_{r=r_0^-}\\ &\quad = -(E_1 - E)\int_0^{r_0}\{F_{KE_1}(r,\lambda)F_{KE}(r,\lambda) + G_{KE_1}(r,\lambda)G_{KE}(r,\lambda)\}dr.\end{aligned} \tag{15.10}$$

Taking the limit $E_1 \to E$, we have

$$\begin{aligned}&\lim_{E_1\to E}\left.\frac{F_{kE_1}(r,\lambda)G_{KE}(r,\lambda) - G_{KE_1}(r,\lambda)F_{KE}(r,\lambda)}{E_1 - E}\right|_{r=r_0^-}\\ &\quad = \{G_{KE}(r_0,\lambda)\}^2\frac{\partial}{\partial E}A_K(E,\lambda)\\ &\quad = -\int_0^{r_0}\{F^2_{KE}(r,\lambda) + G^2_{KE}(r,\lambda)\}dr < 0.\end{aligned} \tag{15.11}$$

Thus, when $|E| \ge M$ we have

$$\begin{aligned}&A_K(E,\lambda) = A_K(M,\lambda) - c_1^2k^2 + \cdots, \quad \text{when } E \gtrsim M,\\ &A_K(E,\lambda) = A_K(-M,\lambda) + c_2^2k^2 + \cdots, \quad \text{when } E \lesssim -M,\end{aligned} \tag{15.12}$$

where c_1^2 and c_2^2 are non-negative numbers, and the momentum k is defined as follows:

$$k = (E^2 - M^2)^{1/2}. \tag{15.13}$$

Similarly, from the boundary condition that the radial functions $F_{KE}(r,\lambda)$ and $G_{KE}(r,\lambda)$ for $|E| \le M$ tend to zero at infinity, we obtain by integrating Eq. (15.9) in the region $r_0 \le r < \infty$

$$\{G_{KE}(r_0,\lambda)\}^2 \frac{\partial}{\partial E}\left(\frac{F_{KE}(r,\lambda)}{G_{KE}(r,\lambda)}\right)\bigg|_{r=r_0^+}$$
$$= \int_{r_0}^{\infty} \{F_{KE}^2(r,\lambda) + G_{KE}^2(r,\lambda)\} dr > 0. \tag{15.14}$$

Thus, as the energy E increases, the ratio $F_{KE}(r,\lambda)/G_{KE}(r,\lambda)$ at r_0^- decreases monotonically, but the ratio $F_{KE}(r,\lambda)/G_{KE}(r,\lambda)$ at r_0^+ increases monotonically for $|E| \le M$. This is called the generalized Sturm-Liouville theorem [297].

3 The Number of Bound States

Now, we solve Eq. (15.5) for the energy $|E| \le M$. In the region $0 \le r < r_0$, when $\lambda = 0$ we have

$$\begin{aligned} F_{KE}(r,0) &= e^{-i(K-1/2)\pi/2}\sqrt{(M+E)\pi k_1 r/2}\, J_{K-\frac{1}{2}}(ik_1 r), \\ G_{KE}(r,0) &= e^{-i(K-3/2)\pi/2}\sqrt{(M-E)\pi k_1 r/2}\, J_{K+\frac{1}{2}}(ik_1 r), \end{aligned} \tag{15.15}$$

where $J_m(x)$ is the Bessel function, and

$$k_1 = (M^2 - E^2)^{1/2}. \tag{15.16}$$

When $\lambda = 0$ the ratio at $r = r_0^-$ is

$$\begin{aligned} A_K(E,0) &= -i\left(\frac{M+E}{M-E}\right)^{1/2} \frac{J_{K-\frac{1}{2}}(ik_1 r_0)}{J_{K+\frac{1}{2}}(ik_1 r_0)} \\ &= \begin{cases} -\frac{2M(2K+1)}{k_1^2 r_0} \sim -\infty, & \text{when } E \sim M, \\ -\frac{2K+1}{2Mr_0}, & \text{when } E \sim -M. \end{cases} \end{aligned} \tag{15.17}$$

In the region $r_0 < r < \infty$, due to the cutoff potential we have $V(r) = 0$ and also

$$\begin{aligned} F_{KE}(r,\lambda) &= e^{i(K+1/2)\pi/2}\sqrt{(M+E)\pi k_1 r/2}\, H^{(1)}_{K-\frac{1}{2}}(ik_1 r), \\ G_{KE}(r,\lambda) &= e^{i(K+3/2)\pi/2}\sqrt{(M-E)\pi k_1 r/2}\, H^{(1)}_{K+\frac{1}{2}}(ik_1 r), \end{aligned} \tag{15.18}$$

where $H_m^{(1)}(x)$ is the Hankel function of the first kind. The ratio at $r = r_0^+$ that does not depend on λ is given by

$$\frac{F_{KE}(r,\lambda)}{G_{KE}(r,\lambda)}\bigg|_{r=r_0^+} = -i\left(\frac{M+E}{M-E}\right)^{1/2} \frac{H^{(1)}_{K-\frac{1}{2}}(ik_1 r_0)}{H^{(1)}_{K+\frac{1}{2}}(ik_1 r_0)}$$

$$= \begin{cases} \frac{2Mr_0}{2K-1}, & \text{when } E \sim M \text{ and } K \geq 1, \\ -2Mr_0 \log(k_1 r_0) \sim \infty, & \text{when } E \sim M \text{ and } K = 1/2, \\ \frac{k_1^2 r_0}{2M(2K-1)} \sim 0, & \text{when } E \sim -M \text{ and } K \geq 1, \\ \frac{-k_1^2 r_0 \log(k_1 r_0)}{2M} \sim 0, & \text{when } E \sim -M \text{ and } K = 1/2. \end{cases} \tag{15.19}$$

It is evident from Eqs. (15.17) and (15.19) that as the energy E increases from $-M$ to M, there is no overlap between two variant ranges of the ratio at two sides of r_0 when $\lambda = 0$ (no potential) except for $K = 1/2$ where there is a half bound state at $E = M$. The half bound state will be discussed in next section.

As λ increases from zero to one, the potential $V(r, \lambda)$ changes from zero to the given potential $V(r)$, and $A_K(E, \lambda)$ changes, too. If $A_K(M, \lambda)$ decreases across the value $2Mr_0/(2K - 1)$ as λ increases, an overlap between the variant ranges of the ratios at two sides of r_0 appears. Since the ratio $A_K(E, \lambda)$ of two radial functions at r_0^- decreases monotonically as the energy E increases, and the ratio at r_0^+ increases monotonically, the overlap means that there must be one and only one energy where the matching condition (15.8) is satisfied, namely, a bound state appears.

As λ increases, $A_K(M, \lambda)$ decreases to $-\infty$, jumps to ∞, and then decreases again across the value $2Mr_0/(2K - 1)$, so that another bound state appears. Note that when r_0 is a zero point of the wavefunction $G_{KE}(r, \lambda)$, $A_K(E, \lambda)$ goes to infinity. It is not a singularity.

On the other hand, as λ increases, if $A_K(-M, \lambda)$ decreases across zero, an overlap between the variant ranges of the ratios at two sides of r_0 disappears so that a bound state disappears. Therefore, each time $A_K(M, \lambda)$ decreases across the value $2Mr_0/(2K - 1)$ as λ increases, a new overlap between the variant ranges of the ratios at two sides of r_0 appears such that a scattering state of a positive energy becomes a bound state, and each time $A_K(-M, \lambda)$ decreases across zero, an overlap between the variant ranges of the ratio at two sides of r_0 disappears such that a bound state becomes a scattering state of a negative energy. Conversely, each time $A_K(M, \lambda)$ increases across the value $2Mr_0/(2K - 1)$, an overlap between the variant ranges disappears such that a bound state becomes a scattering state of a positive energy, and each time $A_K(-M, \lambda)$ increases across zero, a new overlap between the variant ranges appears such that a scattering state of a negative energy becomes a bound state. Now, the number n_K of bound states with the parameter K is equal to the sum (or subtraction) of four times as λ increases from zero to one: the times that $A_K(M, \lambda)$ decreases across the value $2Mr_0/(2K - 1)$, minus the times that $A_K(M, \lambda)$ increases across the value $2Mr_0/(2K - 1)$, minus the times that $A_K(-M, \lambda)$ decreases across zero, plus the times that $A_K(-M, \lambda)$ increases across zero.

When $K = 1/2$, the value $2Mr_0/(2K - 1)$ becomes infinity. We may check the times that $A_K(M, \lambda)^{-1}$ increases (or decreases) across zero to replace the times that $A_K(M, \lambda)$ decreases (or increases) across infinity.

4 The Relativistic Levinson Theorem

We turn to discuss the phase shifts of the scattering states. Solving Eq. (15.5) in the region $r_0 < r < \infty$ for the energy $|E| > M$, we have

$$\begin{aligned} f_{KE}(r,\lambda) &= B(E)\left(\frac{\pi kr}{2}\right)^{1/2} \\ &\quad \cdot \left\{\cos\delta_K(E,\lambda) J_{K-\frac{1}{2}}(kr) - \sin\delta_K(E,\lambda) N_{K-\frac{1}{2}}(kr)\right\}, \\ g_{KE}(r,\lambda) &= \left(\frac{\pi kr}{2}\right)^{1/2} \\ &\quad \cdot \left\{\cos\delta_K(E,\lambda) J_{K+\frac{1}{2}}(kr) - \sin\delta_K(E,\lambda) N_{K+\frac{1}{2}}(kr)\right\}, \end{aligned} \tag{15.20}$$

where $N_m(x)$ denotes the Neumann function, and $B(E)$ is defined as

$$B(E) = \begin{cases} (\frac{E+M}{E-M})^{1/2}, & \text{when } E > M, \\ -(\frac{|E|-M}{|E|+M})^{1/2}, & \text{when } E < -M. \end{cases} \tag{15.21}$$

The asymptotic form of the solution (15.20) at $r \to \infty$ is given by

$$\begin{aligned} f_{KE}(r,\lambda) &\sim B(E)\cos(kr - K\pi/2 + \delta_K(E,\lambda)), \\ g_{KE}(r,\lambda) &\sim \sin(kr - K\pi/2 + \delta_K(E,\lambda)). \end{aligned} \tag{15.22}$$

Substituting Eq. (15.20) into the matching condition (15.8), we obtain the formula for the phase shift $\delta_K(E,\lambda)$:

$$\begin{aligned} \tan\delta_K(E,\lambda) &= \frac{J_{K+\frac{1}{2}}(kr_0)}{N_{K+\frac{1}{2}}(kr_0)} \frac{A_K(E,\lambda) - B(E)\frac{J_{K-\frac{1}{2}}(kr_0)}{J_{K+\frac{1}{2}}(kr_0)}}{A_K(E,\lambda) - B(E)\frac{N_{K-\frac{1}{2}}(kr_0)}{N_{K+\frac{1}{2}}(kr_0)}} \\ &= \frac{J_{K-\frac{1}{2}}(kr_0)}{N_{K-\frac{1}{2}}(kr_0)} \frac{A_K(E,\lambda)^{-1} - \frac{1}{B(E)}\frac{J_{K+\frac{1}{2}}(kr_0)}{J_{K-\frac{1}{2}}(kr_0)}}{A_K(E,\lambda)^{-1} - \frac{1}{B(E)}\frac{N_{K+\frac{1}{2}}(kr_0)}{N_{K-\frac{1}{2}}(kr_0)}}. \end{aligned} \tag{15.23}$$

The phase shift $\delta_K(E,\lambda)$ is determined up to a multiple of π due to the period of the tangent function. We use the convention that the phase shifts for the free particles $V(r) = 0$ vanish

$$\delta_K(E,0) = 0. \tag{15.24}$$

Under this convention, the phase shifts $\delta_K(E)$ are determined completely as λ increases from zero to one:

$$\delta_K(E) \equiv \delta_K(E,1). \tag{15.25}$$

The phase shifts $\delta_K(\pm M,\lambda)$ are the limits of the phase shifts $\delta_K(E,\lambda)$ as E tends to $\pm M$. At the sufficiently small k, $k \ll 1/r_0$, when $E > M$, we have

$$\tan \delta_K(E,\lambda) \sim -\frac{\pi (kr_0/2)^{2K-1}}{(K+1/2)!(K-1/2)!} \cdot \frac{A_K(M,\lambda)(kr_0/2)^2 - Mr_0(K+1/2)}{A_K(M,\lambda) - c_1^2k^2 - \frac{2Mr_0}{2K-1}[1+\frac{(kr_0)^2}{(2K-1)(2K-3)}]}, \quad (15.26)$$

when $K > 3/2$,

$$\tan \delta_K(E,\lambda) \sim -\frac{\pi}{2}\left(\frac{kr_0}{2}\right)^2 \cdot \frac{A_K(M,\lambda)(kr_0/2)^2 - 2Mr_0}{A_K(M,\lambda) - c_1^2k^2 - Mr_0[1-\frac{(kr_0)^2}{2}\log(kr_0)]}, \quad (15.27)$$

when $K = 3/2$,

$$\tan \delta_K(E,\lambda) \sim (kr_0)\frac{A_K(M,\lambda)(kr_0/2)^2 - 3Mr_0/2}{A_K(M,\lambda) - c_1^2k^2 - 2Mr_0[1-(kr_0)^2]}, \quad (15.28)$$

when $K = 1$,

$$\tan \delta_K(E,\lambda) \sim \frac{\pi}{2\log(kr_0)} \frac{A_K(M,\lambda)^{-1} + c_1^2k^2 - k^2r_0/(4M)}{A_K(M,\lambda)^{-1} + c_1^2k^2 + [2Mr_0\log(kr_0)]^{-1}}, \quad (15.29)$$

when $K = 1/2$. When $E < -M$ we have for a sufficient k

$$\tan \delta_K(E,\lambda) \sim -\frac{\pi (kr_0/2)^{2K+1}}{(K+1/2)!(K-1/2)!} \cdot \frac{A_K(-M,\lambda) + (2K+1)/(2Mr_0)}{A_K(-M,\lambda) + c_2^2k^2 + \frac{k^2r_0}{2M(2K-1)}}, \quad (15.30)$$

when $K \geq 1$,

$$\tan \delta_K(E,\lambda) \sim -\pi\left(\frac{kr_0}{2}\right)^2 \frac{A_K(-M,\lambda) + 1/(Mr_0)}{A_K(-M,\lambda) + c_2^2k^2 - \frac{k^2r_0\log(kr_0)}{2M}}, \quad (15.31)$$

when $K = 1/2$. The asymptotic forms (15.12) have been used in deriving above formulas. In addition to the leading terms, we include the next leading terms in some of Eqs. (15.26)–(15.29), (15.30) and (15.31) to be used only for the critical case where the leading terms are canceled to each other.

First, from Eqs. (15.26)–(15.29), (15.30) and (15.31) we see that, except for some critical cases, $\tan \delta_K(E,\lambda)$ tends to zero as E goes to $\pm M$, i.e., $\delta_K(\pm M,\lambda)$ are always equal to the multiple of π. In other words, if the phase shift $\delta_K(E,\lambda)$ for a sufficiently small k is expressed as a positive or negative acute angle plus $n\pi$, its limit $\delta_K(M,\lambda)$ or $\delta_K(-M,\lambda)$ is equal to $n\pi$. This means that $\delta_K(M,\lambda)$ or $\delta_K(-M,\lambda)$ changes discontinuously when $\delta_K(E,\lambda)$ changes through the value $(n+1/2)\pi$.

Second, from Eq. (15.25) we have

$$\begin{aligned}
\left.\frac{\partial \delta_K(E,\lambda)}{\partial A_K(E,\lambda)}\right|_E &= -\left(\frac{E+M}{E-M}\right)^{1/2} \\
&\quad \cdot \frac{2[\cos\delta_K(E,\lambda)]^2}{\pi k r_0[N_{K+\frac{1}{2}}(kr_0)A_K(E,\lambda) - B(E)N_{K-\frac{1}{2}}(kr_0)]^2} \\
&\le 0, \quad E > M \\
\left.\frac{\partial \delta_K(E,\lambda)}{\partial A_K(E,\lambda)}\right|_E &= \left(\frac{|E|-M}{|E|+M}\right)^{1/2} \\
&\quad \cdot \frac{2[\cos\delta_K(E,\lambda)]^2}{\pi k r_0[N_{K+\frac{1}{2}}(kr_0)A_K(E,\lambda) - B(E)N_{K-\frac{1}{2}}(kr_0)]^2} \\
&\ge 0, \quad E < -M.
\end{aligned} \tag{15.32}$$

Equation (15.32) shows that, as the ratio $A_K(E,\lambda)$ decreases, the phase shift $\delta_K(E,\lambda)$ for $E > M$ increases monotonically, but $\delta_K(E,\lambda)$ for $E < -M$ decreases monotonically. In terms of the monotonic properties we are able to determine the jump of the phase shifts $\delta_K(\pm M,\lambda)$.

We first consider the scattering states of a positive energy with a sufficiently small momentum k. As $A_K(E,\lambda)$ decreases, if $\tan\delta_K(E,\lambda)$ changes sign from positive to negative, the phase shift $\delta_K(M,\lambda)$ jumps by π. Note that in this case if $\tan\delta_K(E,\lambda)$ changes sign from negative to positive, the phase shift $\delta_K(M,\lambda)$ keeps invariant. Conversely, as $A_K(E,\lambda)$ increases, if $\tan\delta_K(E,\lambda)$ changes sign from negative to positive, the phase shift $\delta_K(M,\lambda)$ jumps by $-\pi$. Therefore, as λ increases from zero to one, each time the $A_K(M,\lambda)$ decreases from near and larger than the value $2Mr_0/(2K-1)$ to smaller than that value, the denominator in Eq. (15.32) changes sign from positive to negative and the rest factor keeps positive, so that the phase shift $\delta_K(M,\lambda)$ jumps by π. We have shown in the previous section that each time the $A_K(M,\lambda)$ decreases across the value $2Mr_0/(2K-1)$, a scattering state of a positive energy becomes a bound state. Conversely, each time the $A_K(M,\lambda)$ increases across that value, the phase shift $\delta_K(M,\lambda)$ jumps by $-\pi$, and a bound state becomes a scattering state of a positive energy.

Then, we consider the scattering states of a negative energy with a sufficiently small k. As $A_K(E,\lambda)$ decreases, if $\tan\delta_K(E,\lambda)$ changes sign from negative to positive, the phase shift $\delta_K(-M,\lambda)$ jumps by $-\pi$. However, in this case if $\tan\delta_K(E,\lambda)$ changes sign from positive to negative, the phase shift $\delta_K(-M,\lambda)$ keeps invariant. Conversely, as $A_K(E,\lambda)$ increases, if $\tan\delta_K(E,\lambda)$ changes sign from positive to negative, the phase shift $\delta_K(-M,\lambda)$ jumps by π. Therefore, as λ increases from zero to one, each time the $A_K(-M,\lambda)$ decreases from a small and positive number to a negative one, the denominator in Eqs. (15.30) and (15.31) changes sign from positive to negative and the rest factor keeps negative, so that the phase shift $\delta_K(-M,\lambda)$ jumps by $-\pi$. In the previous section we have shown that each time the $A_K(-M,\lambda)$ decreases across zero, a bound state becomes a scattering state of a negative energy. Conversely, each time the $A_K(-M,\lambda)$ increases across zero, the phase shift $\delta_K(-M,\lambda)$ jumps by π, and a scattering state of a negative energy becomes a bound state. Therefore, we obtain the Levinson theorem for the Dirac equation in D dimensions for non-critical cases:

$$\delta_K(M) + \delta_K(-M) = n_K \pi. \tag{15.33}$$

It is obvious that the Levinson theorem (15.33) holds for both positive and negative K in the non-critical cases.

For the special case $K = 1/2$ and $E \sim M$, where the value $2Mr_0/(2K-1)$ is infinity. Since $\{A_K(E,\lambda)\}^{-1}$ increases as $A_K(E,\lambda)$ decreases, we can study the variance of $\{A_K(E,\lambda)\}^{-1}$ in this case instead. For the energy $E > M$ where the momentum k is sufficiently small, when $\{A_K(M,\lambda)\}^{-1}$ increases from negative to positive as λ increases, both the numerator and denominator in Eqs. (15.26)–(15.29) change signs, but not simultaneously. The numerator changes sign first, and then the denominator changes. The front factor in Eqs. (15.26)–(15.29) is negative so that $\tan\delta_K(E,\lambda)$ first changes from negative to positive when the numerator changes sign, and then changes from positive to negative when the denominator changes sign. It is in the second step that the phase shift $\delta_K(M,\lambda)$ jumps by π. Similarly, each time $\{A_K(M,\lambda)\}^{-1}$ decreases across zero as λ increases, $\delta_K(M,\lambda)$ jumps by $-\pi$.

For $\lambda = 0$ and $K = 1/2$, the numerator in Eqs. (15.26)–(15.29) is equal to zero, and the phase shift $\delta_K(M,0)$ is defined to be zero. In this case there is a half bound state at $E = M$. If $\{A_K(M,\lambda)\}^{-1}$ increases ($A_K(M,\lambda)$ decreases) as λ increases from zero, the front factor in Eqs. (15.26)–(15.29) is negative, the numerator first becomes positive, and then the denominator changes sign from negative to positive, such that the phase shift $\delta_K(M,\lambda)$ jumps by π and simultaneously the half bound state becomes a bound state with $E < M$.

Now, we turn to study the critical cases. First, we study the critical case for $E = M$, where the ratio $A_K(M,1)$ is equal to the value $2Mr_0/(2K-1)$. It is easy to obtain the following solution of $E = M$ in the region $r_0 < r < \infty$, satisfying the radial equations (15.5) and the matching condition (15.7) at r_0:

$$f_{KM}(r,1) = 2Mr^{-K+1}, \qquad g_{KM}(r,1) = (2K-1)r^{-K}. \tag{15.34}$$

It is a bound state when $K > 3/2$, but called a half bound state when $K \leq 3/2$. A half bound state is not a bound state, because its wavefunction is finite but not square integrable.

For definiteness, we assume that in the critical case, as λ increases from a number near and less than one and finally reaches one, $A_K(M,\lambda)$ decreases and finally reaches, but not across, the value $2Mr_0/(2K-1)$. In this case, when $\lambda = 1$ a new bound state of $E = M$ appears for $K > 3/2$, but does not appear for $K \leq 3/2$. We should check whether or not the phase shift $\delta_K(M,1)$ increases by an additional π as λ increases and reaches one.

It is evident from the next leading terms in the denominator of Eqs. (15.26)–(15.29) that the denominator for $K \geq 3/2$ has changed sign from positive to negative as $A_K(M,\lambda)$ decreases and finally reaches the value $2Mr_0/(2K-1)$, i.e., the phase shift $\delta_K(M,\lambda)$ jumps by an additional π at $\lambda = 1$. Simultaneously, a new bound state of $E = M$ appears for $K > 3/2$, but only a half bound state appears for $K = 3/2$, so that the Levinson theorem (15.33) holds for the critical case with $K > 3/2$, but it has to be modified for the critical case when a half bound state occurs at $E = M$ and $K = 3/2$:

$$\delta_K(M) + \delta_K(-M) = (n_K + 1)\pi. \tag{15.35}$$

For $K = 1$, the $\tan\delta_K(E, 1)$ tends to infinity as $\{A_K(M, \lambda)\}^{-1}$ increases and finally reaches $2Mr_0$, i.e., the phase shift $\delta_K(M, \lambda)$ jumps by $\pi/2$. Simultaneously, only a new half bound state of $E = M$ for $K = 1$ appears, so that the Levinson theorem (15.33) has to be modified for the critical case when a half bound state occurs at $E = M$ and $K = 1$:

$$\delta_K(M) + \delta_K(-M) = \left(n_K + \frac{1}{2}\right)\pi. \tag{15.36}$$

For $K = 1/2$ the next leading term with $\log(kr_0)$ in the denominator of Eqs. (15.26)–(15.29) dominates so that the denominator keeps negative (does not change sign!) as $\{A_K(M, \lambda)\}^{-1}$ increases and finally reaches zero, namely, the phase shift $\delta_K(M, \lambda)$ does not jump, no matter whether the rest part in Eq. (15.29) keeps positive or has changed to negative. Simultaneously, only a new half bound state of $E = M$ for $K = 1/2$ appears, so that the Levinson theorem (15.33) holds for the critical case with $K = 1/2$.

This conclusion holds for the critical case where $A_K(M, \lambda)$ increases and finally reaches, but not across, the value $2Mr_0/(2K - 1)$. Therefore, for the critical case when a half bound state occurs at $E = M$ and $K \leq 3/2$ the Levinson theorem has to be modified as follows

$$\delta_K(M) + \delta_K(-M) = \left(n_K + K - \frac{1}{2}\right)\pi. \tag{15.37}$$

Second, we study the critical case for $E = -M$, where the ratio $A_K(-M, 1)$ is equal to zero. It is easy to obtain the following solution of $E = -M$ in the region $r_0 < r < \infty$, satisfying the radial equations (15.5) and the matching condition (15.7) at r_0:

$$f_{KM}(r, \lambda) = 0, \qquad g_{KM}(r, \lambda) = r^{-K}. \tag{15.38}$$

It is a bound state when $K \geq 1$, but a half bound state when $K = 1/2$.

For definiteness, once again we assume that in the critical case, as λ increases from a number near and less than one and finally reaches one, $A_K(-M, \lambda)$ decreases and finally reaches zero, so that when $\lambda = 1$ the energy of a bound state decreases to $E = -M$ for $K \geq 1$, but a bound state becomes a half bound state for $K = 1/2$. We should check whether or not the phase shift $\delta_K(-M, 1)$ decreases by π as λ increases and reaches one.

For the energy $E < -M$ where the momentum k is sufficiently small, one can see from the next leading terms in the denominator of Eqs. (15.30) and (15.31) that the denominator does not change sign as $A_K(-M, \lambda)$ decreases and finally reaches zero, namely, the phase shift $\delta_K(-M, \lambda)$ does not jump by an additional $-\pi$ at $\lambda = 1$. Simultaneously, the energy of a bound state decreases to $E = -M$ for $K \geq 1$, but a bound state becomes a half bound state for $K = 1/2$, so that the

Levinson theorem (15.33) holds for the critical case with $K \geq 1$, but it has to be modified when a half bound state occurs at $E = -M$ and $K = 1/2$:

$$\delta_K(M) + \delta_K(-M) = (n_K + 1)\pi. \tag{15.39}$$

Combining Eqs. (15.33), (15.37), (15.39) and their corresponding forms for the negative K, we obtain the relativistic Levinson theorem in D dimensions.

5 Discussions on General Case

Now, we discuss the general case where the potential $V(r)$ has a tail at $r \geq r_0$. Let r_0 be so large that only the leading term in $V(r)$ is concerned:

$$V(r) \sim br^{-n}, \quad r \geq r_0, \tag{15.40}$$

where b is a non-vanishing constant and n is a positive constant, not necessary to be an integer. Substituting it into Eq. (15.5) and changing the variable r to ξ:

$$\xi = \begin{cases} kr = r\sqrt{E^2 - M^2}, & \text{when } |E| > M, \\ \kappa r = r\sqrt{M^2 - E^2}, & \text{when } |E| \leq M, \end{cases} \tag{15.41}$$

we obtain the radial equations in the region $r_0 \leq r < \infty$:

$$\begin{aligned} \frac{d}{d\xi} g_{KE}(\xi) + \frac{K}{\xi} g_{KE}(\xi) &= \left(\frac{E}{|E|}\sqrt{\frac{E-M}{E+M}} - \frac{b}{\xi^n} k^{n-1} \right) f_{KE}(\xi), \\ -\frac{d}{dr} f_{KE}(\xi) + \frac{K}{r} f_{KE}(\xi) &= \left(\frac{E}{|E|}\sqrt{\frac{E+M}{E-M}} - \frac{b}{\xi^n} k^{n-1} \right) g_{KE}(\xi), \end{aligned} \tag{15.42}$$

for $|E| > M$, and

$$\begin{aligned} \frac{d}{d\xi} g_{KE}(\xi) + \frac{K}{\xi} g_{KE}(\xi) &= \left(-\sqrt{\frac{M-E}{M+E}} - \frac{b}{\xi^n} \kappa^{n-1} \right) f_{KE}(\xi), \\ -\frac{d}{dr} f_{KE}(\xi) + \frac{K}{r} f_{KE}(\xi) &= \left(\sqrt{\frac{M+E}{M-E}} - \frac{b}{\xi^n} \kappa^{n-1} \right) g_{KE}(\xi), \end{aligned} \tag{15.43}$$

for $|E| \leq M$. As far as the Levinson theorem is concerned, we are only interested in the solutions with the sufficiently small k and κ. If $n \geq 3$, in comparison with the first term on the right hand side of Eq. (15.42) or Eq. (15.43), the potential term with a factor k^{n-1} (or κ^{n-1}) is too small to affect the phase shift at the sufficiently small k and the variant range of the ratio $f_{jE}(r, \lambda)/g_{jE}(r, \lambda)$ at r_0^+. Therefore, the proof given in the previous sections is valid for those potential with a tail so that the Levinson theorem (15.33) holds.

When $n = 2$ and $b \neq 0$, we will only keep the leading terms for the small parameter k (or κ) in solving Eq. (15.42) (or Eq. (15.43)). First, we calculate the solutions with the energy $E \sim M$. Let

$$\alpha = (K^2 - K + 2Mb + 1/4)^{1/2} \neq K - \frac{1}{2}. \tag{15.44}$$

If $\alpha^2 < 0$, there is an infinite number of bound states. We will not discuss this case as well as the case with $\alpha = 0$ here. When $\alpha^2 > 0$, we take $\alpha > 0$ for convenience. Some formulas given in the previous sections will be changed.

When $E \le M$ we have

$$\begin{aligned} f_{KE}(r,\lambda) &= e^{i(\alpha+1)\pi/2} 2M(\pi\kappa r/2)^{1/2} H_\alpha^{(1)}(i\kappa r), \\ g_{KE}(r,\lambda) &= e^{i(\alpha+1)\pi/2} \kappa(\pi\kappa r/2)^{1/2} \\ &\quad \cdot \left\{ -\frac{d}{d(\kappa r)} H_\alpha^{(1)}(i\kappa r) + \frac{K-1/2}{\kappa r} H_\alpha^{(1)}(i\kappa r) \right\}. \end{aligned} \tag{15.45}$$

Hence, the ratio at $r = r_0^+$ for $E = M$ is

$$\left. \frac{f_{KE}(r,\lambda)}{g_{KE}(r,\lambda)} \right|_{r=r_0^+} = \frac{2Mr_0}{K+\alpha-1/2}, \qquad E = M. \tag{15.46}$$

When $E > M$ we have

$$\begin{aligned} f_{KE}(r,\lambda) &= 2M(\pi kr/2)^{1/2} \{\cos\eta_\alpha(E,\lambda) J_\alpha(kr) - \sin\eta_\alpha(E,\lambda) N_\alpha(kr)\}, \\ g_{KE}(r,\lambda) &= k(\pi kr/2)^{1/2} \\ &\quad \cdot \left\{ \cos\eta_\alpha(E,\lambda) \left(-\frac{d}{d(kr)} J_\alpha(kr) + \frac{K-1/2}{kr} J_\alpha(kr) \right) \right. \\ &\quad \left. - \sin\eta_\alpha(E,\lambda) \left(-\frac{d}{d(kr)} N_\alpha(kr) + \frac{K-1/2}{kr} N_\alpha(kr) \right) \right\}. \end{aligned} \tag{15.47}$$

When (kr) tends to infinity, the asymptotic forms of the solutions are given by

$$\begin{aligned} f_{KE}(r,\lambda) &\sim 2M\cos(kr - \alpha\pi/2 - \pi/4 + \eta_\alpha(E,\lambda)), \\ g_{KE}(r,\lambda) &\sim k\sin(kr - \alpha\pi/2 - \pi/4 + \eta_\alpha(E,\lambda)). \end{aligned} \tag{15.48}$$

In comparison with the solution (15.22) we obtain the phase shift $\delta_K(E,\lambda)$ for $E > M$:

$$\delta_K(E,\lambda) = \eta_\alpha(E,\lambda) + (K - \alpha - 1/2)\pi/2, \qquad E > M. \tag{15.49}$$

From the matching condition (15.8), for the sufficiently small k we obtain

$$\begin{aligned} \tan\eta_\alpha(E,\lambda) &\sim \frac{-\pi(kr_0/2)^{2\alpha}}{\Gamma(\alpha+1)\Gamma(\alpha)} \left(\frac{K-\alpha-1/2}{K+\alpha-1/2} \right) \\ &\quad \cdot \frac{A_K(M,\lambda) - 2Mr_0/(K-\alpha-1/2)}{A_K(M,\lambda) - 2Mr_0/(K+\alpha-1/2)}. \end{aligned} \tag{15.50}$$

Therefore, as λ increases from zero to one, each time the $A_K(M,\lambda)$ decreases from near and larger than the value $2Mr_0/(K+\alpha-1/2)$ to smaller than that value, the denominator in Eq. (15.50) changes sign from positive to negative and the rest factor keeps positive, so that $\eta_\alpha(M,\lambda)$ jumps by π. Simultaneously, from Eq. (15.46) a new overlap between the variant ranges of the ratio at two sides of r_0 appears such that a scattering state of a positive energy becomes a bound state. Conversely, each time the $A_K(M,\lambda)$ increases across that value, $\eta_\alpha(M,\lambda)$ jumps by $-\pi$, and a bound state becomes a scattering state of a positive energy.

Second, we calculate the solutions with the energy $E \sim -M$. Let

$$\beta = (K^2 + K - 2Mb + 1/4)^{1/2} \neq K + \frac{1}{2}. \tag{15.51}$$

Similarly, we only discuss the cases with $\beta^2 > 0$, and take $\beta > 0$.

When $E \geq -M$ we have

$$\begin{aligned} f_{KE}(r,\lambda) &= -e^{i(\beta+1)\pi/2}\kappa(\pi\kappa r/2)^{1/2} \\ &\quad \cdot \left\{\frac{d}{d(\kappa r)}H_\beta^{(1)}(i\kappa r) + \frac{K+1/2}{\kappa r}H_\beta^{(1)}(i\kappa r)\right\}, \\ g_{KE}(r,\lambda) &= e^{i(\beta+1)\pi/2}2M(\pi\kappa r/2)^{1/2}H_\beta^{(1)}(i\kappa r). \end{aligned} \tag{15.52}$$

Hence, the ratio at $r = r_0^+$ for $E = -M$ is

$$\left.\frac{f_{KE}(r,\lambda)}{g_{KE}(r,\lambda)}\right|_{r=r_0^+} = -\frac{K - \beta + 1/2}{2Mr_0}, \qquad E = -M. \tag{15.53}$$

When $E < -M$ we have

$$\begin{aligned} f_{KE}(r,\lambda) &= -k(\pi kr/2)^{1/2} \\ &\quad \cdot \left\{\cos\eta_\beta(E,\lambda)\left(\frac{d}{d(kr)}J_\beta(kr) + \frac{K+1/2}{kr}J_\beta(kr)\right)\right. \\ &\quad \left. - \sin\eta_\beta(E,\lambda)\left(\frac{d}{d(kr)}N_\beta(kr) + \frac{K+1/2}{kr}N_\beta(kr)\right)\right\}, \\ g_{KE}(r,\lambda) &= 2M(\pi kr/2)^{1/2} \\ &\quad \cdot \{\cos\eta_\beta(E,\lambda)J_\beta(kr) - \sin\eta_\beta(E,\lambda)N_\beta(kr)\}. \end{aligned} \tag{15.54}$$

When (kr) tends to infinity, the asymptotic forms for the solutions are:

$$\begin{aligned} f_{KE}(r,\lambda) &\sim k\sin(kr - \beta\pi/2 - \pi/4 + \eta_\beta(E,\lambda)), \\ g_{KE}(r,\lambda) &\sim 2M\cos(kr - \beta\pi/2 - \pi/4 + \eta_\beta(E,\lambda)). \end{aligned} \tag{15.55}$$

In comparison with the solution (15.22) we obtain the phase shift $\delta_K(E,\lambda)$ for $E < -M$:

$$\delta_K(E,\lambda) = \eta_\beta(E,\lambda) + (K - \beta + 1/2)\pi/2, \qquad E < -M. \tag{15.56}$$

From the matching condition (15.8), for the sufficiently small k we obtain

$$\tan\eta_\alpha(E,\lambda) \sim \frac{-\pi(kr_0/2)^{2\beta}}{\Gamma(\beta+1)\Gamma(\beta)} \cdot \frac{A_K(-M,\lambda) + \frac{K+\beta+1/2}{2Mr_0}}{A_K(-M,\lambda) + \frac{K-\beta+1/2}{2Mr_0}}. \tag{15.57}$$

Therefore, as λ increases from zero to one, each time the $A_K(-M,\lambda)$ decreases from near and larger than the value $-(K - \beta + 1/2)/(2Mr_0)$ to smaller than that value, the denominator in Eq. (15.57) changes sign from positive to negative and the rest factor keeps negative, so that $\eta_\beta(-M,\lambda)$ jumps by $-\pi$. Simultaneously, from Eq. (15.53) an overlap between the variant ranges of the ratio at two sides of r_0 disappears such that a bound state becomes a scattering state of a negative energy. Conversely, each time the $A_K(-M,\lambda)$ increases across that value, $\eta_\beta(-M,\lambda)$ jumps by π, and a scattering state of a negative energy becomes a bound state.

In summary, we obtain the modified relativistic Levinson theorem for non-critical cases when the potential has a tail (15.40) with $n=2$ at infinity:

$$\delta_K(M)+\delta_K(-M)=n_K\pi+(2K-\alpha-\beta)\pi/2. \tag{15.58}$$

We will not discuss the critical cases in detail. In fact, the modified relativistic Levinson theorem (15.58) holds for the critical cases of $\alpha>1$ and $\beta>1$. When $0<\alpha<1$ or $0<\beta<1$, $\eta_\alpha(M,1)$ or $\eta_\beta(-M,1)$ in the critical case will not be multiple of π, respectively, so that Eq. (15.58) is violated for those critical cases.

Furthermore, for the potential (15.40) with a tail at infinity, when $n>2$, even if it contains a logarithm factor, for any arbitrarily small positive ϵ, one can always find a sufficiently large r_0 such that $|V(r)|<\epsilon/r^2$ in the region $r_0<r<\infty$. Thus, from Eqs. (15.44) and (15.51) we have for the sufficiently small ϵ

$$\begin{aligned}\alpha&=(K^2-K\pm 2M\epsilon+1/4)^{1/2}\sim K-\frac{1}{2},\\ \beta&=(K^2+K\mp 2M\epsilon+1/4)^{1/2}\sim K+\frac{1}{2}.\end{aligned} \tag{15.59}$$

Hence, Eq. (15.58) coincides with Eq. (15.33). In this case the Levinson theorem (15.33) still holds for the non-critical case.

6 Friedel Theorem

The Levinson theorem is closely related to the Friedel theorem [452] in three dimensions described by

$$\Delta N=\frac{2}{\pi}\sum_{l=0}^{\infty}(2l+1)\delta_l(E_F). \tag{15.60}$$

It states that the change of the number of states ΔN around a potential barrier can be expressed as the Friedel sum rule [453], which sets up the relation between ΔN and the phase shifts at the Fermi energy. It provides a powerful method in calculating some of the important properties of electron structure. Recently, this theorem has been generalized to the system of Dirac fermions [454, 455] in two and three dimensions as well as in arbitrary dimensions [456].

As shown in Ref. [456], the generalized Friedel theorem for Dirac fermions in D dimensions is given by

$$\Delta N=\begin{cases}\frac{1}{\pi}\sum_{\pm K}\sum_{l=0}^{\infty}2^d\binom{D+l-2}{l}\\ \quad\cdot([\delta_K(E_F)-\delta_K(M)+\delta_K(-E_F')-\delta_K(-M)]\\ \quad+\epsilon_K(-1)^{|K|}\frac{\pi}{2}[\sin^2\delta_K(-M)-\sin^2\delta_K(M)]), & D=2d+1,\\ \frac{1}{\pi}\sum_{\pm K}\sum_{l=0}^{\infty}2^{d-1}\binom{D+l-2}{l}\\ \quad\cdot[\delta_K(E_F)-\delta_K(M)+\delta_K(-E_F')-\delta_K(-M)], & D=2d,\end{cases} \tag{15.61}$$

where $\delta_K(E_F)[\delta_K(-E'_F)]$ and $\delta_K(\pm M)$ are the phase shifts of scattering states at Fermi energy $E_\lambda = E_F(E_\lambda = -E'_F)$ for Dirac fermions (antifermions) and the critical points $E_\lambda = \pm M$ of zero momentum with $\pm M$ being effective mass of fermion (antifermion). The notation $\sum_{\pm K}$ denotes the summation over the quantum numbers $K = \pm|K|$, and $\epsilon_K = +1$ (-1) for $K > 0$ $(K < 0)$. The total angular momentum $K = \pm(2l + D - 1)/2$.

7 Comparison Theorem

Before ending this Chapter, we are going to give a brief sketch of the comparison theorem for Dirac equation in arbitrary dimensions due to its interest. For the Dirac problems, this theorem has been discussed by some authors [457–461].

We consider two different attractive potentials $V_a(r)$ and $V_b(r)$ satisfying $V_a(r) < V_b(r)$. If we write the corresponding pairs of radial wavefunction as $\{F_a, G_a\}$ and $\{F_b, G_b\}$, then in terms of Eqs. (15.1) we have

$$\frac{dG_a(r)}{dr} + \frac{K}{r}G_a(r) = [E_a - V_a(r) - M]F_a(r), \tag{15.62}$$

$$-\frac{dF_a(r)}{dr} + \frac{K}{r}F_a(r) = [E_a - V_a(r) + M]G_a(r), \tag{15.63}$$

$$\frac{dG_b(r)}{dr} + \frac{K}{r}G_b(r) = [E_b - V_b(r) - M]F_b(r), \tag{15.64}$$

$$-\frac{dF_b(r)}{dr} + \frac{K}{r}F_b(r) = [E_b - V_b(r) + M]G_b(r). \tag{15.65}$$

The difference between Eq. (15.62) multiplied by $F_b(r)$ and Eq. (15.65) multiplied by $G_a(r)$ leads to the following formula

$$\begin{aligned}\frac{d}{dr}[G_a(r)F_b(r)] &= [-M - V_a(r) + E_a]F_a(r)F_b(r)\\ &\quad + [-M + V_b(r) - E_b]G_a(r)G_b(r).\end{aligned} \tag{15.66}$$

In a similar way, the difference between Eq. (15.64) multiplied by $F_a(r)$ and Eq. (15.63) multiplied by $G_b(r)$ allows us to obtain

$$\begin{aligned}\frac{d}{dr}[F_a(r)G_b(r)] &= [-M - V_b(r) + E_b]F_a(r)F_b(r)\\ &\quad + [-M + V_a(r) - E_a]G_a(r)G_b(r).\end{aligned} \tag{15.67}$$

In terms of Eqs. (15.66) and (15.67) we have

$$\begin{aligned}&[(E_a - E_b) - (V_a(r) - V_b(r))][F_a(r)F_b(r) + G_a(r)G_b(r)]\\ &= -\frac{d}{dr}[F_a(r)G_b(r) - G_a(r)F_b(r)].\end{aligned} \tag{15.68}$$

By integrating this equation with respect to argument $r \in [0, \infty)$ and considering the boundary conditions $F(0) = G(0) = 0$ and $F(\infty) = G(\infty) = 0$, we have

$$\int_0^\infty (E_a - E_b)[F_a(r)F_b(r) + G_a(r)G_b(r)]dr$$
$$= \int_0^\infty (V_a(r) - V_b(r))[F_a(r)F_b(r) + G_a(r)G_b(r)]dr. \tag{15.69}$$

Therefore, two integrals have the same sign for the node-free wavefunctions $\{F(r), G(r)\}$. It should be emphasized that the potentials and eigenvalues are both real. On the contrary, if the potential parameters stray into a region such that the corresponding energy level becomes complex, then Eq. (15.69) will no longer lead to the expected result $V_a(r) < V_b(r) \Rightarrow E_a < E_b$ since the complex energy levels cannot be well ordered.

Recently, Hall has supposed that the potential $V = V(r,a)$ depends smoothly on a parameter a and restated this theorem [460, 461]. For the real attractive central potential $V(r,a)$ and the corresponding discrete Dirac eigenvalue $E(a) = E_{Kn}(a)$, one has

$$\begin{aligned} E'(a) \geq 0, \quad \text{for } \frac{\partial V}{\partial a} \geq 0, \\ E'(a) \leq 0, \quad \text{for } \frac{\partial V}{\partial a} \leq 0. \end{aligned} \tag{15.70}$$

Moreover, if we suppose that E^1_{Kn} and E^2_{Kn} are Dirac eigenvalues corresponding to two distinct attractive central potentials $V_a(r)$ and $V_b(r)$, then we have

$$E^{(a)}_{Kn} \leq E^{(b)}_{Kn}, \quad \text{for } V_a(r) \leq V_b(r). \tag{15.71}$$

Finally, it should be pointed out that those attractive central potentials $V_a(r)$ and $V_b(r)$ are time-independent.

8 Concluding Remarks

In this Chapter we have uniformly studied the Levinson theorem for the Dirac equation in $(D+1)$ dimensions by the Sturm-Liouville theorem. Thus, the Levinson theorem have been established by proving the number of bound states to be equal to the sum of the phase shifts of the scattering states at $E = \pm M$. The general case where the potential $V(r)$ has a tail at $r > r_0$ has also been analyzed. Finally, we have discussed the Friedel and comparison theorems for the radial Dirac equation in arbitrary dimensions D.

Chapter 16
Generalized Hypervirial Theorem for Dirac Equation

1 Introduction

The recurrence relations for matrix elements are very useful in quantum calculations, but their direct computation are generally very cumbersome. An interesting and useful recurrence formula is the Blanchard's relation for arbitrary non-relativistic matrix elements of the form $\langle n_1 l_1 | r^\lambda | n_2 l_2 \rangle$, where the symbol $|nl\rangle$ stands for non-relativistic hydrogenic radial eigenstate.

As shown in Chap. 10, we have generalized the hypervirial theorem for the non-relativistic equation case in arbitrary dimensions D. In this Chapter we attempt to present the generalized hypervirial theorem for relativistic Dirac equation case.

This Chapter is organized as follows. In Sect. 2 we are going to derive the generalized hypervirial theorem for Dirac equation in off-diagonal case. Section 3 is devoted to the diagonal case. We summarize our conclusions in Sect. 4.

2 Relativistic Recurrence Relation

Based on the study in Chap. 4, when the factor $r^{(1-D)/2}$ is built into each radial function $\Psi_{K,[j]}(\mathbf{x}, t)$ we may write out the relativistic radial Hamiltonian[1]

$$H_a = -i\alpha_r\left(\frac{d}{dr} + \frac{\beta}{r}K_a\right) + M\beta + V(r), \tag{16.1}$$

where

$$\begin{aligned} \alpha_r &= \frac{1}{r}\vec{\alpha}\cdot\vec{r} = \begin{pmatrix} 0 & -1 \\ -1 & 0 \end{pmatrix}, \qquad \beta = \begin{pmatrix} 1 & 0 \\ 0 & -1 \end{pmatrix}, \\ K_a &= \mp[j_a + (D-2)/2] \end{aligned} \tag{16.2}$$

[1] Note that the first derivative d/dr comes from the radial momentum operator $p_r = -i\hbar[d/dr + (D-1)/2r]$ as given in Eq. (3.53). Due to the introduction of the factor $r^{(1-D)/2}$, the second term $(D-1)/2r$ in p_r is canceled each other.

S.-H. Dong, *Wave Equations in Higher Dimensions*,
DOI 10.1007/978-94-007-1917-0_16,

with the property $\alpha_r \beta = -\beta \alpha_r$. For example, we have in three dimensions

$$K = \kappa = \begin{cases} -(l+1), & j = l + \frac{1}{2}, \\ l, & j = l - \frac{1}{2}, \end{cases} \tag{16.3}$$

and $K = 0$ in one dimension.

As a result, the reduced radial Dirac Hamiltonian H can be written as a matrix form

$$H_a = \begin{bmatrix} M + V(r) & i(\frac{d}{dr} - \frac{K_a}{r}) \\ i(\frac{d}{dr} + \frac{K_a}{r}) & V(r) - M \end{bmatrix} \tag{16.4}$$

with the property

$$H_a \psi(r) = E_a \psi(r), \tag{16.5}$$

where we have introduced purely radial wavefunction

$$\psi(r) = \begin{pmatrix} G(r) \\ iF(r) \end{pmatrix}. \tag{16.6}$$

For bound states, they may be normalized by the relation

$$(\psi(r), \psi(r)) = \int_0^\infty [G^2(r) + F^2(r)] dr = 1. \tag{16.7}$$

We use inner products without the radial measure r^{D-1} because the factor $r^{(1-D)/2}$ is already built into each radial function as mentioned above.

The corresponding radial equations are written as

$$\begin{aligned} \frac{dG(r)}{dr} + \frac{K}{r} G(r) &= [E - V(r) + M] F(r), \\ -\frac{dF(r)}{dr} + \frac{K}{r} F(r) &= [E - V(r) - M] G(r). \end{aligned} \tag{16.8}$$

It is found that this set of radial equations are slightly different from those given in Chap. 4 due to the exchange between M and $-M$. Such a modification does not affect the final result at all.

By using the definition of the Hamiltonian H_a given in Eq. (16.1), we have

$$H_2 f(r) - f(r) H_1 = -i\alpha_r \left(f'(r) + \frac{\beta}{r} \Delta_{21}^- f(r) \right), \tag{16.9}$$

where $\Delta_{21}^{\mp} = K_2 \mp K_1$. Thus, acting this relation on the eigenfunctions yields

$$(E_2 - E_1) \langle n_2 K_2 | f(r) | n_1 K_1 \rangle = -i\alpha_r \langle n_2 K_2 | f'(r) + \frac{\beta}{r} \Delta_{21}^- f(r) | n_1 K_1 \rangle, \tag{16.10}$$

where the matrix elements of radial functions are calculated as follows:

$$\begin{aligned} \langle n_2 K_2 | f(r) | n_1 K_1 \rangle &= \int f(r) [F_2^*(r) F_1(r) + G_2^*(r) G_1(r)] dr, \\ \langle n_2 K_2 | \beta f(r) | n_1 K_1 \rangle &= \int f(r) [F_2^*(r) F_1(r) - G_2^*(r) G_1(r)] dr. \end{aligned} \tag{16.11}$$

In a similar way, one is able to obtain the following relations

$$H_2\xi = -\left\{ f''(r) + f'(r)\frac{d}{dr} - \frac{\beta}{r^2}\Delta_{21}^- f \right.$$
$$+ \frac{\beta}{r}\Delta_{21}^-\left(f' + f\frac{d}{dr}\right) - \frac{\beta}{r}K_2\left[f' + \frac{\Delta_{21}^-}{r}\beta f(r)\right]\Bigg\}$$
$$- i\alpha_r(V(r) - M\beta)\left\{ f'(r) + \frac{\Delta_{21}^-}{r}\beta f(r)\right\} \tag{16.12}$$

and

$$\xi H_1 = -\left\{ f'(r) - \frac{\Delta_{21}^-}{r}\beta f(r)\right\}\frac{d}{dr}$$
$$- \frac{1}{r}\beta K_1\left\{ f'(r) - \frac{\Delta_{21}^-}{r}\beta f(r)\right\}$$
$$- i\alpha_r\left\{ f'(r) + \frac{\Delta_{21}^-}{r}\beta f(r)\right\}(M\beta + V(r)), \tag{16.13}$$

where

$$\xi = H_2 f(r) - f(r)H_1. \tag{16.14}$$

The difference between (16.12) and (16.13) leads to

$$H_2\xi - \xi H_1 = -f''(r) + \frac{\Delta_{21}^-}{r^2}\beta f(r)$$
$$+ \frac{(\Delta_{21}^-)^2}{r^2} f(r) - \frac{2\beta}{r}\Delta_{21}^- f(r)\frac{d}{dr}$$
$$+ \frac{2K_1}{r}\beta f'(r) + 2i\alpha_r M\beta\left\{ f'(r) + \frac{\Delta_{21}^-}{r}\beta f(r)\right\}, \tag{16.15}$$

from which we have

$$(E_2 - E_1)^2\langle n_2K_2|f(r)|n_1K_1\rangle$$
$$= \langle n_2K_2|H_2\xi - \xi H_1|n_1K_1\rangle$$
$$= \langle n_2K_2| - f''(r) + \frac{\Delta_{21}^-}{r^2}\beta f(r) + \frac{(\Delta_{21}^-)^2}{r^2} f(r) - \frac{2\beta}{r}\Delta_{21}^- f(r)\frac{d}{dr}$$
$$+ \frac{2K_1}{r}\beta f'(r) + 2i\alpha_r M\beta\left\{ f'(r) + \frac{\Delta_{21}^-}{r}\beta f(r)\right\}|n_1K_1\rangle, \tag{16.16}$$

where the first derivative d/dr in (16.16) can be substituted by the Hamiltonian given in (16.1).

On the other hand, the addition between (16.12) and (16.13) allows us to obtain the following off-diagonal matrix elements

$$(E_2^2 - E_1^2)\langle n_2K_2|f(r)|n_1K_1\rangle$$
$$= \langle n_2K_2|H_2\xi + \xi H_1|n_1K_1\rangle$$

$$= \langle n_2K_2|-f''(r) - 2f'(r)\frac{d}{dr} + \frac{\Delta_{21}^-}{r^2}\beta f(r) + \frac{\Delta_{21}^-\Delta_{21}^+}{r^2} f(r)$$
$$- 2i\alpha_r\left\{f'(r) + \frac{\Delta_{21}^-}{r}\beta f(r)\right\}V(r)|n_1K_1\rangle. \tag{16.17}$$

Due to the presence of Dirac matrices, we also need to calculate off-diagonal matrix elements of expressions involving $\alpha_r f(r)$ and $\beta f(r)$, i.e.,

$$H_2(-i\alpha_r f(r)) = -\left\{f'(r) + f(r)\frac{d}{dr}\right\}$$
$$+ \frac{K_2}{r}\beta f(r) + i\alpha_r(M\beta - V(r))f(r) \tag{16.18}$$

and

$$(-i\alpha_r f(r))H_1 = -f(r)\left(\frac{d}{dr} + \frac{K_1}{r}\beta\right) - i\alpha_r(M\beta + V(r))f(r). \tag{16.19}$$

Summing and subtracting them lead to

$$(E_2 + E_1)\langle n_2K_2|-i\alpha_r f(r)|n_1K_1\rangle$$
$$= \langle n_2K_2|H_2(-i\alpha_r f(r)) - i\alpha_r f(r)H_1|n_1K_1\rangle$$
$$= \langle n_2K_2|-f'(r) - 2f(r)\frac{d}{dr} + \frac{\Delta_{21}^-}{r}\beta f(r) - 2i\alpha_r V(r)f(r)|n_1K_1\rangle, \tag{16.20}$$

$$(E_2 - E_1)\langle n_2K_2|-i\alpha_r f(r)|n_1K_1\rangle$$
$$= \langle n_2K_2|H_2(-i\alpha_r f(r)) + i\alpha_r f(r)H_1|n_1K_1\rangle$$
$$= \langle n_2K_2|-f'(r) + \frac{\Delta_{21}^+}{r}\beta f(r) + 2i\alpha_r\beta M(r)f(r)|n_1K_1\rangle. \tag{16.21}$$

In addition, one has

$$H_2\beta f(r) + \beta f(r)H_1 = -i\alpha_r\left\{\frac{\Delta_{21}^-}{r}f(r) + \beta f'(r)\right\}$$
$$+ 2(M + \beta V(r))f(r), \tag{16.22}$$

from which we have

$$(E_2 + E_1)\langle n_2K_2|\beta f(r)|n_1K_1\rangle$$
$$= \langle n_2K_2|-i\alpha_r\left\{\frac{\Delta_{21}^-}{r}f(r) + \beta f'(r)\right\}$$
$$+ 2(M + \beta V(r))f(r)|n_1K_1\rangle. \tag{16.23}$$

Equations (16.10)–(16.23) are the basic results of our problem. Even though we might consider, as in the non-relativistic case, radial functions of the form $f(r) = r^\lambda$ and insert the explicit expression for the Coulomb potential, we do not show them for simplicity.

3 Diagonal Case

Let us study the diagonal case $\Delta_{21}^{-}=0$, i.e., $K_2=K_1$. In this case, it is shown from Eq. (16.10) that

$$(E_2-E_1)\langle n_2K_1|f(r)|n_1K_1\rangle=-i\alpha_r\langle n_2K_1|f'(r)|n_1K_1\rangle. \tag{16.24}$$

Similarly, Eqs. (16.16) and (16.17) are simplified as

$$\begin{aligned}&(E_2-E_1)^2\langle n_2K_1|f(r)|n_1K_1\rangle\\&\quad=\langle n_2K_1|H_2\xi-\xi H_1|n_1K_1\rangle\\&\quad=\langle n_2K_2|-f''(r)+\frac{2K_1}{r}\beta f'(r)+2i\alpha_r M\beta f'(r)|n_1K_1\rangle, \qquad (16.25)\\&(E_2^2-E_1^2)\langle n_2K_1|f(r)|n_1K_1\rangle\\&\quad=\langle n_2K_1|H_2\xi+\xi H_1|n_1K_1\rangle\\&\quad=\langle n_2K_1|-f''(r)-2f'(r)\frac{d}{dr}-2i\alpha_r f'(r)V(r)|n_1K_1\rangle. \qquad (16.26)\end{aligned}$$

Likewise, Eqs. (16.20) and (16.21) become

$$\begin{aligned}&(E_2+E_1)\langle n_2K_1|-i\alpha_r f(r)|n_1K_1\rangle\\&\quad=\langle n_2K_1|-f'(r)-2f(r)\frac{d}{dr}-2i\alpha_r V(r)f(r)|n_1K_1\rangle, \qquad (16.27)\\&(E_2-E_1)\langle n_2K_1|-i\alpha_r f(r)|n_1K_1\rangle\\&\quad=\langle n_2K_1|-f'(r)+\frac{2K_2}{r}\beta f(r)+2i\alpha_r\beta M(r)f(r)|n_1K_1\rangle, \qquad (16.28)\\&(E_2+E_1)\langle n_2K_1|\beta f(r)|n_1K_1\rangle\\&\quad=\langle n_2K_1|-i\alpha_r\beta f'(r)+2(M+\beta V(r))f(r)|n_1K_1\rangle. \qquad (16.29)\end{aligned}$$

Finally, let us consider a very special case $f(r)=1$ and $V(r)=-Z\alpha/r$. If so, we have

$$(E_2+E_1)\langle n_2K_1|\beta|n_1K_1\rangle=\langle n_2K_1|2\left(M-\beta\frac{Z\alpha}{r}\right)|n_1K_1\rangle. \tag{16.30}$$

Before ending this section, we want to give a few useful remarks on the radial Hamiltonian H given in (16.4). First, it is found that this Hamiltonian is hermitian if we consider $(d/dr)^\dagger=-d/dr$, i.e., $H^\dagger=H$. Therefore, the eigenvalues of the Hamiltonian are reals. Second, it might be noted that there are other two forms of the radial Hamiltonian. The first one is given by

$$H=-i\alpha_r\left(\frac{d}{dr}+\frac{K}{r}\beta\right)+M\beta+V(r),\quad \alpha_r=\sigma_2=\begin{pmatrix}0&-i\\i&0\end{pmatrix}, \tag{16.31}$$

whose matrix form becomes

$$H=\begin{bmatrix}M+V(r)&-(\frac{d}{dr}-\frac{K}{r})\\(\frac{d}{dr}+\frac{K_a}{r})&V(r)-M\end{bmatrix} \tag{16.32}$$

satisfying

$$H\psi(r) = E\psi(r), \quad \psi(r) = \begin{pmatrix} G(r) \\ F(r) \end{pmatrix}. \tag{16.33}$$

The second is the formalism used by Davydov [462]

$$H = \alpha_r p_r + i\alpha_r \frac{\beta}{r} K + M\beta + V, \quad \alpha_r = \sigma_2, \tag{16.34}$$

satisfying

$$H\Psi(r) = E\Psi(r), \tag{16.35}$$

where

$$\Psi(r) = r^{(1-D)/2} \begin{pmatrix} F(r) \\ G(r) \end{pmatrix}, \tag{16.36}$$

and p_r is given by Eq. (3.53).

4 Conclusions

In this Chapter we have generalized hypervirial theorem for the relativistic Dirac equation both in off-diagonal case and in diagonal case. The key issue is to construct the radial Dirac Hamiltonian. Three different expressions of the Hamiltonian have been given.

Chapter 17
Kaluza-Klein Theory

1 Introduction

The model of higher-dimensional space-time is a powerful ingredient to be needed to unify the interactions of various fields in nature. Before starting out the higher-dimensional Kaluza-Klein theory, let us first consider the five-dimensional Kaluza-Klein theory. Almost ninety years ago Kaluza put forward the issue that our universe has more than four dimensions [9–11]. Kaluza introduced an extra compactified dimension to unify two fundamental forces of gravitation and electromagnetism in our world. Kaluza began by taking as action the five-dimensional pseudo-Riemannian manifold, and he imposed a so-called cylinder condition: the components of the metric g_{IJ} should not depend on the space like 5th dimension, i.e., he neglected the effect of the 5th or scalar potential, and set all derivatives of four-dimensional quantities with respect to the 5th coordinate to zero. The 5×5 metric introduced by him is given by

$$g_{IJ}^{(5)} = \begin{pmatrix} g_{\mu\nu}^{(4)} & \alpha A_\nu \\ \alpha A_\mu & 2V \end{pmatrix} \tag{17.1}$$

with $\partial_5 g_{IJ} = 0$. The $g_{\mu\nu}^{(4)}$, A_μ, V and $\alpha = \sqrt{2G_N}$ represent the metric of the four-dimensional space-time, the gauge field from electrodynamics, the dilaton and the coupling constant related to the Newton constant G_N, respectively. He evaluated the Ricci tensor for linearized fields

$$R_{\mu\nu} = \partial_i \Gamma^i_{\mu\nu}, \qquad R_{5\nu} = -\alpha \partial^\mu F_{\mu\nu}, \qquad R_{55} = -\Box V. \tag{17.2}$$

The Kaluza's theory was, up to the point where he introduced the particle, a vacuum theory. It did not contain an additional term in the action besides the five-dimensional Ricci scalar: the field equations are given by setting the Ricci tensor (17.2) equal to zero.

This theory can be separated into further sets of equations, one of which is equivalent to Einstein field equations, another set equivalent to Maxwell electromagnetic field equations and the final part an extra scalar field now termed the "radion". It

S.-H. Dong, *Wave Equations in Higher Dimensions*,

DOI 10.1007/978-94-007-1917-0_17,

should be noted that such a splitting of five-dimensional space-time into the Einstein's equations and Maxwell's equations in four dimensions was first discovered in 1914 by Gunnar Nordström, in the context of his theory of gravity, but subsequently forgotten [8].

Five years later, shortly after the discovery of the Schrödinger equation, in 1926 Oskar Klein improved and extended Kaluza's treatment, and revealed a very interesting geometrical interpretation of gauge transformations [10, 11]. He proposed that the fourth spatial dimension is curled up in a circle of very small radius, so that a particle moving a short distance along that axis would return to where it began. The distance a particle can travel before reaching its initial position is said to be the size of the dimension. This extra dimension is a compact set, and the phenomenon of having a space-time with compact dimensions is referred to as compactification.[1] The extension of extra dimension turned out to be comparable to the Planck length. In Klein's theory, he proposed a more fruitful interpretation of the five-dimensional metric

$$g_{IJ}^{(5)} = \begin{pmatrix} g_{\mu\nu}^{(4)} + VA_\mu A_\nu & VA_\nu \\ VA_\mu & V \end{pmatrix}. \tag{17.3}$$

He then varies the action and recovers the full field equations of general relativity, with the energy-momentum tensor of the electro-magnetic fields, and the source-free Maxwell's equations

$$G_{\mu\nu} = \kappa T_{\mu\nu}, \qquad \nabla_\mu F^{\mu\nu} = 0. \tag{17.4}$$

So Kaluza's vacuum theory with the cylinder condition contains the full field equations. However, this vacuum theory cannot produce non-singular rotation symmetric particle solutions.

After that, this theory was developed by some authors. For example, Einstein investigated this theory most intensively from 1938 to 1943. This was performed by himself and his coauthors Bergmann and Bargmann in the classical Kaluza-Klein theory with constant dilaton [463–466]. One of their primary objectives was to find a nonsingular particle solution. Unfortunately, in the full theory this search got frustrated. In 1943, Einstein and Pauli argued that solitons cannot exist in this theory, a result that Einstein may have been disappointed with [466]. Since there are problems with this theory it predicts the existence of a massless scalar, which has not been seen so the theory is not phenomenologically acceptable even if it works as the first model of unified theory. Extra dimensions were then abandoned from a phenomenological point of view and looked upon as mathematical tools. When the string theory was invented in the 1980s [467], the extra dimensions were considered to have some importance from the physical point of view [468, 469]. String theory

[1] Up to now, there exist two main types of compactifications of an extra dimension. The first is a circular compactification. With this extra dimension the entire space can be denoted $M_4 \times S^1$, where M_4 is the regular Minkowski space and S^1 is the circle with a radius R. It is chosen to be small enough to avoid detection. Another type is an orbifold compactification by imposing the discrete Z_2 symmetry, where $y \to -y$ on the circle.

attempts to address the question of merging quantum mechanics with general relativity into a consistent theory. In the original formulation of those string models, the size of the extra dimensions where supposed to be compactified to manifolds of small radii with sizes is about the order of the Planck length 10^{-36} m, such that they remain hidden to the experiment, thus explaining the reason why we see only four dimensions. It is believed that the relevant energy scale where quantum gravity effects would become important is given by the Planck mass defined through the fundamental constant as

$$M_P c^2 = \sqrt{\frac{\hbar c^5}{8\pi G_N}} \sim 2.4 \times 10^{18}\ \text{GeV}. \tag{17.5}$$

Clearly, this is much above any energy scale we have ever tested. If this is the correct description of nature, then extra dimensions do exist but they have little influence in physics at the electroweak scale. Nevertheless, the possibility of extra dimensions at the weak scale was put forwarded and a new research direction started.

Up to now, two main types of extra dimensions by regarding the fields that live on them can be clarified as large extra dimensions and TeV flat extra dimensions. The first case corresponds to extra dimensions where only gravity can propagate, i.e., gravity has only been tested up to scales $\sim$1 μm so these types of dimensions must have a size of a micron or less. The second corresponds to those there are constraints coming from collider and indirect tests that force this kind of extra dimensions to be at the TeV scale or more. Another type is the warped extra dimensions [470–472], of which a more famous model is Arkani-Hamed, Dimopoulos and Dvali (ADD) model.[2] Unfortunately, current search shows no signals of extra dimensions [473–477]. The extra dimensions has evolved from a single idea to a new general paradigm with some authors applying it as a tool to address the large number of key issues that remain unanswerable within the standard model context. However, the search for extra dimensions is not over yet. On the contrary, it has only just started. This is because its discovery would produce a fundamental change in how we view the universe and also some surprising results.

Since the solitons were described in the Kaluza theory in the 1980s, the modern versions of Kaluza-Klein theory have allowed the 5th coordinate to play an important physical role. Space-time-matter theory dates from 1992, and is motivated by the old idea of Einstein to give a geometrical description of matter [478–481]. This is realized by embedding the 4-dimensional Einstein equations with sources in the apparently-empty 5-dimensional Ricci-flat equations, so there is an effective or induced energy-momentum tensor in which properties of matter like the density and pressure are given in terms of the 5th potential and derivatives of the 4-dimensional

[2]The large extra dimension scenario of ADD was proposed as a potential solution to the hierarchy problem, i.e., the question of why the reduced Planck scale $\bar{M}_{\rm pl} \simeq 2.4 \cdot 10^{18}$ is so much larger than the weak scale $\sim$1 TeV. They propose that we live on an assumed to be rigid $4D$ hypersurface (also named a wall or brane). The gravity is allowed to propagate in a $(4+D)$-dimensional n-torus T^n, whose radii are equal to R.

space-time potentials with respect to the 5th coordinate. Membrane theory appearing in 1998 is motivated by explaining the strength of particle interactions compared to gravity, or alternatively the smallness of particle masses compared to the Planck value [482–484]. This is done by embedding 4-dimensional space-time as a singular hypersurface in a 5-dimensional manifold, so particle interactions are confined to a sheet while gravity is diluted by propagating into the bulk of the 5th dimension.[3] Both space-time-matter theory and membrane theory are in agreement with observations. We suggest the reader refer to those review papers to recognize their developments [485, 486].

Even in the absence of a completely satisfying theoretical physics framework, the idea of exploring extra, compactified, dimensions is of considerable interest in the experimental physics and astrophysics communities. A variety of predictions, with real experimental consequences, can be made in the case of large extra dimensions or warped models. For example, on the simplest of principles, one might expect to have standing waves in the extra compactified dimension(s). If the radius of a spatial extra dimension is given by R, then the invariant mass of such standing waves would be $M_n = nh/Rc$, where n is an integer, h the Planck's constant and c the speed of light. This set of possible mass values is often called the Kaluza-Klein tower. Similarly, in thermal quantum field theory a compactification of the Euclidean time dimension leads to the Matsubara frequencies and thus to a discretized thermal energy spectrum. On the other hand, examples of experimental pursuits include work by the Collider Detector at Fermilab (CDF) collaboration, which has re-analyzed particle collider data for the signature of effects associated with large extra dimensions/warped models. Brandenberger and Vafa have speculated that in the early universe, cosmic inflation causes three of the space dimensions to expand to cosmological size while the remaining dimensions of space remained microscopic.

Until now, the higher-dimensional generalizations of this theory to include weak and strong interactions have attracted much attention for many particle physicists in the past few years [487–491]. For instance, to unify gravity with the strong and electroweak forces, the symmetry group of standard model, SU(3) × SU(2) × U(1) was used. However, in order to convert this interesting geometrical construction into a true model of reality founders on a number of issues, then the fermions must be introduced in nonsupersymmetric models. Nevertheless, Kaluza-Klein theory remains an important milestone in theoretical physics and is often embedded in more sophisticated theories. The revival of interest in Kaluza-Klein theory arose in the first instance from work in string theories [492, 493], and then from the usefulness

[3] In modern geometry the extra 5th dimension can be understood as a circle group U(1) since the electromagnetism can be formulated essentially as a gauge theory on a fiber bundle, the circle bundle, with gauge group U(1). If one is able to understand this geometrical interpretation, it is relatively straightforward to replace U(1) by a general Lie group. Such generalizations are often called Yang-Mills theories. If a distinction is drawn between them, then the Yang-Mills theories occur on a flat space-time, while Kaluza-Klein treats a more general case of curved space-time. The base space of Kaluza-Klein theory need not be four-dimensional space-time; it can be any pseudo-Riemannian manifold, or even a supersymmetric manifold or orbifold.

of extra spatial dimensions in the construction of $N = 8$ supergravity theory [494, 495]. In these contributions, it would have been possible to regard the extra spatial dimensions as a mathematical device. Although the order of the compactification scale of the additional dimensions has not been confirmed and are also of considerable interest recently, larger extra dimensions were invoked in order to provide a breakthrough of hierarchy problem in some approaches [483, 496, 497]. It is believed that the research on higher-dimensional space-time is valuable and become a focus in the physical community, therefore the theory needs to be explored deeply, extensively and in various directions.

Due to limited space, many other topics are not being covered, including brane intersecting models, cosmology of models with extra dimensions both in flat and warped bulk backgrounds; Kaluza-Klein dark matter; an extended discussion on black hole physics; as well as many others. The interested reader that would like to go beyond the present note can consult any of the excellent reviews [469, 485, 498].

As what follows, we shall review the development of the high-dimensional Kaluza-Klein theory. In particular, $(4 + D)$-dimensional Kaluza-Klein theories and the particle spectrum of Kaluza-Klein theory for fermions are to be discussed.

2 $(4 + D)$-Dimensional Kaluza-Klein Theories

As mentioned above, in order to unify gravitation, not only with electromagnetism but also with weak and strong interactions, it is necessary to generalize the five-dimensional theory to a higher-dimensional theory so as to obtain a non-Abelian gauge group [485].

For the moment, remember that gravity is a geometric property of the space. Then, the first thing to notice is that in a higher-dimensional world, where Einstein gravity is assumed to hold, the gravitational coupling does not necessarily coincide with the well known Newton constant G_N. To explain this more clearly, let us assume that there are N extra space-like[4] dimensions which are compactified into circles of the same radius R so the space is factorized as an $M_4 \times K$ with $K = T^N$ manifold. The higher-dimensional gravity action can be written as

$$S = -\frac{1}{16\pi G_*} \int d^{4+N}x \sqrt{|g_{(4+N)}|} R_{(4+N)}, \tag{17.6}$$

where G_* is the fundamental gravity coupling, $g_{(4+N)}$ the metric in the whole $(4 + N)$-dimensional spaces.[5] We assume that the extra dimensions are flat without curvature, thus the metric has the form

$$ds^2 = g_{\mu\nu}(x) dx^\mu dx^\nu - \delta_{ab} dy^a dy^b, \tag{17.7}$$

[4]The choice of a time-like extra dimension shall lead to the tachyon. Tachyons are well known to be very dangerous in most theories. This seems to imply that we should only pick the space-like solution.

[5]In fact in $(4 + D)$ dimensions a gauge field can be decomposed into a $4D$ gauge Kaluza-Klein tower plus D distinct scalar towers.

where the metric $g_{\mu\nu}$ depends only on the four-dimensional coordinates x^μ for $\mu = 0, 1, 2, 3$ and $\delta_{ab} dy^a dy^b$ denotes the line element on bulk, $a \in [1, N]$. Thus, it is easy to see that $\sqrt{|g_{(4+N)}|} = \sqrt{|g_4|}$ and $R_{(4+N)} = R_4$ so that we might integrate out the extra dimensions in Eq. (17.6) to obtain an effective $4D$ action

$$S = -\frac{V_n}{16\pi G_*} \int d^4x \sqrt{|g_4|} R_4, \tag{17.8}$$

where V_n denotes the volume of the extra space. This equation is precisely the standard gravity action in $4D$ if one makes the following identification

$$G_N = G_* / V_n, \tag{17.9}$$

which is given by a volumetric scaling of the truly fundamental gravity scale.

We denote by y^n the coordinates of the compact manifold K. An isometry of K is a coordinate transformation $y \to y_1$ which leaves the form of the metric $\tilde{g}_{mn}$ for K invariant:

$$y \to y_1 : \tilde{g}^1_{mn}(y_1) = \tilde{g}_{mn}(y_1). \tag{17.10}$$

The general infinitesimal isometry is given by

$$I + i\varepsilon^a s_a : y^n \to y^{n'} = y^n + \varepsilon^a \zeta^n_a(y), \tag{17.11}$$

where s_a are the generators. The infinitesimal parameters ε^a are independent of y, and the Killing vectors ζ^n_a, which are associated with the independent infinitesimal isometries, obey the Lie algebra

$$\zeta^m_b \partial_m \zeta^n_c - \zeta^m_c \partial_m \zeta^n_b = -C^a_{bc} \zeta^n_a, \tag{17.12}$$

where C^a_{bc} are the structure constants with the following property

$$[s_a, s_b] = i C^c_{ab} s_c. \tag{17.13}$$

For example, the D-dimensional sphere S^D has isometry group SO$(D+1)$.

Now, let us consider the non-Abelian gauge transformations [499]. The ground-state metric for the compactified $(4+D)$-dimensional theory may be written as

$$\bar{g}_{AB} = \mathrm{diag}^{(0)}\{\eta_{\mu\nu}, -\tilde{g}_{mn}(y)\} \tag{17.14}$$

where $\eta_{\mu\nu} = (1, -1, -1, -1)$ is the metric of Minkowski space M_4 and $\tilde{g}_{mn}(y)$ the metric of the compact manifold. The non-Abelian gauge fields of the theory may be expressed by the expansion about the ground state

$$\bar{g}_{AB} = \begin{pmatrix} g_{\mu\nu}(x) - \tilde{g}_{mn}(y) B^m_\mu B^n_\nu & B^n_\mu \\ B^m_\nu & -\tilde{g}_{mn}(y) \end{pmatrix}, \quad B^n_\mu \equiv \zeta^n_a(y) A^a_\mu(x). \tag{17.15}$$

Non-Abelian gauge transformations come from considering the effect on the components $\bar{g}_{\mu n}$ of the metric of the infinitesimal isometry with x-dependent parameters:

$$y^n \to y^n + \zeta^n_a(y)\varepsilon^a(x). \tag{17.16}$$

It shall be found that

$$A^a_\mu \to A^{a'}_\mu = A^a_\mu + \partial_\mu \varepsilon^a(x) + C^c_{ab}\varepsilon^b(x) A^c_\mu. \tag{17.17}$$

If taking $C_{ab}^{c} = g f_{ab}^{c}$ and $s_a = g S_a$ one finds that this transformation is just the usual Yang-Mills gauge transformation. Consequently, we have

$$[S_a, S_b] = i f_{ab}^{c} S_c. \tag{17.18}$$

Thus, non-Abelian gauge transformations are generated by x-dependent infinitesimal isometries of the compact manifold K.

3 Particle Spectrum of Kaluza-Klein Theories for Fermions

In this section we are going to give a review of the particle spectrum of the Kaluza-Klein theories for fermions [500]. We consider, for example, a massless spinor particle ψ in $(4 + D)$ dimensions. The Dirac equation can be written as

$$i(\gamma_\mu \partial_\mu + \Gamma_\alpha e_\alpha^m \nabla_m)\psi = 0 \tag{17.19}$$

with

$$\nabla_m = \partial_m - i\omega_m, \quad \omega_m = \frac{1}{2}\omega_m^{\alpha\beta} M_{\alpha\beta}, \; M_{\alpha\beta} = \frac{1}{4} i[\Gamma_\alpha, \Gamma_\beta], \tag{17.20}$$

where ω is an antisymmetric tensor and $M_{\alpha\beta}$ the spinor representation of the tangent space group SO(D) of the compact manifold K. The γ_μ are $(2^{2+D/2} \times 2^{2+D/2})$ gamma matrices satisfying

$$\{\gamma_\mu, \gamma_\nu\} = \gamma_\mu \gamma_\nu + \gamma_\nu \gamma_\mu = 2\delta_{\mu\nu}, \quad \mu, \nu = 0, 1, 2, 3. \tag{17.21}$$

The gamma matrices of the compact space satisfy

$$\{\Gamma_\alpha, \Gamma_\beta\} = 2\delta_{\alpha\beta}, \quad \alpha, \beta = 4, \ldots, 3 + D \tag{17.22}$$

and

$$\{\Gamma_\alpha, \gamma_\mu\} = 0. \tag{17.23}$$

Accordingly, we have

$$\nabla_m \psi = \left(\partial_m - \frac{1}{2} i \omega_m^{\alpha\beta} M_{\alpha\beta} \right) \psi. \tag{17.24}$$

Obviously, $M = -i\Gamma_\alpha e_\alpha^m \nabla_m$ plays the role of the mass operator since its eigenvalues give the particle masses in four dimensions. It may be shown that the operator M has no zero eigenvalue [501, 502]. Note that the observed fermions are in fact zero modes of the Dirac operator, and that their small non-zero masses arise from physics at a much lower energy scale. Thus, we must change some assumptions made in deriving Lichnerowitz's theorem. One possibility, which has been explored by Destri *et al.* [503] and by Wu and Zee [504], is to introduce torsion on the internal manifold K. Even so their explorations have shown that this possible escape route is also unattractive. This is because there is the arbitrariness of precisely how the torsion should be introduced.

Next, suppose that D is even, i.e., $D = 2N$. Define Γ matrices as follows

$$\Gamma = \Gamma_1 \Gamma_2 \ldots \Gamma_D = i^N \sigma_3 \times \sigma_3 \times \cdots \times \sigma_3, \tag{17.25}$$

which anticommutes with all matrices $\Gamma_1, \Gamma_2, \ldots, \Gamma_D$. The gamma matrices for full tangent space group SO(1, 3 + D) are given by

$$\bar{\Gamma}_A = \begin{cases} \gamma_\mu \times I, & A = \mu = 0, 1, 2, 3, \\ i\gamma_5 \times \Gamma_{A-3}, & A = 4, \ldots, D+3. \end{cases} \tag{17.26}$$

In full $(4 + D)$-dimensional space the "chirality" χ is defined by

$$\chi = \bar{\Gamma}_0\bar{\Gamma}_1\bar{\Gamma}_2\ldots\bar{\Gamma}_{D+3} = -i(-1)^N\gamma_5 \times \Gamma \tag{17.27}$$

and χ anticommutes with all matrices $\bar{\Gamma}_0, \bar{\Gamma}_1, \bar{\Gamma}_2, \ldots, \bar{\Gamma}_{D+3}$ since γ_5 and Γ are the pseudo-orthogonal group SO(1, 3) and SO(D), respectively. Therefore, χ commutes with all generators M^{AB} of SO(1, 3 + D) and can be used to label inequivalent spinor representations. Since

$$\chi^2 = -(\gamma_5)^2 \times \Gamma_2 = \pm 1, \tag{17.28}$$

where ± 1 correspond odd and even N, respectively. The corresponding eigenvalues are given by

$$\chi = \begin{cases} \pm i, & N = \text{even}, \\ \pm 1, & N = \text{odd}. \end{cases} \tag{17.29}$$

Let us consider a special case. For odd N, one has $\chi = \pm 1$, we have spinor representations with $\gamma_5 = +1, \Gamma = -i$ or $\gamma_5 = -1, \Gamma = +i$. Thus, fermions with left-handed physical chirality have internal chirality $+i$, and satisfy a different Dirac equation from the right-handed fermions. Thus, the zero modes might have different quantum numbers. Actually, the only possibility of this happening is for odd N. This is because for even N the complex conjugate of the spinor representation with $\chi = +i$ is equivalent to the spinor representation with $\chi = -1$. It is known from Eq. (17.27) that for even N we have

$$\gamma_5 = \pm 1, \qquad \Gamma = \mp 1 \tag{17.30}$$

for $\chi = +i$, while

$$\gamma_5 = \pm 1, \qquad \Gamma = \pm 1 \tag{17.31}$$

for $\chi = -i$. The complex conjugate of the $\gamma_5 = 1$ spinor is the $\gamma_5 = -1$ spinor as usual in SO(1, 3). But the complex conjugate of $\Gamma = -1$ spinor of SO($2N$) is equal to itself for even N.

Even so we still have to arrange that the zero modes of the Dirac operator with $\Gamma = +i$ form a complex representation of the symmetry group. That is to say, we must arrange that the $\Gamma = +i$ zero modes transform differently from $\Gamma = -i$ zero modes. Unfortunately, the possibility of achieving this, even in $(4n+2)$ dimensions, is severely constrained by a theorem of Atiyah and Hirzebruch [505], which requires that in the absence of elementary gauge fields, the zero modes of the Dirac operator form a real representation (in any even number of dimensions). The detailed discussed was done by Witten [500] by introducing gauge fields in order to arrange for the zero modes to form a complex representation.

The most comprehensive study of complete fermion spectrum, not just the zero modes, has been carried out by Schellekens [506, 507]. He has expressed the eigenvalues of the fermion mass-squared operator M^2, in the presence of a general instanton background gauge field configuration on a symmetric coset space G/H, in terms of the Casimir invariants of G and H. For massless fermions the problem is that given a fermion transforming according to some representation of $H \subset G_{YM}$ to determine in which representation of G the zero modes occur. This has been done in general for the hyperspheres S^D and the complex project planes CP^N. In the former case when D is even, and the fermion is in an irreducible representation of $H = \mathrm{SO}(D)$, then the zero modes form an irreducible representation of $G = \mathrm{SO}(D+1)$. This generalizes a previously known result [508] for $D = 4$.

Watamura [509, 510] has also provided a general treatment of compactification in the case $K = CP^N$ in the presence of a monopole U(1) gauge field, and has shown how for a particular choice of the monopole charge, the fermionic zero modes belong to the fundamental representation of $G = \mathrm{SU}(N+1)$.

4 Warped Extra Dimensions

Up to now we have been working in the simplest picture where the energy density on the brane does not affect the space time curvature, but rather it has been taken as a perturbation on the flat extra space. However, for large brane densities this may not be the case. The first approximation to the problem can be done by considering a five-dimensional model where branes are located at the two ends of a closed 5th dimension. Hence, the 5th dimension would be a slice of an Anti-de Sitter space with flat branes at its edges. As a result, one can keep the branes flat paying the price of curving the extra dimension. Such curved extra dimensions are usually referred as warped extra dimensions. Historically, the possibility was first mentioned by Rubakov and Shaposhnikov in Refs. [511–514], who suggested the cosmological constant problem be understood under this light: the matter fields vacuum energy on the brane could be canceled by the bulk vacuum, leaving an almost zero cosmological constant for the brane observer even though no specific model was given there. It was actually Gogberashvili [515] who provided the first exact solution for a warped metric. Nevertheless, these models are best known after Randall and Sundrum (RS) who linked the solution to the hierarchy problem [470]. Later developments suggested that the warped metrics could even provide an alternative to compactification for the extra dimensions [470].

5 Conclusions

It is hard to address too many interesting topics of the area in detail, as we would have liked, without facing trouble with the limiting space of this book. In this Chapter we have reviewed the development of the high-dimensional Kaluza-Klein theory.

The contents are concerned with traditional five-dimensional Kaluza-Klein theory, $(4+D)$-dimensional Kaluza-Klein theories and the particle spectrum of the Kaluza-Klein theory for fermions.

One of the most direct ways to observe the extra spatial dimensions may be in the cosmological context. Here, Kaluza-Klein theory can give a satisfactory late-time cosmology, and also suggests various mechanisms for achieving the desired cosmological inflation at early times, though none of these has so far proved entirely compelling when detailed calculations have been performed.

In summary, it seems quite likely that even if the original pure Kaluza-Klein theory cannot be sustained, extra spatial dimensions will play an important role in the eventual unified theory of interactions, and in understanding early cosmology.

Part V
Conclusions and Outlooks

Chapter 18
Conclusions and Outlooks

1 Conclusions

We are now in the position to draw conclusions for this work. Motivated by the physical laws in higher dimensions leading to insight concerning those in lower dimensions, we have introduced wave equations in higher dimensions and put the mathematical and physical concepts and techniques like the wave equations and group theory related to the higher dimensions at the reader's disposal. In some sense we have provided a comprehensive description of the wave equations including the non-relativistic Schrödinger equation, relativistic Dirac and Klein-Gordon equations in arbitrary dimensions and their wide applications in quantum mechanics.

We have introduced the fundamental theory about the SO(N) group which has been used in the successive Chaps. 3–5 including the non-relativistic Schrödinger equation, relativistic Dirac and Klein-Gordon equations. As important applications in non-relativistic quantum mechanics, we have applied our theories proposed in Part II to study some quantum systems such as the harmonic oscillator, Coulomb potential, wavefunction ansatz method, Levinson theorem, generalized hypervirial theorem, exact and proper quantization rules and Langer modification, position-dependent mass Schrödinger equation. As for as those important generalized applications to relativistic quantum mechanics in higher dimensions, we have studied the Levinson theorem and generalized hypervirial theorem for the Dirac equation, the Klein-Gordon equation and Kaluza-Klein theory. A number of previous results are summarized and some new materials are presented.

2 Outlooks

Considering the interest in higher dimensional quantum physics, some new appearing fields could be studied. For example, the superstring theory, supergravity shall become interesting and challenging. On the other hand, the searching of the extra dimensions both in theory and in experiment also becomes exciting.

S.-H. Dong, *Wave Equations in Higher Dimensions*,
DOI 10.1007/978-94-007-1917-0_18, © Springer Science+Business Media B.V. 2011

Appendix A
Introduction to Group Theory

In this Appendix we are going to outline some basic definitions of the groups based on textbooks by Weyl, Wybourne, Hamermesh, Miller and others [136–138, 169].

Definition A **group** G is a set of elements $\{e, f, g, h, k, \ldots\}$ together with a binary operation which associates with any ordered pair of elements f, g in G a third element fg. The binary operation named the group multiplication is subject to the following four requirements:

- Closure: if f, g are in G then fg is also in G,
- Identity element: there exists an identity element e in G (a unit) such that $ef = fe = f$ for any $f \in G$,
- Inverses: for every $f \in G$ there exists an inverse element $f^{-1} \in G$ such that $ff^{-1} = f^{-1}f = e$,
- Associative law: the identity $f(hk) = (fh)k$ is satisfied for all elements $f, h, k \in G$.

As far as those basic concepts such as Abelian group, subgroup, homomorphism, isomorphism, representation, irreducible representation and commutation relation has been reviewed in Chap. 2. Accordingly, as what follows we want to give a few typical examples to indicate the variety of mathematical objects with the structure of groups. For more examples, we suggest the reader refer to textbooks by Miller [138, 169].

Example 1 For the real numbers R, we consider the **addition** as the group product. The product of two elements a, b is their sum $a + b$. We may take 0 as an **identity element**. The **inverse** of an element c is its negative $-c$. The set odd real number R constructs an infinite Abelian group. Among the subgroups of R are the integers, the even integers and the group consisting of the element 0 alone.

Example 2 For a given group G, two elements $\{0, 1\} \in G$, the group multiplication is given by $0 \cdot 0 = 0$, $0 \cdot 1 = 1 \cdot 0 = 1$, $1 \cdot 1 = 0$. We choose 0 as the **identity element**. This is an Abelian group with order $n = 2$. It has only two subgroups, $\{0\}$ and $\{0, 1\}$.

S.-H. Dong, *Wave Equations in Higher Dimensions*,

DOI 10.1007/978-94-007-1917-0,

Sometimes it is very convenient to present the law of combination in the form of a group table, i.e., we label the rows and columns of a square array in terms of the elements of the group. In the box in the nth row and mth column, we record the product of the element labeling the nth row by the element labeling the mth column. An explicit example for the group of order 2 is given as

	e	a
e	e	a
a	a	e

(A.1)

For the group of order 3, one has

	e	a	b
e	e	a	b
a	a	b	e
b	b	e	a

(A.2)

where $b = a^2$.

There are two distinct structures for abstract groups of order 4:

(i)

	e	a	b	c
e	e	a	b	c
a	a	e	c	b
b	b	c	e	a
c	c	b	a	e

(A.3)

where $a^2 = b, ab = c = a^3, a^4 = b^2 = e$. This group corresponds a cyclic group $\{a, a^2, a^3, a^4 = e\}$.

(ii)

	e	a	b	c
e	e	a	b	c
a	a	b	c	e
b	b	c	e	a
c	c	e	a	b

(A.4)

where $a^2 = b^2 = c^2 = e$, $ab = c$, $ac = b$, $bc = a$. This group is called the four-group V.

1 Subgroups

If we select from the elements of the group G a subset H, we use the notation $H \in G$ to symbolize that H is contained in G. If the subset H forms a group, then we say H is a subgroup of the group G. There are two trivial subgroups for every group: the group consisting of the identity element alone and the whole group G itself. These two subgroups are said to be improper subgroups. The problem of finding all other proper subgroups of a given group G has become one of the main problems of the group theory. It should be noted that H forms a group under the same law of combination as G. Testing a subset H to see whether H is a subgroup requires that

- the product of any pair of elements of subset H be in H;
- H contain the inverse of each of its elements.

A good example refers to the rational numbers. As we know, the rational numbers form a group G under addition. The positive rational numbers form a group G_1 under multiplication, but G_1 is not a subgroup of G even though the elements of G_1 are a subset of the elements of G. This is because G_1 are not satisfied with the second condition. For a finite group only the first requirement is needed. However, both requirements are needed for infinite groups.

If H is a subgroup of G, and K is a subgroup of H, then K is a subgroup of G. This transitive relation leads to the idea of sequences of subgroups, each containing all those preceding it in the sequence. Each group contains all the successive subgroups, i.e., $G \in H \in K \in \cdots$; and in this case, all these groups are isomorphic: $G \approx H \approx K \approx \cdots$. Therefore, a group can be isomorphic with one of its proper subgroups, but this is impossible for groups of finite order. The symmetry groups S_n, which are the permutations of degree n, are important because they actually exhaust the possible structures of finite groups as shown by Cayley's theorem:

Cayley's theorem *Every group G of order n is isomorphic with a subgroup of the symmetry group* S_n.

Thus, S_{n-1} is isomorphic with the n subgroups of S_n. The permutations associated with each group element can be obtained by looking at the group table. For example, in the four-group V given above, to find the permutation corresponding to, say the element a, we write in the top line the symbols $(e\ a\ b\ c)$ and enter below them the symbols as they appear in the row where we multiply by a on the left, i.e., $(a\ e\ c\ b)$ so

$$a \to \pi_a = \begin{pmatrix} e & a & b & c \\ a & e & c & b \end{pmatrix}. \tag{A.5}$$

If we label the elements (e, a, b, c) as $(1, 2, 3, 4)$, the permutations corresponding to (e, a, b, c) are given by

$$\begin{aligned} \pi_e &= \begin{pmatrix} 1 & 2 & 3 & 4 \\ 1 & 2 & 3 & 4 \end{pmatrix}, \\ \pi_a &= \begin{pmatrix} 1 & 2 & 3 & 4 \\ 2 & 1 & 4 & 3 \end{pmatrix} = (12)(34), \\ \pi_b &= \begin{pmatrix} 1 & 2 & 3 & 4 \\ 3 & 4 & 1 & 2 \end{pmatrix} = (13)(24), \\ \pi_c &= \begin{pmatrix} 1 & 2 & 3 & 4 \\ 4 & 3 & 2 & 1 \end{pmatrix} = (14)(23), \end{aligned} \tag{A.6}$$

where (12) or (34), (13), (24), (14), (23) is called a transposition. The permutation groups formed in this way have two special features:

- They are subgroups of order n of the symmetry group S_n.

- It is seen that, except for the identity e, the permutations leave no symbol unchanged since π_b takes a_i into ba_i, which is equal to a_i only if b is the identity e.

Permutations in S_n with these two properties are called regular permutations.

2 Cosets

Lagrange's theorem *The order of a subgroup of a finite group is a divisor of the order of the group.*

Suppose two sets of h distinct elements each, H and aH contained in G. If G has not been exhausted, choosing some element b of G which is not contained in either H or aH. Thus, the set bH will again generate h new elements of the group G. Continuing this process, we can exhibit the group G as the sum of a finite number of distinct sets of h elements each

$$G = H + aH + bH + \cdots + mH. \tag{A.7}$$

Thus the order g of the group G is a multiple of the order h of its subgroup H, i.e.,

$$g = mh, \tag{A.8}$$

where m is the index of the subgroup H under the group G. This means that h is a divisor of g so that the orders of all elements of a finite group must be divisors of the order of the group. The sets of elements of the form aH in (A.7) are called left cosets of H in G. Of course, we are able to multiply H on the right to yield the corresponding right cosets as follows

$$G = H + Ha_1 + Hb_1 + \cdots + Hm_1. \tag{A.9}$$

It is seen from Eq. (A.8) that a group, whose order is a prime number, has no proper subgroups and is necessarily cyclic. Such a group can be generated from any of its elements other than the identity element.

Lagrange's theorem can be used to find the possible structures of groups for a given order. Here we present an explicit example, say order 6, to show its advantage. Since the order of the group is 6, then the order of each of its elements is a divisor of 6, i.e., 1, 2, 3 or 6. If the group contains an element a of order 6, then the group is the cyclic group $\{a, a^2, a^3, a^4, a^5, a^6 = e\}$. To find other possible structures, let us assume that the group contains no element of order 6, but has an element a of order 3. Thus this group contains the subgroup $H \in \{a, a^2, a^3 = e\}$. If the group also contains another element b, then it contains six different elements $\{e, a, a^2, b, ba, ba^2\}$. The element b has order 2 or 3. After analyzing its order carefully, it is found that the order of element b cannot be 3 and it must be 2. This is because $b^3 = e$ if the order of b is 3, then element b^2 must be one of the six elements listed above. Obviously, we cannot have $b^2 = e$. Assuming that $b^2 = b, ba$ or ba^2 implies that $b = e, a$ or a^2, which contradicts the assumption that b is different from these elements. Moreover, if we assume that $b^2 = a$ or a^2 implies $ba = e$ or

$ba^2 = e$, respectively. Both of them contradict our assumptions $b^3 = e$. As a result, the order of element b must be 2. Consequently, we are able to construct the group table as follows [138]:

$$\begin{array}{c|cccccc} e & a & a^2 & b & ba & ba^2 \\ \hline a & a^2 & e & ba^2 & b & ba \\ a^2 & e & a & ba & ba^2 & b \\ b & ba & ba^2 & e & a & a^2 \\ ba & ba^2 & b & a^2 & e & a \\ ba^2 & b & ba & a & a^2 & e \end{array} \tag{A.10}$$

with $a^3 = b^2 = (ab)^2 = e$. This group is isomorphic with permutation group S_3.

3 Conjugate Classes

An element $b \in G$ is said to be conjugate to element a if we may find an element $v \in G$ such that

$$vav^{-1} = b. \tag{A.11}$$

In order to separate the group into classes of elements which are conjugate to one another we shall use an equivalence relation to separate a set into classes. An important feature of all elements in the same class is that they have the same order. For example, the distinct classes in S_4 are given by

- e;
- (12), (13), (14), (23), (24), (34);
- (12)(34), (13)(24), (14)(23);
- (123), (132), (124), (142), (134), (143), (234), (243);
- (1234), (1243), (1324), (1342), (1423), (1432).

The permutations of S_n operate on a total of n symbols. Assume that we resolve the permutations into independent cycles and let the number of 1-cycle be v_1, of 2-cycle be $v_2, \ldots$, of n-cycles be v_n. Due to the conservation of the total number n of symbols, we have the following relation

$$\sum_{i=1}^{n} i v_i = n. \tag{A.12}$$

If we let

$$\begin{aligned}
\sum_{i=1}^{n} v_i &= \lambda_1, \\
\sum_{i=2}^{n} v_i &= \lambda_2, \\
\sum_{i=3}^{n} v_i &= \lambda_3, \\
&\vdots \\
v_n &= \lambda_n,
\end{aligned} \tag{A.13}$$

then we have

$$\sum_{i=1}^{n} \lambda_i = n, \quad \lambda_1 \geq \lambda_2 \geq \lambda_3 \geq \cdots \geq \lambda_n \geq 0. \tag{A.14}$$

The splitting-up of n into a sum of n integers as in (A.14) is called a partition $[\lambda_1, \lambda_2, \ldots, \lambda_n]$ of n. Therefore, the problem to find the number of conjugate classes in S_n is reduced to the problem to partition n. Conversely, there is a corresponding cycle structure for a given partition as in (A.14)

$$\begin{aligned}
v_1 &= \lambda_1 - \lambda_2, \\
v_2 &= \lambda_2 - \lambda_3, \\
&\vdots \\
v_n &= \lambda_n.
\end{aligned} \tag{A.15}$$

Here we list the partitions for a few simple S_n and the total number r of classes

$$\begin{aligned}
&S_1: \quad [1]; \quad r = 1, \\
&S_2: \quad [2], [1^2]; \quad r = 2, \\
&S_3: \quad [3], [21], [1^3]; \quad r = 3, \\
&S_4: \quad [4], [31], [2^2], [21^2], [1^4]; \quad r = 5,
\end{aligned} \tag{A.16}$$

where $[1^2] = [11]$ and also others. These distinct classes can also be shown by Young table.

4 Invariant Subgroups

Starting with a subgroup $H \subset G$, we can form a set of elements aHa^{-1} where $a \in G$. This set of elements is again a subgroup of G and is said to be a conjugate subgroup of H in G. By choosing various elements a from G, we are able to obtain different conjugate subgroups. It may happen that for all a,

$$aHa^{-1} = H, \tag{A.17}$$

which means that all the conjugate subgroups of H in G are identical with H. Thus, we say that H is an invariant subgroup. On the other hand, it is shown from (A.17) that this equation can be written as

$$aH = Ha, \tag{A.18}$$

which leads to another definition of an invariant subgroup: the subgroup H is invariant in G if the left and right cosets formed with any element $a \in G$ are the same. This implies that the subgroup H commutes with all elements of the group G.

A group which has no invariant subgroups is said to be simple. Otherwise, a group is said to be semisimple if none of its invariant subgroups are Abelian. All the subgroups of an Abelian group are clearly invariant. There are four special properties for invariant subgroups:

- The product of two cosets of an invariant subgroups is again a coset, i.e., $(aH)(bH) = (ab)H$;
- Multiplying any coset of H by H yields the coset;
- For a coset aH, we find the coset which is its inverse, i.e., $(a^{-1}H)(aH) = a^{-1}aHH = H$;
- The invariant subgroup contains elements of G in complete class, i.e., for any class of G, H contains either all or none of the elements in the class.

When we consider the cosets of H as elements and define product as the result of coset multiplication, the cosets of the invariant subgroup form a group which is called the factor group G/H, whose order is the index of H in G.

Appendix B
Group Representations

Before starting with group representations [138] we sketch a basic definition related with the equivalent matrices. The so-called equivalent matrices mean that the matrix representations of the linear operator T in different bases are equivalent. If a linear operator T is defined in the space L, the mapping S induces a linear operator T' in L':

$$T' = STS^{-1}. \tag{B.1}$$

The operator T' is well defined: S^{-1} takes vectors $\mathbf{x}' \in L'$ into vectors $\mathbf{x} \in L$, operator T transforms vectors $\mathbf{x} \in L$ into vectors $\mathbf{y} \in L$, and finally S takes vectors $\mathbf{y} \in L$ into vectors $\mathbf{y}' \in L'$. The final result is that T' takes vectors $\mathbf{x}' \in L'$ into vectors $\mathbf{y}' \in L'$. This means that the operator T' is the transform of T by the operator S. In the matrices representatives, the S is referred as a similarity transformation.

By mapping an arbitrary group G homomorphically on a group of operators $D(G)$ in the vector space L, we say that the operator group $D(G)$ is a matrix representation of group G in the representation space L. The operator corresponding to the element $R \in G$ is denoted by $D(R)$. Given the elements R and S of group G, then we have

$$D(RS) = D(R)D(S), \tag{B.2}$$

$$D(R^{-1}) = [D(R)]^{-1} \tag{B.3}$$

and

$$D(E) = 1. \tag{B.4}$$

The corresponding matrix representations of group G are given by

$$D_{ij}(E) = \delta_{ij}, \quad i, j \in [1, n], \tag{B.5}$$

$$D_{ij}(RS) = \sum_k D_{ik}(R)D_{kj}(S) \equiv D_{ik}(R)D_{kj}(S). \tag{B.6}$$

Sometimes we use $D_{ij}^{(\mu)}(R)$ to distinguish those different representations.

S.-H. Dong, *Wave Equations in Higher Dimensions*,
DOI 10.1007/978-94-007-1917-0,

If the homomorphic mapping of G on $D(G)$ reduces to an isomorphism, then we call the representation as the faithful representations. Thus, the order of the group of matrices $D(G)$ is equal to the order g of group G.

1 Characters

If the matrices $D(R)$ will be replaced by their transforms by some matrix S, then the matrices

$$D'(R) = SD(R)S^{-1} \tag{B.7}$$

also provide a representation of group G, which is equivalent to the representation $D(R)$. Note that equivalent representations have the same structure even though the matrices look different.

By taking the sum of the diagonal elements of the matrix, we find that the trace of a matrix $D(R)$ is invariant under a transformation of the coordinate axes. The trace is defined by

$$\chi(R) = \sum_j D_{jj}(R). \tag{B.8}$$

We see that equivalent representations have the same set of characters. For convenience, we use $\chi^{(\mu)}(R)$ to denote the character of R in the μ representation. When we describe a group by listing the characters of its elements in a given representation, the same number character is assigned to all the elements in a given class since the conjugate elements in the group G always have the same character. If we label the classes of the group G by $K_i, i \in [1, \nu]$, the representation will be described by the set of characters χ_i, where ν is the number of the classes in G.

2 Construction of Representations

In physics we start not from an abstract group, but from a group of transformations of the configuration space of a physical system. One of our problems is to determine how to go about constructing representations of group G. Another is to see what connection representations have with physics.

For a transformation T belonging to the group of transformations G, the representations can be constructed by $x' = Tx$. Suppose an associated linear operator O_T acting on the functions $\psi(x)$:

$$\psi'(x') \equiv O_T\psi(x') = \psi(x), \quad x' = Tx. \tag{B.9}$$

This means that the transformed function $\psi' \equiv O_T\psi$ takes the same value at the image point x' that the original function ψ had at the object point x. In fact, Eq. (B.9) can also be written as

$$O_T\psi(Tx) = \psi(x), \quad \text{or} \quad O_T\psi(x) = \psi(T^{-1}x). \tag{B.10}$$

Based on this, we shall find that the operators also satisfy the same relations as the group elements, i.e., $O_{ST} = O_S O_T$ and $O_{S^{-1}} = (O_S)^{-1}$. Therefore, if we can find a representation for the operators, we automatically obtain a representation of the group G. Here, we give a useful remark on (B.10). This formula implies that O_T operating on ψ replaces x by $T^{-1}x$. If $O_T\psi$ is identical with ψ, i.e., $O_T\psi(x) \equiv \psi(x)$ so that $\psi(Tx) \equiv \psi(x)$. In this case the function ψ is invariant under the operator O_T or under the transformation T.

Let us give a simple example to show how this method works [138]. Consider the symmetry group C_i with two elements E and I. Given any function $\psi(x)$, from (B.10) we have

$$\psi(x) = O_E\psi(Ex) = O_E\psi(x), \qquad \Psi(-x) = O_E\psi(-Ex) = O_E\psi(-x), \tag{B.11}$$

which means that O_E is an identity operator. Similarly, we have

$$\psi(x) = O_I\psi(Ix) = O_I\psi(-x), \qquad \psi(-x) = O_I\psi(-Ix) = O_I\psi(x), \tag{B.12}$$

which means that the operator I changes the sign of x in ψ.

As a result, $O_E\psi(\pm x)$, $O_I\psi(\pm x)$ can be expressed by linear combinations of $\psi(\pm x)$ as follows:

$$\begin{cases} O_E\psi(x) = \psi(x) + 0\cdot\psi(-x), \\ O_E\psi(-x) = 0\cdot\psi(x) + \psi(-x); \end{cases} \qquad \begin{cases} O_I\psi(x) = 0\cdot\psi(x) + \psi(-x), \\ O_I\psi(-x) = \psi(x) + 0\cdot\psi(-x), \end{cases} \tag{B.13}$$

which implies that the operators transform the functions $\psi(\pm x)$ among themselves. Therefore, by taking $\psi(x) = g_1$, $\psi(-x) = g_2$, we may write

$$O_R g_i = \sum_{k=1}^{2} g_k D_{ki}(R), \quad i = 1, 2. \tag{B.14}$$

When compared with (B.13), we find that

$$D(E) = \begin{pmatrix} 1 & 0 \\ 0 & 1 \end{pmatrix}, \qquad D(I) = \begin{pmatrix} 0 & 1 \\ 1 & 0 \end{pmatrix}, \tag{B.15}$$

which give a two-dimensional representation of the group C_i:

$$I^2 = E, \qquad [D(I)]^2 = \begin{pmatrix} 0 & 1 \\ 1 & 0 \end{pmatrix}\begin{pmatrix} 0 & 1 \\ 1 & 0 \end{pmatrix} = \begin{pmatrix} 1 & 0 \\ 0 & 1 \end{pmatrix} = D(E). \tag{B.16}$$

The general procedure to construct representations can be summarized as follows. We start from any set of linearly independent functions and apply all the operators O_R corresponding to elements R of the transformation group G to each of the functions. Then we get a set of functions which can all be expressed linearly in terms of n of them, say ψ_i, $i \in [1, n]$. In mathematical language, we may express it as

$$O_R\psi_\nu = \sum_{\mu=1}^{n} \psi_\mu D_{\mu\nu}(R), \quad \nu = 1, \ldots, n, \tag{B.17}$$

from which we may get a proper homomorphism of G on $D(G)$, i.e.,

$$D_{\sigma\nu}(SR) = \sum_{\mu=1}^{n} D_{\sigma\mu}(S) D_{\mu\nu}(R). \tag{B.18}$$

3 Reducible and Irreducible Representations

In general, if we may find a basis in which all matrices $D(R)$ of an n-dimensional representation can be brought to the form

$$D(R) = \left[\begin{array}{c|c} D^{(1)}(R) & A^{(1)}(R) \\ \hline 0 & D^{(2)}(R) \end{array}\right], \tag{B.19}$$

where $D^{(1)}(R)$ are m-by-m matrices, the $D^{(2)}(R)$ are $(n-m)$-by-$(n-m)$ matrices, $A^{(1)}(R)$ is a rectangular matrix with m rows and $(n-m)$ columns and 0 represents a matrix with $(n-m)$ rows and m columns all of whose elements are 0, then we may say that the representation $D(R)$ is reducible. Successively, we may transform the basis in the m dimensional space of $D^{(1)}$ and try to bring all matrices $D^{(1)}(R)$ to the following form

$$D^{(1)}(R) = \left[\begin{array}{c|c} D^{(3)}(R) & A^{(2)}(R) \\ \hline 0 & D^{(4)}(R) \end{array}\right], \tag{B.20}$$

where $D^{(3)}(R)$ is q-dimensional, and $D^{(4)}$ is $(m-q)$-dimensional and also apply the same procedure to the matrices $D^{(2)}(R)$. The repeated process clearly comes to an end. Finally, we may obtain k sets of matrices $D^{(1)}(R), \ldots, D^{(k)}(R)$, which are irreducible representations of dimension m_i $(n = \sum_{i=1}^{k} m_i)$.

We give an intrinsic criterion of reducibility as follows: If there exists some subspace of dimension $m < n$ which is invariant under all transformations of the group, the representation is reducible. For example, in the case of (B.19) the subspace of the first m components is invariant:

$$D(R) = \left[\begin{array}{c|c} D^{(1)}(R) & A^{(1)}(R) \\ \hline 0 & D^{(2)}(R) \end{array}\right] \begin{bmatrix} \mathbf{x} \\ \cdots \\ 0 \end{bmatrix} = \begin{bmatrix} D^{(1)}(R)\mathbf{x} \\ \cdots \\ 0 \end{bmatrix}. \tag{B.21}$$

If there is no proper subspace which is invariant, the representation is irreducible. If it is possible to find a basis in which all the matrices of the representation have the following form

$$D(R) = \begin{bmatrix} D^{(1)}(R) & 0 \\ 0 & D^{(2)}(R) \end{bmatrix}, \tag{B.22}$$

then we say that this representation is fully reducible. It can be written as $D = D^{(1)}(R) + D^{(2)}(R)$.

4 Schur's Lemmas

There are two main purposes of these lemmas. The first is to find some simpler criteria for irreducibility. The second is to give some restrictions on the number of non-equivalent representations.

Lemma 1 *If $D(R)$ and $D'(R)$ are two irreducible representations of a group G, having different dimensions, then if the matrix A satisfies*

$$D(R)A = AD'(R) \tag{B.23}$$

for all $R \in G$, it follows that $A = 0$.

Lemma 1.1 *If $D(R)$ and $D'(R)$ are irreducible representations of a group G having the same dimensions, and if the matrix A satisfies*

$$D(R)A = AD'(R) \tag{B.24}$$

for all $R \in G$, then either $D(R)$ and $D'(R)$ are equivalent or $A = 0$.

Lemma 2 *If the matrices $D(R)$ are an irreducible representation of a group G, and if*

$$AD(R) = D(R)A \tag{B.25}$$

for all $R \in G$, then $A = \text{constant} \cdot \mathbf{1}$.

That is to say, if a matrix commutes with all the matrices of an irreducible representation, the matrix must be a multiple of the unit matrix **1**.

5 Criteria for Irreducibility

Let us consider an arbitrary representation $D(R)$, which can be expressed in terms of irreducible representations as

$$D(R) = \sum_{\mu} a_{\mu} D^{(\mu)}(R), \tag{B.26}$$

where μ are integers. By taking its trace, one gets the compound character

$$\chi_j = \sum_{\mu} a_{\mu} \chi_j^{\mu}. \tag{B.27}$$

We see that the compound character is a linear combination of simple characters with positive integral coefficients calculated by

$$a_{\mu} = \frac{1}{g} \sum_{j} g_j \chi_j^{(\mu)*} \chi_j. \tag{B.28}$$

On the other hand, based on (B.27) we have

$$\sum_j \chi_j \chi_j^* g_j = \sum_{\mu,\nu} a_\mu a_\nu \sum_j g_j \chi_j^{(\nu)} \chi_j^{(\mu)*} = g \sum_\nu a_\nu^2. \tag{B.29}$$

In particular, if the representation is irreducible, all the coefficients $a_\nu = 0$ except for one which is unity. Therefore, if the original representation is irreducible, its characters must satisfy the following relation

$$\sum_j g_j |\chi_j|^2 = g. \tag{B.30}$$

This gives us a simple criterion for irreducibility.

It should be pointed out that (B.29) and (B.30) are extremely useful tools for calculating the character χ_j. If by some means we find a representation of the group G, we may calculate the compound character χ and evaluate

$$\frac{1}{g} \sum_j g_j |\chi_j|^2 = \sum_\nu a_\nu^2. \tag{B.31}$$

If this quantity is unity, then the representation must be irreducible. In practice, we may use the orthogonality relations for the characters to calculate them. That is, the scalar product with weight factors g_j of any two rows or any two columns is equal to zero

$$\sum_j \chi_j^{(\mu)} \chi_j^{(\nu)*} g_j = g \delta_{\mu\nu}, \qquad \sum_{\mu=1}^{r} \chi_i^{(\mu)} \chi_j^{(\mu)*} = \frac{g}{g_j} \delta_{ij}. \tag{B.32}$$

General theorem *The number of nonequivalent irreducible representations of a group G is equal to the number of classes in the group.*

6 Expansion of Functions in Basis Functions of Irreducible Representations

As we know, any function ψ can be expressed as a sum of functions which can act as base functions in the various irreducible representations

$$\psi = \sum_\mu \sum_{j=1}^{n_\mu} \psi_j^{(\mu)}, \tag{B.33}$$

where the base functions for the μth irreducible unitary representation satisfy the equations

$$O_R \psi_j^{(\mu)} = \sum_k \psi_k^{(\mu)} D_{kj}^{(\mu)}(R). \tag{B.34}$$

We now try to find the condition that a given function must satisfy in order that it may belong to the ith row of a given representation. It is found that a set of functions

$$\psi_l^{(\mu)} = \frac{n_\mu}{g} \sum_R D_{lm}^{(\mu)*}(R) O_R \psi_m^{(\mu)} \tag{B.35}$$

satisfy (B.34) and become a sufficient and necessary condition to make the given function belong to the ith row of a given representation. That is to say, they form a basis for the μth irreducible representation.

We now return to (B.33) and ask how to find the $\psi_j^{(\mu)}$ for a given function ψ, i.e., how do we resolve the given function into a sum of functions, each of which belongs to a particular row of some irreducible representation? This can be realized by introducing a projection operator

$$P_i^{(\nu)} = \frac{n_\nu}{g} \sum_R D_{ii}^{(\nu)*}(R) O_R \tag{B.36}$$

with property

$$P_i^{(\nu)} \psi_j^{(\mu)} = \psi_i^{(\nu)} \delta_{\nu\mu} \delta_{ij}. \tag{B.37}$$

Applying it to (B.33), we obtain the following result

$$\psi_i^{(\nu)} = \frac{n_\nu}{g} \sum_R D_{ii}^{(\nu)*}(R) O_R \psi. \tag{B.38}$$

In analogy to the above, we say that a function belongs to the μth irreducible representation if it is a sum of functions belonging to the various rows of that representation, i.e.,

$$\psi^{(\mu)} = \sum_{i=1}^{n_\mu} \psi_i^{(\mu)}. \tag{B.39}$$

If we sum over i from 1 to n_μ we find that the projection operator becomes

$$P^{(\nu)} = \frac{n_\nu}{g} \sum_R \chi^{(\nu)*}(R) O_R \tag{B.40}$$

with the property

$$P^{(\nu)} \psi^{(\mu)} = \psi^{(\nu)} \delta_{\nu\mu}. \tag{B.41}$$

Therefore, we have

$$\psi = \sum_\mu \psi^{(\mu)}, \quad \psi^{(\mu)} = P^{(\mu)} \psi. \tag{B.42}$$

Appendix C
Fundamental Properties of Lie Groups and Lie Algebras

In this Appendix we are ready to outline some fundamental properties of Lie groups and Lie algebras since they are very helpful for studying quantum systems with central physical potentials.

Up to now, we have dealt exclusively with finite groups. However, the symmetry groups for many physical systems consist of an infinite rather than a finite number of elements. Thus physical problems require that we examine the theory of the representations of groups with an infinite number of elements.

The infinite groups include two main types such as the infinite discrete and continuous groups [138]. The elements of infinite discrete group, R_a, are labeled by a subscript a which runs through the integers $1, 2, \ldots, \infty$. The group manifold is a countable set of "point" R_a. As what follows, we shall pay more attention to continuous groups since the Lie groups are closely related to them.

1 Continuous Groups

We say the group is continuous if some generalized definition of "nearness" or continuity is imposed on the elements of the group manifold. For instance, the set of transformations $x' = ax + b$ forms a group, where two parameters $a, b \in (-\infty, \infty)$. We say that such a group is a two-parameter continuous group. Generally, an r-parameter continuous group has its elements labeled by r continuously varying real parameters $a_i, i \in [1, r]$ so that the elements of the group are $R(a_1, \ldots, a_r) = R(a)$. The continuity is expressed in terms of distances in the parameter space. Two group elements $R(a)$ and $R(a')$ are "near" to each other if the distance $[\sum_{i=1}^{r}(a_i - a_i')^2]^{1/2}$ is very small.

The requirements that the elements $R(a)$ form a continuous group are the same as for finite groups. There must be a set of parameter values a^0 such that

$$\begin{aligned} &R(a^0)R(a) = R(a)R(a^0) = R(a), \\ &R(\bar{a})R(a) = R(a)R(\bar{a}) = R(0), \\ &R(\bar{a}) = [R(a)]^{-1}, \qquad R(c) = R(b)R(a), \end{aligned} \tag{C.1}$$

S.-H. Dong, *Wave Equations in Higher Dimensions*,
DOI 10.1007/978-94-007-1917-0,

where $R(a^0)$ is the identity element of the group, and $R(\bar{a})$ is the element inverse to $R(a)$. The c are real functions of the real parameters a and b

$$c_j = \phi_j(a_1, \ldots, a_r; b_1, \ldots, b_r), \quad j \in [1, r]. \tag{C.2}$$

Even though the requirements are the same as for finite groups, we require that the parameters of a product be analytic functions of the parameters of the factors; i.e., the function in (C.1) shall possess derivatives of all orders with respect to both variables. Also, we require that the $\bar{a}$ in $R(\bar{a})$ be analytic functions of the a. We then get an r-parameter Lie group.

When we say that we have an r-parameter group, this implies that the r parameters are essential. An r-parameter Lie group of transformations is a group of transformations

$$x'_i = g_i(x_1, \ldots, x_n; a_1, \ldots, a_r), \quad i \in [1, n], \tag{C.3}$$

or symbolically

$$x' = g(x; a). \tag{C.4}$$

If the parameters are not essential, then there exist parameter values $a_1 + \varepsilon_1$, $\ldots, a_r + \varepsilon_r$, where the ε's are arbitrarily small quantities which are functions of the a_i with

$$g_i(x; a) = g_i(x; a + \varepsilon) \tag{C.5}$$

for all values of the argument x. The necessary and sufficient condition for the r parameters a_i to be essential is that it shall be impossible to find r functions $\chi_k(a)$ which satisfy

$$\sum_{k=1}^{r} \chi_k(a) \frac{\partial g_i(x; a)}{\partial a_k} = 0, \quad \text{for all } x \text{ and } a, \ i \in [1, n]. \tag{C.6}$$

Starting with (C.4), we can differentiate the x''s with respect to the x's and obtain a set of equations from which the finite set of parameters a can be eliminated. We shall then be left with a finite set of partial differential equations for the x''s which no longer contain any arbitrary elements. Moreover, the general solution of this set of partial differential equations will rely on just r arbitrary constants. We say that the group is finite and continuous. Otherwise, we get an infinite continuous group.

Here we give several typical examples of Lie groups. The first is the one-parameter Abelian group $x' = ax, a \neq 0$. Its identity element is $a = 1$, and inverse element $\bar{a} = 1/a$. Its product element is $c = ba$. The second is the orthogonal group in two dimensions O(2). This group is concerned with those transformations which leave $x^2 + y^2$ invariant. This invariance condition

$$\begin{aligned} &x'^2 + y'^2 = (a_1 x + a_2 y)^2 + (a_3 x + a_4 y)^2 = x^2 + y^2, \\ &a_1^2 + a_3^2 = 1, \qquad a_2^2 + a_4^2 = 1, \qquad a_1 a_2 + a_3 a_4 = 0 \end{aligned} \tag{C.7}$$

imposes three conditions on the four parameters. Thus we have a one-parameter group written as

$$\begin{pmatrix} x' \\ y' \end{pmatrix} = \begin{pmatrix} \cos\varphi & -\sin\varphi \\ \sin\varphi & \cos\varphi \end{pmatrix} \begin{pmatrix} x \\ y \end{pmatrix}, \tag{C.8}$$

where $\varphi \in [0, 2\pi]$ is the angle of rotation about the z-axis. This group is Abelian. The angle of the resultant of two transformations is the sum of the angles of the individual transformations. Finally, we consider the orthogonal group in n dimensions O(n). This group leaves $\sum_{i=1}^{n} x_i^2$ invariant. Thus, we impose $[n + n(n-1)/2]$ conditions on the n^2 parameters, which leave us with $[n(n-1)/2]$ essential parameters.

So far we have considered only real transformations of real variables. As what follows, let us consider a typical example with complex variables. It is the unitary group in two dimensions U(2)

$$\begin{pmatrix} x_1' \\ x_2' \end{pmatrix} = \begin{pmatrix} a_{11} & a_{12} \\ a_{21} & a_{22} \end{pmatrix} \begin{pmatrix} x_1 \\ x_2 \end{pmatrix}, \tag{C.9}$$

where x, a are complex and $\det A \neq 0$. This group is in fact a unitary group with $AA^\dagger = 1$. Similarly, the unitary group in n dimensions U(n), i.e., $r' = Ar$, $AA^\dagger = 1$, imposes $[n + 2n(n-1)/2]$ conditions on the $2n^2$ real parameters, leaving us with n^2 essential real parameters.

2 Infinitesimal Transformations and Lie Algebra

The transformation $x' = f(x; a)$ takes all points of the space from their initial positions x to final positions x'. It is natural to introduce the concept of infinitesimal transformations when we consider the gradual shift of the points of the space. It can also be proved that a one-parameter continuous group is equivalent to a group of translations and must be Abelian. Due to the infinitesimal transformations, we introduce the infinitesimal operators of the group as follows

$$X_j = \sum_{i=1}^{n} w_{ij}(x) \frac{\partial}{\partial x_i}. \tag{C.10}$$

The operator $\mathbf{1} + \sum_j X_j \delta a_j$ is close to the identity operator. On the other hand, the infinitesimal operators X_j have the property that their commutators satisfy

$$[X_\alpha, X_\beta] = w_{i\alpha} \frac{\partial}{\partial x_i} w_{j\beta} \frac{\partial}{\partial x_j} - w_{j\beta} \frac{\partial}{\partial x_j} w_{i\alpha} \frac{\partial}{\partial x_i} = c_{\alpha\beta}^{\gamma} X_\gamma, \tag{C.11}$$

where $c_{\alpha\beta}^{\gamma}$ are called the structure constants of the Lie group with the property $c_{\alpha\beta}^{\gamma} = -c_{\beta\alpha}^{\gamma}$.

Substituting (C.11) into the Jacobi identity

$$[[X_\alpha, X_\beta], X_\gamma] + [[X_\beta, X_\gamma], X_\alpha] + [[X_\gamma, X_\alpha], X_\beta] = 0, \tag{C.12}$$

we find

$$c_{\alpha\beta}^{\mu} c_{\mu\gamma}^{\nu} + c_{\beta\gamma}^{\mu} c_{\mu\alpha}^{\nu} + c_{\gamma\alpha}^{\mu} c_{\mu\beta}^{\nu} = 0. \tag{C.13}$$

Based on this, we have

$$x_i' = \left[1 + \sum_j X_j \delta a_j\right] x_i = x_i + \sum_j w_{ij}(x) \delta a_j. \tag{C.14}$$

Consider an example in order to show how to define the infinitesimal operators. For example, the orthogonal transformations are characterized by the statement that the transpose A^T of the matrix A of the transformation is its inverse $AA^T = \mathbf{1}$. The proper rotations have determinant unity so that the infinitesimal rotations have a matrix of the form $A = \mathbf{1} + B$ where B has all its elements in the neighborhood of zero. The orthogonality condition requires $\mathbf{1} = A\,A^T = (\mathbf{1} + B)(\mathbf{1} + B^T) \sim \mathbf{1} + B + B^T$, which leads to $B + B^T = 0$. Therefore B has a skew-symmetric matrix with the form

$$B = \begin{bmatrix} 0 & \zeta & -\eta \\ -\zeta & 0 & \xi \\ \eta & -\xi & 0 \end{bmatrix}, \tag{C.15}$$

where ξ, η and ζ are constants.

The infinitesimal rotations are given by

$$dx = \zeta y - \eta z, \qquad dy = -\zeta x + \xi z, \qquad dz = \eta x - \xi y, \tag{C.16}$$

and the infinitesimal operators are

$$\begin{aligned} X_1 &= z\frac{\partial}{\partial y} - y\frac{\partial}{\partial z}, \\ X_2 &= x\frac{\partial}{\partial z} - z\frac{\partial}{\partial x}, \\ X_3 &= y\frac{\partial}{\partial x} - x\frac{\partial}{\partial y}, \end{aligned} \tag{C.17}$$

which correspond to the angular momentum operators for three coordinate directions and satisfy familiar commutation relations

$$[X_1, X_2] = X_3, \qquad [X_2, X_3] = X_1, \qquad [X_3, X_1] = X_2. \tag{C.18}$$

This means that the commutators of the infinitesimal operators are linearly expressible in terms of the infinitesimal operators. If we set $iX_\mu = J_\mu$, then we have

$$[J_1, J_2] = iJ_3, \qquad [J_2, J_3] = iJ_1, \qquad [J_3, J_1] = iJ_2. \tag{C.19}$$

We have known that for an r-parameter transformation group there are r linearly independent infinitesimal operators X_j. Linear combinations of these quantities can be formed to give an r-dimensional vector space. If we are considering problems concerned with the structure of the Lie group, we should take only linear combinations with real coefficients. The r-parameter Lie group has associated with it a real r-dimensional vector space of quantities $\sum_j a_j X_j$ which is closed under a multiplication defined by (C.11). This is the Lie algebra of the Lie group. To every Lie algebra there corresponds to a Lie group; the structure constants determine the Lie group locally. To deduce all possible structures of Lie algebra is a formidable mathematical problem. We do not address it here and suggest the reader refer to Ref. [138].

3 Structure of Compact Semisimple Lie Groups and Their Algebras

When we use the term "closed" to describe Lie groups whose parameters vary over a finite range, the group manifold itself is then said to be compact. Generally speaking, a set M is compact if every infinite subset of M contains a sequence which converges to an element of M. The Lie algebra of a compact Lie group is also said to be compact.

Cartan theorem *The necessary and sufficient condition for a Lie algebra to be semisimple is that* $\det g_{\mu\nu} \neq 0$, *where* $g_{\mu\nu} = c^{\beta}_{\mu\alpha} c^{\alpha}_{\nu\beta}$.

A necessary and sufficient condition for a semisimple Lie algebra to be compact is that the matrix $g_{\mu\nu}$ be negative definite. Finally, note that for a compact semisimple Lie algebra we can choose the basis so that $g_{\mu\nu} = -\delta_{\mu\nu}$. The actual analysis of the structure of compact semisimple Lie algebras is beyond the scope of this Appendix.

4 Irreducible Representations of Lie Groups and Lie Algebras

For simplicity, we restrict ourselves to compact Lie groups in order to prove that every representation of a compact is equivalent to a unitary representation and can be fully reducible to a sum of irreducible representations, all of which have finite dimensions. The regular representation contains all irreducible representations.

Theorem *For semisimple Lie groups, every representation of finite degree is fully reducible.*

Casimir theorem *The operator* $C = g^{ij} X_i X_j$ *commutes with all operators of the representation and is a multiple of the unit operator.*

For compact groups $g_{ij} = -\delta_{jj}$ in a suitable basis, so that the Casimir operator becomes

$$C = \sum_j X_j^2. \tag{C.20}$$

For rotation group O(3) with the operators J_μ where $\mu = 1, 2, 3$, the Casimir operator is calculated as

$$C = \sum_i J_i^2, \quad i = 1, 2, 3. \tag{C.21}$$

The number of operators which are required to give a complete set is equal to the rank of the studied algebra. This can be defined as follows: for any element A, we look for all independent solutions of the equation $[A, X] = 0$, which always has at least the one solution $X = A$. We now vary A to reduce the number of independent solutions of this equation to a minimum number l called the rank of the algebra. The l operators of the Casimir type are required to characterize an irreducible representation.

Before ending this Appendix, let us review a few useful commutator identities to simplify the commutation relations. We often use the following well-known relations

$$[AB, C] = A[B, C] + [A, C]B, \qquad [A, BC] = B[A, C] + [A, B]C \tag{C.22}$$

to simplify the commutators involving the products of the operators. On the other hand, it is shown that for arbitrary polynomial functions r, t of the indicated operators,

$$r(A)[t(A), B] = [t(A), r(A)B], \qquad [A, t(B)]r(B) = [Ar(B), t(B)] \tag{C.23}$$

can be effectively used to move the operators in and out of the commutators. In particular, we find that the following formula is very useful

$$[A, B^m] = \sum_{i=0}^{m-1} B^i [A, B] B^{m-i-1}, \tag{C.24}$$

which can be proved by induction on m.

In addition, we have to calculate the operator transformations of the form $e^{-B} A e^B$ in scaling transformations. For this purpose, let us define

$$g(\alpha) = e^{-\alpha B} A e^{\alpha B}. \tag{C.25}$$

By differentiating, we have

$$g(\alpha)' = e^{-\alpha B}[A, B] e^{\alpha B}, \qquad g(\alpha)'' = e^{-\alpha B}[[A, B], B] e^{\alpha B} \tag{C.26}$$

and in general

$$g^m(\alpha) = e^{-\alpha B}[\ldots[[A, B], B], \ldots, B] e^{\alpha B}, \tag{C.27}$$

where the multiple commutator contains B exactly m times. Let us expand the $g(\alpha)$ in Taylor series

$$g(\alpha) = \sum_{m=0}^{\infty} \frac{1}{m!} g^m(0) \alpha^m. \tag{C.28}$$

If setting $\alpha = 1$, we get the following identity

$$e^{-B} A e^B = A + [A, B] + \frac{1}{2!}[[A, B], B] + \frac{1}{3!}[[[A, B], B], B] + \cdots, \tag{C.29}$$

which can be used to obtain some scaling transformations.

Appendix D
Angular Momentum Operators in Spherical Coordinates

In this Appendix we give a brief review of the angular momentum operators in spherical coordinates [462, 516] due to its wide applications in angular momentum theory.

It is well known that the expressions for the components of the angular momentum operator $\mathbf{L} = \mathbf{r} \times \mathbf{p}$ in Cartesian coordinates can be easily written as

$$L_x = yp_z - zp_y = -i\hbar\left(y\frac{\partial}{\partial z} - z\frac{\partial}{\partial y}\right), \tag{D.1}$$

$$L_y = zp_x - xp_z = -i\hbar\left(z\frac{\partial}{\partial x} - x\frac{\partial}{\partial z}\right), \tag{D.2}$$

$$L_z = xp_y - yp_x = -i\hbar\left(x\frac{\partial}{\partial y} - y\frac{\partial}{\partial x}\right), \tag{D.3}$$

from which we can obtain the following commutation relation

$$[L_i, L_j] = i\hbar\varepsilon_{ijk}L_k, \tag{D.4}$$

where the indices i, j, k can be x, y, or z, and the coefficient ε_{ijk} is named Levi-Civita symbol (3.26). Thus we establish the commutation relations for the components of the angular momentum of a spinless particle.

We now want to find the form of these operators in spherical coordinates. By using the following transformation equations

$$x = r\sin\theta\cos\varphi, \qquad y = r\sin\theta\sin\varphi, \qquad z = r\cos\theta, \tag{D.5}$$

then we have

$$r = \sqrt{x^2 + y^2 + z^2}, \qquad \cos\theta = \frac{z}{r}, \qquad \tan\varphi = \frac{y}{x}, \tag{D.6}$$

where $r \in [0, \infty)$, $\theta \in [0, \pi]$, $\varphi \in [0, 2\pi]$.

S.-H. Dong, *Wave Equations in Higher Dimensions*,
DOI 10.1007/978-94-007-1917-0, © Springer Science+Business Media B.V. 2011

Consequently, we have

$$\begin{aligned}
&\frac{\partial r}{\partial x}=\sin\theta\cos\varphi, && \frac{\partial r}{\partial y}=\sin\theta\sin\varphi, && \frac{\partial r}{\partial z}=\cos\theta,\\
&\frac{\partial \theta}{\partial x}=\frac{\cos\theta\cos\varphi}{r}, && \frac{\partial \theta}{\partial y}=\frac{\cos\theta\sin\varphi}{r}, && \frac{\partial \theta}{\partial z}=-\frac{\sin\theta}{r},\\
&\frac{\partial \varphi}{\partial x}=-\frac{\sin\varphi}{r\sin\theta}, && \frac{\partial \varphi}{\partial y}=\frac{\cos\varphi}{r\sin\theta}, && \frac{\partial \varphi}{\partial z}=0.
\end{aligned} \tag{D.7}$$

Making use of these relations, we are able to obtain the following expressions

$$\begin{aligned}
L_x &= i\hbar\left(\sin\varphi\frac{\partial}{\partial\theta}+\frac{\cos\theta}{\sin\theta}\cos\varphi\frac{\partial}{\partial\varphi}\right),\\
L_y &= -i\hbar\left(\cos\varphi\frac{\partial}{\partial\theta}-\frac{\cos\theta}{\sin\theta}\sin\varphi\frac{\partial}{\partial\varphi}\right),\\
L_z &= -i\hbar\frac{\partial}{\partial\varphi},
\end{aligned} \tag{D.8}$$

from which we have

$$L^2=L_x^2+L_y^2+L_z^2=-\hbar^2\left\{\frac{1}{\sin\theta}\frac{\partial}{\partial\theta}\left(\sin\theta\frac{\partial}{\partial\theta}\right)+\frac{1}{\sin^2\theta}\frac{\partial^2}{\partial\varphi^2}\right\}, \tag{D.9}$$

with the following properties

$$[L^2, L_i]=0, \quad i=x,y,z. \tag{D.10}$$

Instead of the operators L_x and L_y, we often use the linear combinations of the operators L_x and L_y as

$$L_\pm = L_x \pm iL_y = \hbar e^{\pm i\varphi}\left[\pm\frac{\partial}{\partial\theta}+i\cot\theta\frac{\partial}{\partial\varphi}\right], \tag{D.11}$$

which are called the ladder operators for the magnetic quantum number m but leave the angular momentum quantum number l unchanged, that is,

$$\begin{aligned}
L_+Y_l^m(\theta,\varphi) &= \hbar\sqrt{l(l+1)-m(m+1)}\,Y_l^{m+1}(\theta,\varphi),\\
L_-Y_l^m(\theta,\varphi) &= \hbar\sqrt{l(l+1)-m(m-1)}\,Y_l^{m-1}(\theta,\varphi),
\end{aligned} \tag{D.12}$$

where $Y_l^m(\theta,\varphi)$ is the spherical harmonics and becomes the common eigenfunction of L^2 and L_z. They correspond to the eigenvalues $l(l+1)\hbar^2$ and $m\hbar$, respectively

$$\begin{aligned}
L^2Y_l^m(\theta,\varphi) &= l(l+1)\hbar^2Y_l^m(\theta,\varphi),\\
L_zY_l^m(\theta,\varphi) &= m\hbar Y_l^m(\theta,\varphi).
\end{aligned} \tag{D.13}$$

If we use the properties of the ladder operators $L_\pm$, then we may express the spherical harmonics as

$$\begin{aligned}
Y_l^m(\theta,\varphi) &= \sqrt{\frac{(l+m)!}{(2l)!(l-m)!}}(L_-)^{l-m}Y_l^l(\theta,\varphi),\\
Y_l^m(\theta,\varphi) &= \sqrt{\frac{(l-m)!}{(2l)!(l+m)!}}(L_+)^{l+m}Y_l^{-l}(\theta,\varphi).
\end{aligned} \tag{D.14}$$

It is well known that these ladder operators $L_\pm$ together with the operator L_z satisfy the following commutation relations

$$[L_z, L_\pm] = \pm L_\pm, \qquad [L_+, L_-] = 2L_z, \tag{D.15}$$

which correspond to an SU(2) group. On the other hand, it is not difficult to see that

$$\begin{aligned} L^2 &= L_+L_- + L_z^2 - L_z \\ &= L_-L_+ + L_z^2 + L_z. \end{aligned} \tag{D.16}$$

Finally, let us calculate the mean values and root-mean-square deviations of the operators L_x and L_y in the state $Y_l^m(\theta, \varphi)$ in terms of the ladder operators $L_\pm$. By inverting formulas (D.11), we have

$$L_x = \frac{1}{2}(L_+ + L_-), \qquad L_y = \frac{1}{2}(L_+ - L_-). \tag{D.17}$$

Thus we see that $L_x Y_l^m(\theta, \varphi)$ and $L_y Y_l^m(\theta, \varphi)$ are linear combinations of the states $Y_l^{m+1}(\theta, \varphi)$ and $Y_l^{m-1}(\theta, \varphi)$. As a result, we are able to obtain their mean values as follows:

$$\begin{aligned} \langle L_x \rangle &= \langle L_y \rangle = 0, \\ \langle L_x^2 \rangle &= \langle L_y^2 \rangle = \frac{\hbar^2}{2}[l(l+1) - m^2], \\ \Delta L_x &= \Delta L_y = \frac{\hbar}{\sqrt{2}}\sqrt{l(l+1) - m^2}. \end{aligned} \tag{D.18}$$

It is well known that these ladder operators $L_\pm$ together with the operator L_z satisfy the following commutation relations:

[illegible]

Appendix E
Confluent Hypergeometric Functions

In this Appendix we shall give some basic properties of the confluent hypergeometric functions for the sake of easy reference. It is well known that the confluent hypergeometric functions are defined by the series

$$ {}_1F_1(\alpha;\beta;z) = 1 + \frac{\alpha}{\beta}\frac{z}{1!} + \frac{\alpha(\alpha+1)}{\beta(\beta+1)}\frac{z^2}{2!} + \frac{\alpha(\alpha+1)(\alpha+2)}{\beta(\beta+1)(\beta+2)}\frac{z^3}{3!} + \cdots, \tag{E.1} $$

where the parameter α is arbitrary, but the parameter β is supposed not zero or a negative integer.

The confluent hypergeometric functions ${}_1F_1(\alpha;\beta;z)$ satisfy the differential equation

$$ z\frac{d^2\phi}{dz^2} + (\beta - z)\frac{d\phi}{dz} - \alpha\phi = 0. \tag{E.2} $$

Substitution of $\phi = z^{1-\beta}\phi_1$ into (E.2) leads to

$$ z\frac{d^2\phi_1}{dz^2} + (2-\beta-z)\frac{d\phi_1}{dz} - (\alpha-\beta+1)\phi_1 = 0. \tag{E.3} $$

Therefore, the general solution of (E.2) can be expressed as

$$ \phi = a_1 {}_1F_1(\alpha;\beta;z) + a_2 z^{1-\beta} {}_1F_1(\alpha-\beta+1;2-\beta;z), \tag{E.4} $$

which implies that the second term, unlike the first, has a singular point at $z=0$. It should be noted that (E.2) is of Laplace's type and its solutions can be represented as contour integrals as shown by Landau *et al.* [2]. The function ${}_1F_1(\alpha;\beta;z)$ is regular at $z=0$ and has the value 1 there; it satisfies the following relationship

$$ {}_1F_1(\alpha;\beta;z) = e^z {}_1F_1(\beta-\alpha;\beta;-z), \tag{E.5} $$

which is called the Kummer transformation. On the other hand, this function ${}_1F_1(\alpha;\beta;z)$ also satisfies the following recurrence relations [462]

$$ \begin{aligned} &(\beta-\alpha){}_1F_1(\alpha-1;\beta;z) + (2\alpha-\beta+z){}_1F_1(\alpha;\beta;z) = \alpha {}_1F_1(\alpha+1;\beta;z), \\ &(\alpha-\beta+1){}_1F_1(\alpha;\beta;z) + (\beta-1){}_1F_1(\alpha;\beta-1;z) = \alpha {}_1F_1(\alpha+1;\beta;z), \\ &\frac{d}{dz}{}_1F_1(\alpha;\beta;z) = \frac{\alpha}{\beta}{}_1F_1(\alpha+1;\beta+1;z). \end{aligned} \tag{E.6} $$

S.-H. Dong, *Wave Equations in Higher Dimensions*,
DOI 10.1007/978-94-007-1917-0,

By successive applications of (E.6), we have

$$\frac{d^n}{dz^n}{}_1F_1(\alpha;\beta;z) = \frac{\Gamma(\beta)\Gamma(\alpha+n)}{\Gamma(\alpha)\Gamma(\beta+n)}{}_1F_1(\alpha+n;\beta+n;z), \tag{E.7}$$

where $\Gamma(x)$ is the Gamma function.

On the other hand, the polynomials ${}_1F_1(-n;m;z)(m \in [0,n])$ are associated with the generalized Laguerre polynomials defined by

$$\begin{aligned} L_n^m(z) &= (-1)^m \frac{(n!)^2}{m!(n-m)!}\,{}_1F_1(-[n-m];m+1;z) \\ &= \frac{n!}{(n-m)!} e^z \frac{d^n}{dz^n}(e^{-z}z^{n-m}) \\ &= (-1)^m \frac{n!}{(n-m)!} e^z z^{-m} \frac{d^{n-m}}{dz^{n-m}}(e^{-z}z^n), \end{aligned} \tag{E.8}$$

from which we may obtain the following relation [192]

$$L_n^m(z) = \frac{\Gamma(n+m+1)}{\Gamma(n+m)}\,{}_1F_1(-n;m+1;z). \tag{E.9}$$

When $m=0$, the polynomials $L_n^m(z)$ are called simply Laguerre polynomials

$$L_n(z) = e^z \frac{d^n}{dz^n}(e^{-z}z^n). \tag{E.10}$$

Based on Eqs. (E.9) and (E.10), we have

$$L_n(z) = \frac{\Gamma(n+1)}{\Gamma(n)}\,{}_1F_1(-n;1;z). \tag{E.11}$$

We are now in the position to indicate the asymptotic behavior of the confluent hypergeometric functions. For small values of z, the asymptotic value of the function ${}_1F_1(\alpha;\beta;z)$ is given immediately by the first terms of the series (E.1). For large values of the $|z|$, we have

$$\begin{aligned} {}_1F_1(\alpha;\beta;z) &= \frac{\Gamma(\beta)}{\Gamma(\alpha)} z^{\alpha-\beta} e^z [1+\mathcal{O}(|z|^{-1})], \qquad \Re e\, z \to \infty, \\ {}_1F_1(\alpha;\beta;z) &= \frac{\Gamma(\beta)}{\Gamma(\beta-\alpha)} (-z)^{-\alpha} [1+\mathcal{O}(|z|^{-1})], \quad \Re e\, z \to -\infty. \end{aligned} \tag{E.12}$$

On the other hand, for bounded values of z and infinitely large values of one of the parameters we have more asymptotic values of the confluent hypergeometric functions ${}_1F_1(\alpha;\beta;z)$

$$\begin{aligned} {}_1F_1(\alpha;\beta;z) &= 1+\mathcal{O}(|\beta|^{-1}), && \text{if } z \text{ and } \alpha \text{ are bounded}, && \beta \to \infty, \\ {}_1F_1(\alpha;\beta;z) &= e^z(1+\mathcal{O}(|\beta|^{-1})), && \text{if } \beta-\alpha \text{ and } z \text{ are bounded}, && \beta \to \infty. \end{aligned} \tag{E.13}$$

It should be noted that the great significance of the confluent hypergeometric functions in physics is connected with the fact that the solutions of many homogeneous differential equations can be expressed in terms of this function. For example, consider the following second-order differential equation

$$(\alpha_0 y+\beta_0)\frac{d^2\phi}{dy^2}+(\alpha_1 y+\beta_1)\frac{d\phi}{dy}+(\alpha_2 y+\beta_2)\phi=0. \tag{E.14}$$

By taking the following substitution

$$\phi=e^{\nu y}\psi,\quad y=\lambda z+\eta, \tag{E.15}$$

we may transform (E.14) into the form

$$(a_0 z+b_0)\frac{d^2\psi}{dz^2}+(a_1 z+b_1)\frac{d\psi}{dz}+(a_2 z+b_2)\psi=0, \tag{E.16}$$

where

$$\begin{aligned}
&a_0=\frac{\alpha_0}{\lambda}, && a_1=A_1, && a_2=\lambda A_2,\\
&b_0=\frac{\alpha_0\eta+\beta_0}{\lambda^2}, && b_1=\frac{\eta A_1+B_1}{\lambda}, && b_2=\eta A_2+B_2,\\
&A_1=2\alpha_0\nu+\alpha_1, && A_2=\alpha_0\nu^2+\alpha_1\nu+\alpha_2,\\
&B_1=\beta_1+2\beta_0\nu, && B_2=\beta_0\nu^2+b_1\nu+b_2.
\end{aligned} \tag{E.17}$$

If we define the parameters ν, η and λ so that

$$\alpha_0\eta+\beta_0=0,\qquad \alpha_0=-\lambda A_1,\qquad A_2=0, \tag{E.18}$$

then (E.16) coincides with (E.2). Therefore, if we select values of the parameters ν, η and λ, which satisfy the constraint condition (E.18), and then use the transformation (E.15), then we find that an arbitrary equation of the type (E.14) reduces to the equation for the confluent hypergeometric functions (E.2).

By substituting

$$\psi=z^{-\beta/2}e^{z/2}w,\qquad \alpha=1/2-k+\eta,\qquad \beta=1+2\eta, \tag{E.19}$$

we may transform (E.2) into the Whittaker equation

$$\frac{d^2w}{dz^2}+\left(-1/4+\frac{k}{z}+\frac{1/4-\eta^2}{z^2}\right)w=0, \tag{E.20}$$

whose solution is denoted by $W_{k,\eta}$. The connection between the confluent hypergeometric functions ${}_1F_1(\alpha;\beta;z)$ and the Whittaker function $W_{k,\eta}$ is defined by (E.19). Many mathematical functions which are used in physics can be expressed in terms of the Whittaker function. The asymptotic value of the Whittaker functions for large values of z and $|\arg z|<\pi$ is given by

$$W_{k,\eta}=e^{-z/2}z^k(1+\mathcal{O}(z^{-1})). \tag{E.21}$$

neous differential equations can be expressed in terms of this function. For example, consider the following second-order differential equation

References

1. Schiff, L.I.: Quantum Mechanics, 3rd edn. McGraw-Hill, New York (1955)
2. Landau, L.D., Lifshitz, E.M.: Quantum Mechanics-Nonrelativistic Theory, 3rd edn. Pergamon, New York (1977)
3. Dong, S.H.: Factorization Method in Quantum Mechanics. Springer, Netherlands (2007)
4. Bagrov, V.G., Gitman, D.M.: Exact Solutions of Relativistic Wave Equations. Kluwer Academic, Dordrecht (1990)
5. ter Haar, D.: Problems in Quantum Mechanics, 3rd edn. Pion, London (1975)
6. Flügge, S.: Practical Quantum Mechanics. Springer, Berlin (1971)
7. Appelquist, T., Chodos, A., Freund, P.: Modern Kaluza-Klein Theories. Addison-Wesley, Reading (1987)
8. Nordstrom, G.: On the possibility of unifying the electromagnetic and the gravitational fields. Z. Phys. **15**, 504–506 (1914)
9. Kaluza, T.: Zum Unitätsproblem der Physik. Sitzungsber. Preuss. Akad. Wiss. Berlin, Math. Phys. Kl. **1**, 966–972 (1921)
10. Klein, O.: Quantentheorie und fünfdimensionale Relativitätstheorie. Z. Phys. **37**(12), 895–906 (1926)
11. Klein, O.: The atomicity of electricity as a quantum theory law. Nature **118**, 516 (1926)
12. Antoniadis, I., Arkani-Hamed, N., Dimopoulos, S., Dvali, G.: New dimensions at a millimeter to a Fermi and superstrings at a TeV. Phys. Lett. B **436**, 257–263 (1998)
13. Louck, J.D., Shaffer, W.H.: Generalized orbital angular momentum and the n-fold degenerate quantum-mechanical oscillator: Part I. The twofold degenerate oscillator. J. Mol. Spectrosc. **4**, 285–297 (1960)
14. Louck, J.D.: Generalized orbital angular momentum and the n-fold degenerate quantum-mechanical oscillator: Part II. The n-fold degenerate oscillator. J. Mol. Spectrosc. **4**, 298–333 (1960)
15. Louck, J.D.: Generalized orbital angular momentum and the n-fold degenerate quantum-mechanical oscillator: Part III. Radial integrals. J. Mol. Spectrosc. **4**, 334–341 (1960)
16. Appel, P., Kampé de Fériet, J.: Fonctions Hypergéométrique et Hypersphériques Polynomes d'Hermite, p. 202. Gauthier-Villars, Paris (1926)
17. Fock, V.: Zur Theorie des Wasserstoffatoms. Z. Phys. **98**, 145–154 (1935)
18. Bargmann, V.: Zur Theorie des Wasserstoffatoms. Bemerkungen zur gleichamigen Arbeit von V. Foch. Z. Phys. **99**, 576–582 (1936)
19. Sommerfeld, A.: Partial Differential Equation in Physics, p. 227. Academic Press, New York (1949)
20. Erdélyi, A.: Higher Transcendental Functions, vol. 2, Bateman Manuscript Project, p. 232. McGraw-Hill, New York (1953)
21. Bander, M., Itzykson, C.: Group theory and the hydrogen atom (I). Rev. Mod. Phys. **38**, 330–345 (1966)

S.-H. Dong, *Wave Equations in Higher Dimensions*,
DOI 10.1007/978-94-007-1917-0,

22. Shaffer, W.H.: Degenerate modes of vibration and perturbations in polyatomic molecules. Rev. Mod. Phys. **16**, 245–259 (1944)
23. Baker Jr., G.A.: Degeneracy of the n-dimensional, isotropic, harmonic oscillator. Phys. Rev. **103**, 1119–1120 (1956)
24. Alliluev, S.P.: On the relations between accidental degeneracy and hidden symmetry of a system. Sov. Phys. JETP **6**, 156–159 (1958)
25. Rasmussen, W.O., Salamó, S.: An algebraic approach to Coulomb scattering in N dimensions. J. Math. Phys. **20**, 1064–1067 (1979)
26. Al-Jaber, S.M.: Quantization of angular momentum in the N-dimensional space. Nuovo Cimento B **110**, 993–995 (1995)
27. Al-Jaber, S.M.: On the radial-part equation of the wavefunction in N dimensions. Nuovo Cimento B **112**, 761–765 (1997)
28. Al-Jaber, S.M.: The fine structure of the N-dimensional hydrogen atom. Nuovo Cimento B **113**, 651–657 (1998)
29. Al-Jaber, S.M.: Hydrogen atom in N dimensions. Int. J. Theor. Phys. **37**, 1289–1298 (1998)
30. de Broglie, L., Bohm, D., Hillion, P., Halbwachs, F., Takabayasi, T., Vigier, J.P.: Rotator model of elementary particles considered as relativistic extended structures in Minkowski space. Phys. Rev. **129**, 438–450 (1963)
31. Granzow, K.: N-dimensional total orbital angular-momentum operator. II. Explicit representations. J. Math. Phys. **5**(10), 1474–1477 (1964)
32. Bergmann, D., Frishman, Y.: A relation between the hydrogen atom and multidimensional harmonic oscillators. J. Math. Phys. **6**(12), 1855–1856 (1965)
33. Čížek, J., Paldus, J.: An algebraic approach to bound states of simple one-electron systems. Int. J. Quant. Chem. **12**, 875–896 (1977)
34. Kostelecky, V.A., Nieto, M.M., Truax, D.R.: Supersymmetry and the relationship between the Coulomb and oscillator problems in arbitrary dimensions. Phys. Rev. D **32**, 2627–2633 (1985)
35. Balantekin, A.B.: Accidental degeneracies and supersymmetric quantum mechanics. Ann. Phys. **164**(2), 277–287 (1985)
36. Kostelecky, V.A., Russell, N.: Radial Coulomb and oscillator systems in arbitrary dimensions. J. Math. Phys. **37**, 2166–2181 (1996)
37. Zeng, G.J., Su, K.L., Li, M.: Most general and simplest algebraic relationship between energy eigenstates of a hydrogen atom and a harmonic oscillator of arbitrary dimensions. Phys. Rev. A **50**(5), 4373–4375 (1994)
38. Lévai, G., Kónya, B., Papp, Z.: Unified treatment of the Coulomb and harmonic oscillator potentials in D dimensions. J. Math. Phys. **39**, 5811 (1998)
39. Shea, R.W., Aravind, P.K.: Degeneracies of the spherical well, harmonic oscillator and hydrogen atom in arbitrary dimensions. Am. J. Phys. **64**(4), 430 (1996)
40. Jafarizadeh, M.A., Goudarzi, H.: Degeneracy of Schrödinger equation with potential $1/r$ in d-dimensions. Indian J. Phys. B **72**(1), 35–44 (1998)
41. Kirchberg, A., Länge, J.D., Pisani, P.A.G., Wipf, A.: Algebraic solution of the supersymmetric hydrogen atom in d dimensions. Ann. Phys. **303**(2), 359–388 (2003)
42. Witwit, M.R.M.: The eigenvalues of the Schrödinger equation for spherically symmetric states for various types of potentials in two, three and N dimensions, by using perturbative and non-perturbative methods. J. Phys. A, Math. Gen. **24**(19), 4535 (1991)
43. Imbo, T., Pagnamenta, A., Sukhatme, U.: Energy eigenstates of spherically symmetric potentials using the shifted $1/N$ expansion. Phys. Rev. D **29**, 1669–1681 (1984)
44. Imbo, T., Sukhatme, U.: Shifted $1/N$ expansions for energy eigenvalues of the Schrödinger equation. Phys. Rev. D **28**, 418–420 (1983)
45. Imbo, T., Sukhatme, U.: Improved wavefunctions for large-N expansions. Phys. Rev. D **31**, 2655–2658 (1985)
46. Varshni, Y.P.: Eigenenergies of the $r^2+\lambda r^2/(1+gr^2)$ potential obtained by the shifted $1/N$ expansion. Phys. Rev. A **36**, 3009–3014 (1987)
47. Roychoudhury, R.K., Varshni, Y.P.: Shifted $1/N$ expansion and exact solutions for the potential $V(r)=-Z/r+gr+\lambda r^2$. J. Phys. A, Math. Gen. **21**, 3025 (1988)

48. Roychoudhury, R.K., Varshni, Y.P.: Rotating oscillator-shifted $1/N$ expansion and supersymmetric considerations. Phys. Rev. A **37**, 2309–2313 (1988)
49. Christiansen, H., Epele, L.N., Fanchiotte, H., Garcia Canal, C.A.: Improved shifted $1/N$ expansion. Phys. Rev. A **40**, 1760–1764 (1989)
50. Dutt, R., Mukherji, U., Varshni, Y.P.: Shifted large-N expansion for the bound states of the Hellmann potential. Phys. Rev. A **34**, 777–784 (1986)
51. Papp, E.: Quasiclassical approach to the shifting parameter of the $1/N$ method. Phys. Rev. A **36**, 3550–3555 (1987)
52. Atag, S.: Application of the shifted $1/N$ expansion to the rotational-vibrational states of the H_2^+ molecule. Phys. Rev. A **37**, 2280–2283 (1988)
53. Chatterjee, A.: Large-N expansions in quantum mechanics. Phys. Rep. **186**, 249–370 (1990)
54. Hakobyan, Ye.M., Pogosyan, G.S., Sissakian, A.N.: Generalized D-dimensional oscillator: interbasis expansions. Phys. At. Nucl. **61**, 1762–1767 (1998)
55. Papp, E.: q analogs of the radial Schrödinger equation in N space dimensions. Phys. Rev. A **52**, 101–106 (1995)
56. Mlodinov, L.D., Papanicolaou, N.: SO(2, 1) algebra and the large N expansion in quantum mechanics. Ann. Phys. **128**, 314–334 (1980)
57. Gerry, C.C., Togeas, J.B.: A large-N phase integral approximation of Coulomb-type systems using SO(2, 1) coherent states. J. Phys. A, Math. Gen. **19**, 3797 (1986)
58. Yáñez, R.J., Van Assche, W., Dehesa, J.S.: Position and momentum information entropies of the D-dimensional harmonic oscillator and hydrogen atom. Phys. Rev. A **50**, 3065–3079 (1994)
59. Wódkiewicz, K.: Fermi pseudopotential in arbitrary dimensions. Phys. Rev. A **43**, 68–76 (1991)
60. Romera, E., Sánchez-Moreno, P., Dehesa, J.S.: Uncertainty relation for Fisher information of D-dimensional single-particle systems with central potentials. J. Math. Phys. **47**, 103504 (2006)
61. Bender, C.M., Boettcher, S.: Dimensional expansion for the Ising limit of quantum field theory. Phys. Rev. D **48**, 4919–4923 (1993)
62. Bender, C.M., Milton, K.A.: Scalar Casimir effect for a D-dimensional sphere. Phys. Rev. D **50**, 6547–6555 (1994)
63. Romeo, A.: Multidimensional extension of a Wentzel-Kramers-Brillouin improvement for spherical quantum billiard zeta functions. J. Math. Phys. **36**, 4005 (1995)
64. Nieto, M.M.: Existence of bound states in continuous $0 < D < \infty$ dimensions. Phys. Lett. A **293**, 10–16 (2002)
65. Gönül, B., Özer, O., Kocak, M., Tutcu, D., Cancelik, Y.: Supersymmetry and the relationship between a class of singular potentials in arbitrary dimensions. J. Phys. A, Math. Gen. **34**, 8271 (2001)
66. Gómez, F.J., Sesma, J.: Bound states and resonances in sombrero potentials. Phys. Lett. A **286**, 395–400 (2001)
67. Camblong, H.E., Epele, L.N., Franchiotti, H., Canal García, C.A.: Renormalization of the inverse square potential. Phys. Rev. Lett. **85**, 1590–1593 (2000)
68. Nouri, S.: Generalized coherent states for the d-dimensional Coulomb problem. Phys. Rev. A **60**, 1702–1705 (1999)
69. Mustafa, O., Habib Mazharimousavi, S.: Quantum particles trapped in a position-dependent mass barrier; a d-dimensional recipe. Phys. Lett. A **358**, 259–261 (2006)
70. Gönül, B., Koçak, M.: Explicit solutions for N-dimensional Schrödinger equation with position-dependent mass. J. Math. Phys. **47**, 102101 (2006)
71. Chang, L.N., Minic, D., Okamura, N., Takeuchi, T.: Exact solution of the harmonic oscillator in arbitrary dimensions with minimal length uncertainty relations. Phys. Rev. D **65**, 125027 (2002)
72. Burgbacher, F., Lämmerzahl, C., Macias, A.: Is there a stable hydrogen atom in higher dimensions? J. Math. Phys. **40**, 625 (1999)
73. Zhao, W.Q.: Relation between dimension and angular momentum for radially symmetric potential in N-dimensional space. Commun. Theor. Phys. **46**, 429 (2006)

74. López-Rosa, S., Manzano, D., Dehesa, J.S.: Complexity of D-dimensional hydrogenic systems in position and momentum spaces. Physica A **388**, 3273 (2009)
75. Quesne, C.: First-order intertwining operators and position-dependent mass Schrödinger equations in d dimensions. Ann. Phys. **321**, 1221–1239 (2006)
76. Kuru, S., Teğmen, A., Vercin, A.: Intertwined isospectral potentials in an arbitrary dimension. J. Math. Phys. **42**(8), 3344 (2001)
77. Friedberg, R., Lee, T.D., Zhao, W.Q.: Convergent iterative solutions for a Sombrero-shaped potential in any space dimension and arbitrary angular momentum. Ann. Phys. **321**(8), 1981–2015 (2006)
78. Hosoya, H.: Hierarchical structure of the atomic orbital wavefunctions of D-dimensional atom. J. Phys. Chem. **101**, 418–421 (1997)
79. Dunn, M., Watson, D.K.: Continuation of the Schrödinger equation for higher angular-momentum states to D dimensions and interdimensional degeneracies. Few-Body Syst. **21**, 187–209 (1996)
80. Dunn, M., Watson, D.K.: Continuation of the wave function for higher angular momentum states to D-dimensions: I. The generalized Schwartz expansion. Ann. Phys. **251**, 266–318 (1996)
81. Weyl, H.: The Classical Groups. Princeton University Press, Princeton (1939)
82. Gu, X.Y., Duan, B., Ma, Z.Q.: Quantum three-body system in D dimensions. J. Math. Phys. **43**, 2895 (2002)
83. Gu, X.Y., Ma, Z.Q., Duan, B.: Interdimensional degeneracies for a quantum three-body system in D dimensions. Phys. Lett. A **307**, 55–59 (2003)
84. Gu, X.Y., Ma, Z.Q., Sun, J.Q.: Interdimensional degeneracies for a quantum N-body system in D dimensions. Europhys. Lett. **64**, 586 (2003)
85. Gu, X.Y., Ma, Z.Q., Sun, J.Q.: Quantum four-body system in D dimensions. J. Math. Phys. **44**, 3763 (2003)
86. Duan, B., Gu, X.Y., Ma, Z.Q.: Energy levels of P-wave states for a D-dimensional helium atom. Phys. Lett. A **322**, 96–104 (2004)
87. Nieto, M.M.: Hydrogen atom and relativistic pi-mesic atom in N-space dimensions. Am. J. Phys. **47**, 1067 (1979)
88. Joseph, A.: Self-adjoint ladder operators (I). Rev. Mod. Phys. **39**, 829–837 (1967)
89. Coulson, C.A., Joseph, A.: Self-adjoint ladder operators. II. Rev. Mod. Phys. **39**, 838–849 (1967)
90. Joseph, A.: Self-adjoint ladder operators. III. Rev. Mod. Phys. **40**, 845–871 (1968)
91. Gu, X.Y., Ma, Z.Q., Dong, S.H.: Exact solutions to the Dirac equation for a Coulomb potential in $D+1$ dimensions. Int. J. Mod. Phys. E **11**(4), 335–346 (2002)
92. Bollini, C.G., Giambiagi, J.J.: Generalized Klein-Gordon equations in d dimensions from supersymmetry. Phys. Rev. D **32**(12), 3316–3318 (1985)
93. Chatterjee, A.: Large-N solution of the Klein-Gordon equation. J. Math. Phys. **27**, 2331 (1986)
94. Roychoudhury, R.K., Varshni, Y.P.: Shifted $1/N$ expansion and scalar potential in the Dirac equation. J. Phys. A, Math. Gen. **20**, L1083 (1987)
95. Atag, S.: Large-N iterative solution of the Dirac equation. J. Math. Phys. **30**, 696 (1989)
96. Panja, M.M., Dutt, R.: Shifted large-N expansion for the energy levels of relativistic particles. Phys. Rev. A **38**, 3937–3943 (1988)
97. Mustafa, O., Sever, R.: Approach to the shifted $1/N$ expansion for the Klein-Gordon equation. Phys. Rev. A **43**, 5787–5789 (1991)
98. Mustafa, O., Sever, R.: Shifted $1/N$ expansion for the Klein-Gordon equation with vector and scalar potentials. Phys. Rev. A **44**, 4142–4144 (1991)
99. Lin, D.H.: The path integration of a relativistic particle on a D-dimensional sphere. J. Phys. A, Math. Gen. **30**, 3201 (1997)
100. Dong, S.H., Gu, X.Y., Ma, Z.Q., Yu, J.: The Klein-Gordon equation with a Coulomb potential in D dimensions. Int. J. Mod. Phys. E **12**, 555–565 (2003)
101. Saad, N., Hall, R.L., Ciftci, H.: The Klein-Gordon equation with the Kratzer potential in d dimensions. Cent. Eur. J. Phys. **6**, 717–729 (2008)

102. Oyewumi, K.J., Akinpelu, F.O., Agboola, A.D.: Exactly complete solutions of the pseudo-harmonic potential in N-dimensions. Int. J. Theor. Phys. **47**(4), 1039–1057 (2008)
103. Hall, R.L., Aliyu, M.D.: Comparison theorems for the Klein-Gordon equation in d dimensions. Phys. Rev. A **78**, 052115 (2008)
104. Dong, S.H.: The Dirac equation with a Coulomb potential in D dimensions. J. Phys. A, Math. Gen. **36**, 4977 (2003)
105. Dong, S.H., Sun, G.H., Popov, D.: Group theory approach to the Dirac equation with a Coulomb plus scalar potential in $D+1$ dimensions. J. Math. Phys. **44**, 4467–4479 (2003)
106. Ma, Z.Q., Dong, S.H., Gu, X.Y., Yu, J.: The Klein-Gordon equation with a Coulomb plus scalar potential in D dimensions. Int. J. Mod. Phys. E **13**, 597–610 (2004)
107. Wang, L.Y., Gu, X.Y., Ma, Z.Q., Dong, S.H.: Exact solutions to D-dimensional Schrödinger equation with a pseudoharmonic oscillator. Found. Phys. Lett. **15**, 569–576 (2002)
108. Dong, S.H., Sun, G.H.: The Schrödinger equation with a Coulomb plus inverse-square potential in D dimensions. Phys. Scr. **70**, 94–97 (2004)
109. Levinson, N.: On the uniqueness of the potential in a Schrödinger equation for a given asymptotic phase. K. Dan. Vidensk. Selsk. Mat. Fys. Medd. **25**(9), 1–29 (1949)
110. Ma, Z.Q.: The Levinson theorem. J. Phys. A, Math. Gen. **39**(48), R625 (2006)
111. Gu, X.Y., Ma, Z.Q., Dong, S.H.: Levinson theorem for the Dirac equation in $D+1$ dimensions. Phys. Rev. A **67**, 062715 (2003)
112. Dong, S.H., Ma, Z.Q.: Nonrelativistic Levinson's theorem in D dimensions. Phys. Rev. A **65**, 042717 (2002)
113. Dong, S.H., Lozada-Cassou, M.: Generalized hypervirial and recurrence relations for radial matrix elements in arbitrary dimensions. Mod. Phys. Lett. A **20**, 1533–1540 (2005)
114. Dong, S.H., Chen, C.Y., Lozada-Cassou, M.: Generalized hypervirial and Blanchard's recurrence relations for radial matrix elements. J. Phys. B, At. Mol. Opt. Phys. **38**, 2211–2220 (2005)
115. Lin, D.H.: Friedel theorem for Dirac fermions in D dimensions. Phys. Rev. A **74**, 032109 (2006)
116. Lin, D.H.: Friedel sum rule, Levinson theorem, and the Atiyah-Singer index. Phys. Rev. A **75**, 032115 (2007)
117. Cotăescu, I.I.: Remarks on the quantum modes of the scalar field on AdS_{d+1} spacetime. Phys. Rev. D **60**, 107504 (1999)
118. Cotăescu, I.I., Cotăescu, I.I.: Geometric models of $(d+1)$-dimensional relativistic rotating oscillators. J. Math. Phys. **41**(11), 7290 (2000)
119. Saharian, A.A.: Wightman function and vacuum fluctuations in higher dimensional brane models. Phys. Rev. D **73**, 044012 (2006)
120. Kenmoku, M., Ishimoto, K., Nandi, K.K., Shigemoto, K.: Scalar field contribution to rotating black hole entropy. Phys. Rev. D **73**, 064004 (2006)
121. Chatillon, N., Macesanu, C., Trodden, M.: Brane cosmology in an arbitrary number of dimensions. Phys. Rev. D **74**, 124004 (2006)
122. Saa, A.: N-dimensional Vaidya metric with a cosmological constant in double-null coordinates. Phys. Rev. D **75**, 124019 (2007)
123. Goswami, R., Joshi, P.S.: Spherical gravitational collapse in N dimensions. Phys. Rev. D **76**, 084026 (2007)
124. Bouaziz, D., Bawin, M.: Singular inverse square potential in arbitrary dimensions with a minimal length: application to the motion of a dipole in a cosmic string background. Phys. Rev. A **78**, 032110 (2008)
125. Elizalde, E., Odintsov, S.D., Saharian, A.A.: Repulsive Casimir effect from extra dimensions and Robin boundary conditions: from branes to pistons. Phys. Rev. D **79**, 065023 (2009)
126. Chow, D.D.K., Cvetič, M., Lü, H., Pope, C.N.: Extremal black hole/CFT correspondence in (gauged) supergravities. Phys. Rev. D **79**, 084018 (2009)
127. Rogatko, M., Szyplowska, A.: Massive fermion emission from higher dimensional black holes. Phys. Rev. D **79**, 104005 (2009)
128. Belhai, A., Díaz, P., Seguí, A.: Magnetic and electric black holes in arbitrary dimensions. Phys. Rev. D **80**, 044015 (2009)

129. Guo, Z.Q., Ma, B.Q.: Fermion families from two layer warped extra dimensions. J. High Energy Phys. **0808**, 065 (2008)
130. Konoplya, R.A.: Quasinormal behavior of the D-dimensional Schwarzschild black hole and the higher order WKB approach. Phys. Rev. D **68**, 024018 (2003)
131. Melko, R., Mann, R.B.: Studies of the Schroedinger-Newton equations in D dimensions. arXiv:gr-qc/0011004v1 (2000)
132. Kunz, J., Maison, D., Navarro-Léida, F., Viebahn, J.: Rotating Einstein-Maxwell-Dilaton black holes in D dimensions. Phys. Lett. B **639**, 95 (2006)
133. Canfora, F., Giacomini, A., Zerwekh, A.R.: Kaluza-Klein theory in the limit of large number of extra dimensions. Phys. Rev. D **80**, 084039 (2009)
134. Kunstatter, G., Lee, H.C., Leivo, H.P.: Gauge invariance of the one-loop effective potential in $M^d \times S^1$ Kaluza-Klein theory. Phys. Rev. D **33**(4), 1018–1026 (1986)
135. Shiraishi, K.: Multicentered solution for maximally charged dilaton black holes in arbitrary dimensions. J. Math. Phys. **34**(4), 1480 (1993)
136. Weyl, H.: The Theory of Groups and Quantum Mechanics, 2nd (English) edn. Dutton, New York (1931)
137. Wybourne, B.G.: Classical Groups for Physicists. Wiley, New York (1974)
138. Hamermesh, M.: Group Theory and Its Applications to Physical Problems. Addison-Wesley, New York (1964)
139. Ma, Z.Q.: Group Theory in Physics. World Scientific, Singapore (1998) (in Chinese)
140. Ma, Z.Q.: Group Theory for Physicists. World Scientific, Singapore (2007)
141. Racah, G.: Theory of complex spectra. I. Phys. Rev. **61**, 186–197 (1942)
142. Racah, G.: Theory of complex spectra. II. Phys. Rev. **62**, 438–462 (1942)
143. Racah, G.: Theory of complex spectra. III. Phys. Rev. **63**, 367–382 (1943)
144. Wigner, E.P.: Group Theory and Its Application to Quantum Mechanics of Atomic Spectra. Academic Press, New York (1959)
145. Pauli, W.: On the hydrogen spectrum from the standpoint of the new quantum mechanics. Z. Phys. **36**, 336–363 (1926)
146. Schrödinger, E.: Quantisierung als Eigenwertproblem. Ann. Phys. (Leipz.) **80**, 437–490 (1926)
147. Gell-Mann, M., Ne'eman, Y.: The Eightfold Way. Benjamin, New York (1964)
148. Kibler, M.R.: A group-theoretical approach to the periodic table of chemical elements: old and new developments. In: Rouvray, D.H., King, R.B. (eds.) The Mathematics of the Periodic Table. Nova Science Publishers, New York (2005). arXiv:quant-ph/0503039
149. Yurke, B., McCall, S.L., Klauder, J.R.: SU(2) and SU(1, 1) interferometers. Phys. Rev. A **33**, 4033–4053 (1986)
150. Brif, C., Vourdas, A., Mann, A.: Analytic representations based on SU(1, 1) coherent states and their applications. J. Phys. A, Math. Gen. **29**, 5873 (1996)
151. Wodkiewicz, K., Eberly, J.H.: Coherent states, squeezed fluctuations, and the SU(2) and SU(1, 1) groups in quantum-optics applications. J. Opt. Soc. Am. B **2**, 458–466 (1985)
152. Lai, W.K., Buzek, V., Knight, P.L.: Nonclassical fields in a linear directional coupler. Phys. Rev. A **43**, 6323–6336 (1991)
153. Leonhardt, U.: Quantum statistics of a lossless beam splitter: SU(2) symmetry in phase space. Phys. Rev. A **48**, 3265–3277 (1993)
154. Gilmore, R.: Lie Groups, Lie Algebras, and Some of Their Applications. Wiley, New York (1974)
155. Moshinsky, M.: Group Theory and the Many-Body Problem. Gordon & Breach, New York (1968)
156. Barut, A.O., Böhm, A.: Dynamical groups and mass formula. Phys. Rev. **139**, B1107–B1112 (1965)
157. Paldus, J.: Group theoretical approach to the configuration interaction and perturbation theory calculations for atomic and molecular systems. J. Chem. Phys. **61**, 5321–5330 (1974)
158. Paldus, J.: In: Eyring, H., Henderson, D. (eds.) Theoretical Chemistry: Advances and Perspectives, vol. 2, p. 131. Academic Press, New York (1976)

159. Frank, A., Van Isacker, P.: Algebraic Methods in Molecular and Nuclear Structure Physics. Wiley, New York (1994)
160. Iachello, F., Levine, R.D.: Algebraic Theory of Molecules. Oxford University Press, Oxford (1995)
161. Fernández, F., Castro, A.: Algebraic Methods in Quantum Chemistry and Physics. Mathematical Chemistry Series. CRC Press, Boca Raton (1996)
162. Johnston, D.F.: Group theory in solid state physics. Rep. Prog. Phys. **23**, 66–153 (1960)
163. Klauder, J.R., Sudarshan, E.C.G.: Fundamentals of Quantum Optics. Benjamin, New York (1968)
164. Traux, D.R.: Baker-Campbell-Hausdorff relations and unitarity of SU(2) and SU(1, 1) squeeze operators. Phys. Rev. D **31**, 1988–1991 (1985)
165. Barut, A.O., Phillips, E.C.: Matrix elements of representations of non-compact groups in a continuous basis. Commun. Math. Phys. **8**, 52–65 (1968)
166. Perelomov, A.M.: Some remarks on boson pair creation in alternating external field. Phys. Lett. A **39**, 165–166 (1972)
167. Prakash, G.S., Agarwal, G.S.: Mixed excitation- and deexcitation-operator coherent states for the SU(1, 1) group. Phys. Rev. A **50**, 4258–4263 (1994)
168. Gerry, C.C.: Remarks on the use of group theory in quantum optics. Opt. Express **8**, 76–85 (2001)
169. Miller Jr., W.: Symmetry Groups and Their Applications. Academic Press, New York/San Francisco/London (1972)
170. Paz, G.: On the connection between the radial momentum operator and the Hamiltonian in n dimensions. Eur. J. Phys. **22**, 337 (2001)
171. Kolsrud, M.: On the solution of Dirac's equation with Coulomb potential. Phys. Norv. **2**, 43–50 (1966)
172. Hylleraas, E.A.: Matematisk og teoretisk fysikk, vol. IV. Grøndahl, Oslo (1952). English translation: Mathematical and Theoretical Physics, vol. II, Part IV. Wiley-Interscience, New York (1970). Chap. 5
173. Martin, P.C., Glauber, R.J.: Relativistic theory of radiative orbital electron capture. Phys. Rev. **109**, 1307–1325 (1958)
174. Auvil, P.R., Brown, L.M.: The relativistic hydrogen atom: a simple solution. Am. J. Phys. **46**, 679 (1978)
175. Waldenstrøm, S.: On the Dirac equation for the hydrogen atom. Am. J. Phys. **47**, 1098 (1979)
176. Waldenstrøm, S.: Addendum: "On the Dirac equation for the hydrogen atom" [Am. J. Phys. 47, 1098 (1979)]. Am. J. Phys. **48**, 684 (1980)
177. Dong, S.H., Ma, Z.Q.: Exact solutions to the Dirac equation with a Coulomb potential in $2+1$ dimensions. Phys. Lett. A **312**, 78–83 (2003)
178. Lapidus, I.R.: Relativistic one-dimensional hydrogen atom. Am. J. Phys. **51**, 1036 (1983)
179. Bjorken, J.D., Drell, S.D.: Relativistic Quantum Mechanics. McGraw-Hill, New York (1964)
180. Fronsdal, C.: Group theory and application to particle physics (notes by the lecturer). In: Elementary Particle Physics and Field Theory 1. Lectures from Brandeis Summer Institute, 1962. Benjamin, New York (1963)
181. Salam, A.: The formalism of Lie groups. In: Theoretical Physics, p. 173. International Atomic Energy Agency, Vienna (1963)
182. Ma, Z.Q.: In: Dai, Y.B., Hao, B.L. Su, Z.B. (eds.) Yang-Baxter Equation and Quantum Enveloping Algebras. Advanced Series on Theoretical Physical Science, vol. 1. A Collaboration Between World Scientific and Institute of Theoretical Physics. World Scientific, Singapore (1993). Chap. 1
183. Georgi, H.: Lie Algebras in Particle Physics. Benjamin, New York (1982)
184. Dong, S.H., Hou, X.W., Ma, Z.Q.: Relativistic Levinson theorem in two dimensions. Phys. Rev. A **58**, 2160–2167 (1998)
185. de Lange, O.L.: An operator analysis for the Schrödinger, Klein-Gordon, and Dirac equations with a Coulomb potential. J. Math. Phys. **30**, 858 (1989)
186. Faisal, F.H.M., Radozycki, T.: Three-dimensional relativistic model of a bound particle in an intense laser field. II. Phys. Rev. A **48**, 554–557 (1993)

187. Dong, S., Dong, S.H.: Schrödinger equation with a Coulomb field in $2+1$ dimensions. Phys. Scr. **66**, 342–344 (2002)
188. Spector, H.N., Lee, J.: Relativistic one-dimensional hydrogen atom. Am. J. Phys. **53**, 248 (1985)
189. Moss, R.E.: The hydrogen atom in one dimension. Am. J. Phys. **55**, 397 (1987)
190. Galić, H.: Fun and frustration with hydrogen in a $1+1$ dimension. Am. J. Phys. **56**, 312 (1988)
191. Levy, A.A.: Systematic comparison of the quantization rules of hydrogenoid atoms in the Old Quantum, Schrödinger, Klein-Gordon, and Dirac theories, by means of a common set of three parameters. Am. J. Phys. **53**, 454 (1985)
192. Gradshteyn, I.S., Ryzhik, I.M.: Tables of Integrals, Series, and Products, 5th edn. Pergamon, New York (1994)
193. Infeld, L., Hull, T.E.: The factorization method. Rev. Mod. Phys. **23**, 21–68 (1951)
194. Gur, Y., Mann, A.: Radial coherent states—from the harmonic oscillator to the hydrogen atom. Phys. At. Nucl. **68**, 1700–1708 (2005)
195. Goldman, I.I., Krivchenkov, V.D.: Problems in Quantum Mechanics. Pergamon, Oxford (1961)
196. Post, H.R.: Many-particles systems: II. Proc. Phys. Soc. A **69**, 936–938 (1956)
197. Palma, G., Raff, U.: The one-dimensional harmonic oscillator in the presence of a dipole-like interaction. Am. J. Phys. **71**, 247 (2003)
198. Hurley, J.: One-dimensional three-body problem. J. Math. Phys. **8**, 813 (1967)
199. Calogero, F.: Ground state of a one-dimensional N-body system. J. Math. Phys. **10**, 2197 (1969)
200. Calogero, F.: Solution of the one-dimensional N-body problems with quadratic and/or inversely quadratic pair potentials. J. Math. Phys. **12**, 419 (1971)
201. Camiz, P., Gerardi, A., Marchioro, C., Presutti, E., Scacciatelli, E.: Exact solution of a time-dependent quantal harmonic oscillator with a singular perturbation. J. Math. Phys. **12**, 2040 (1971)
202. Dodonov, V.V., Malkin, I.A., Man'ko, V.I.: Even and odd coherent states and excitations of a singular oscillator. Physica **72**, 597–615 (1974)
203. Dodonov, V.V., Malkin, I.A., Man'ko, V.I.: Green function and excitation of a singular oscillator. Phys. Lett. A **39**, 377–378 (1972)
204. Ballhausen, C.J.: A note on the $V = A/x^2 + Bx^2$ potential. Chem. Phys. Lett. **146**, 449–451 (1988)
205. Ballhausen, C.J.: Step-up and step-down operators for the pseudo-harmonic potential $V = r^2/2 + B/2r^2$ in one and two dimensions. Chem. Phys. Lett. **151**, 428–430 (1988)
206. Friš, J., Mandrosov, V., Smorodinsky, Ya.A., Uhlíř, M., Winternitz, P.: On higher symmetries in quantum mechanics. Phys. Lett. **16**, 354–356 (1965)
207. Popov, D.: Barut-Girardello coherent states of the pseudoharmonic oscillator. J. Phys. A, Math. Gen. **34**, 5283–5296 (2001)
208. Sage, M.L.: The vibrations and rotations of the pseudogaussian oscillator. Chem. Phys. **87**, 431–439 (1984)
209. Dong, S.H., Ma, Z.Q.: Algebraic approach to the pseudoharmonic oscillator in 2D. Int. J. Mod. Phys. E **11**, 155–160 (2002)
210. Dong, S., Dong, S.H., Lozada-Cassou, M.: Algebraic approach to a harmonic oscillator plus an inverse squared potential and its recurrence relation. Phys. Scr. **73**, 173–177 (2006)
211. Schrödinger, E.: Further studies on solving eigenvalue problems by factorization. Proc. R. Ir. Acad. A **46**, 183–206 (1940)
212. Berrondo, M., Palma, A.: The algebraic approach to the Morse oscillator. J. Phys. A, Math. Gen. **13**, 773 (1980)
213. Stahlhofen, A., Bleuler, K.: An algebraic form of the factorization method. Nuovo Cimento B **104**, 447–465 (1989)
214. Adams, B.G., Čižek, J., Paldus, J.: In: Löwdin, P.-O. (ed.) Lie Algebraic Methods and Their Applications to Simple Quantum Systems. Advances in Quantum Chemistry, vol. 19. Academic Press, New York (1988)

215. Tangherlini, F.R.: Schwarzschild field in n dimensions and the dimensionality of space problem. Nuovo Cimento **XXVII**, 636–651 (1963)
216. Dong, S.H., Sun, G.H.: The Schrödinger equation with a Coulomb plus inverse-square potential in D dimensions. Phys. Scr. **70**, 94–97 (2004)
217. Ehrenfest, P.: What way does it become manifest in the fundamental laws of physics that space has three dimensions? Proc. Amst. Acad. **20**, 200–209 (1917)
218. Ehrenfest, P.: Welche Rolle spielt die Dimensionalität des Raumes in den Grundgesetzen der Physik? Ann. Phys. (Leipz.) **61**, 440 (1920)
219. Nieto, M.M., Simmons Jr., L.M.: Eigenstates, coherent states, and uncertainty products for the Morse oscillator. Phys. Rev. A **19**, 438–444 (1979)
220. Aebersold, D., Biedenharn, L.C.: Cautionary remark on applying symmetry techniques to the Coulomb problem. Phys. Rev. A **15**, 441–443 (1977)
221. Bose, S.K., Gupta, N.: Exact solution of nonrelativistic Schrödinger equation for certain central physical potentials. Nuovo Cimento B **114**, 299–328 (1998)
222. Khare, A., Behra, S.N.: Anharmonic oscillator model for first order structural phase transition. Pramana J. Phys. **14**, 327–341 (1980)
223. Emin, D.: Small polarons. Phys. Today **35**, 34 (1982)
224. Emin, D., Holstein, T.: Adiabatic theory of an electron in a deformable continuum. Phys. Rev. Lett. **36**, 323–326 (1976)
225. Coleman, S.: "Aspects of Symmetry" Selected Erice Lectures, p. 234. Cambridge University Press, Cambridge (1988)
226. Kaushal, R.S.: Quantum mechanics of noncentral harmonic and anharmonic potentials in two-dimensions. Ann. Phys. (N.Y.) **206**, 90–105 (1991)
227. Kaushal, R.S.: An exact solution of the Schrödinger wave equation for a sextic potential. Phys. Lett. A **142**, 57–58 (1989)
228. Kaushal, R.S., Parashar, D.: On the quantum bound states for the potential $V(r) = ar^2 + br^{-4} + cr^{-6}$. Phys. Lett. A **170**, 335–338 (1992)
229. Özcelik, S., Simsek, M.: Exact solutions of the radial Schrödinger equation for inverse-power potentials. Phys. Lett. A **152**, 145–150 (1991)
230. Dong, S.H., Ma, Z.Q.: Exact solutions to the Schrödinger equation for the potential $V(r) = ar^2 + br^{-4} + cr^{-6}$ in two dimensions. J. Phys. A, Math. Gen. **31**, 9855 (1998)
231. Dong, S.H., Ma, Z.Q., Esposito, G.: Exact solutions of the Schrödinger equation with inverse-power potential. Found. Phys. Lett. **12**, 465–474 (1999)
232. Dong, S.H.: Exact solutions of the two-dimensional Schrödinger equation with certain central potentials. Int. J. Theor. Phys. **39**, 1119–1128 (2000)
233. Dong, S.H.: A new approach to the relativistic Schrödinger equation with central potential: ansatz method. Int. J. Theor. Phys. **40**, 559–567 (2001)
234. Child, M.S., Dong, S.H., Wang, X.G.: Quantum states of a sextic potential: hidden symmetry and quantum monodromy. J. Phys. A, Math. Gen. **33**(32), 5653 (2000)
235. Znojil, M.: Singular anharmonicities and the analytic continued fractions. J. Math. Phys. **30**, 23 (1989)
236. Znojil, M.: Singular anharmonicities and the analytic continued fractions. II. The potentials $V(r) = ar^2 + br^{-4} + cr^{-6}$. J. Math. Phys. **31**, 108 (1990)
237. Voros, A.: Exact resolution method for general $1D$ polynomial Schrödinger equation. J. Phys. A, Math. Gen. **32**, 5993 (1999)
238. Esposito, G.: Scattering from singular potentials in quantum mechanics. J. Phys. A, Math. Gen. **31**, 9493 (1998)
239. Esposito, G.: Complex parameters in quantum mechanics. Found. Phys. Lett. **11**, 535–547 (1998)
240. Barut, A.O.: Magnetic resonances between massive and massless spin-$1/2$ particles with magnetic moments. J. Math. Phys. **21**, 568 (1980)
241. Song, X.: An effective quark-antiquark potential for both heavy and light mesons. J. Phys. G, Nucl. Part. Phys. **17**, 49 (1980)
242. Dong, S.H.: The *ansatz* method for analyzing Schrödinger's equation with three anharmonic potentials in D dimensions. Found. Phys. Lett. **15**, 385–395 (2002)

243. Dong, S.H.: On the solutions of the Schrödinger equation with some anharmonic potentials: wave function ansatz. Phys. Scr. **65**, 289–295 (2002)
244. Morse, P.M.: Diatomic molecules according to the wave mechanics. II. Vibrational levels. Phys. Rev. **34**, 57–64 (1929)
245. Rouse, C.A.: Series solution of the Schrödinger equation with the Morse potential. Phys. Rev. A **36**, 1–4 (1987)
246. Dong, S.H., Sun, G.H.: The series solutions of the non-relativistic equation with the Morse potential. Phys. Lett. A **314**, 261–266 (2003)
247. Dong, S.H., Hou, X.W., Ma, Z.Q.: Levinson's theorem for the Klein-Gordon equation in two dimensions. Phys. Rev. A **59**, 995–1002 (1999)
248. Dong, S.H., Hou, X.W., Ma, Z.Q.: Levinson's theorem for non-local interactions in two dimensions. J. Phys. A, Math. Gen. **31**, 7501 (1998)
249. Newton, R.G.: Analytic properties of radial wave functions. J. Math. Phys. **1**, 319 (1960)
250. Newton, R.G.: Noncentral potentials: the generalized Levinson theorem and the structure of the spectrum. J. Math. Phys. **18**, 1348 (1977)
251. Newton, R.G.: Nonlocal interactions: the generalized Levinson theorem and the structure of the spectrum. J. Math. Phys. **18**, 1582 (1977)
252. Newton, R.G.: Inverse scattering. I. One dimension. J. Math. Phys. **21**, 493 (1982)
253. Newton, R.G.: Scattering Theory of Waves and Particles, 2nd edn. Springer, New York (1982)
254. Jauch, J.M.: On the relation between scattering phase and bound states. Helv. Phys. Acta **30**, 143–156 (1957)
255. Martin, A.: On the validity of Levinson's theorem for non-local interactions. Nuovo Cimento **7**, 607–627 (1968)
256. Eberly, J.H.: Quantum scattering theory in one dimension. Am. J. Phys. **33**, 771 (1965)
257. Ni, G.J.: The Levinson theorem and its generalization in relativistic quantum mechanics. Phys. Energ. Fort. Phys. Nucl. **3**, 432–449 (1979)
258. Newton, R.G.: Inverse scattering by a local impurity in a periodic potential in one dimension. J. Math. Phys. **24**, 2152 (1983)
259. Newton, R.G.: Remarks on inverse scattering in one dimension. J. Math. Phys. **25**, 2991 (1984)
260. Newton, R.G.: Comment on Normalization of scattering states, scattering phase shifts, and Levinson's theorem. Helv. Phys. Acta **67**, 20 (1994)
261. Ma, Z.Q., Ni, G.J.: Levinson theorem for Dirac particles. Phys. Rev. D **31**, 1482–1488 (1985)
262. Ma, Z.Q.: Levinson's theorem for Dirac particles moving in a background magnetic monopole field. Phys. Rev. D **32**, 2203–2212 (1985)
263. Ma, Z.Q.: Levinson's theorem for Dirac particles with a long-range potential. Phys. Rev. D **32**, 2213–2215 (1985)
264. Liang, Y.G., Ma, Z.Q.: Levinson's theorem for the Klein-Gordon equation. Phys. Rev. D **34**, 565–570 (1986)
265. Iwinski, Z.R., Rosenberg, L., Spruch, L.: Nodal structure of zero-energy wavefunctions: new approach to Levinson's theorem. Phys. Rev. A **31**, 1229–1240 (1985)
266. Iwinski, Z.R., Rosenberg, L., Spruch, L.: Nodal structure and phase shifts of zero-incident-energy wavefunctions: multiparticle single-channel scattering. Phys. Rev. A **33**, 946–953 (1986)
267. Bollé, D., Gesztesy, F., Danneels, C., Wilk, S.F.J.: Threshold behavior and Levinson's theorem for two-dimensional scattering systems: a surprise. Phys. Rev. Lett. **56**, 900–903 (1986)
268. Gibson, W.G.: Levinson's theorem and the second virial coefficient in one, two, and three dimensions. Phys. Rev. A **36**, 564–575 (1987)
269. Clemence, D.P.: Low-energy scattering and Levinson's theorem for a one-dimensional Dirac equation. Inverse Probl. **5**, 269 (1987)
270. van Dijk, W., Kiers, K.A.: Time delay in simple one-dimensional systems. Am. J. Phys. **60**, 520 (1992)
271. Baton, G.: Levinson's theorem in one dimension: heuristics. J. Phys. A, Math. Gen. **18**, 479 (1985)

272. Vidal, F., LeTourneux, J.: Multichannel scattering with nonlocal and confining potentials. I. General theory. Phys. Rev. C **45**, 418–429 (1992)
273. Sassoli de Bianchi, M.S.: Levinson's theorem, zero-energy resonances, and time delay in one-dimensional scattering systems. J. Math. Phys. **35**, 2719 (1994)
274. Martin, A., Debianchi, M.S.: Levinson's theorem for time-periodic potentials. Europhys. Lett. **34**, 639 (1996)
275. Kiers, K.A., van Dijk, W.: Scattering in one dimension: the coupled Schrödinger equation, threshold behaviour and Levinson's theorem. J. Math. Phys. **37**, 6033 (1996)
276. Ma, Z.Q., Dai, A.Y.: Levinson's theorem for non-local interactions. J. Phys. A, Math. Gen. **21**, 2085 (1988)
277. Aktosun, T., Klaus, M., van der Mee, C.: On the Riemann-Hilbert problem for the one-dimensional Schrödinger equation. J. Math. Phys. **34**, 2651 (1993)
278. Aktosun, T., Klaus, M., van der Mee, C.: Factorization of scattering matrices due to partitioning of potentials in one-dimensional Schrödinger-type equations. J. Math. Phys. **37**, 5897 (1996)
279. Aktosun, T., Klaus, M., van der Mee, C.: Wave scattering in one dimension with absorption. J. Math. Phys. **39**, 1957 (1998)
280. Aktosun, T., Klaus, M., van der Mee, C.: On the number of bound states for the one-dimensional Schrödinger equation. J. Math. Phys. **39**, 4249 (1998)
281. Ma, Z.Q.: Comment on "Levinson's theorem for the Dirac equation". Phys. Rev. Lett. **76**, 3654 (1996)
282. Poliatzky, N.: Poliatzky replies "Comment on 'Levinson's theorem for the Dirac equation' ". Phys. Rev. Lett. **76**, 3655 (1996)
283. Poliatzky, N.: Levinson's theorem for the Dirac equation. Phys. Rev. Lett. **70**, 2507–2510 (1993)
284. Rosenberg, L., Spruch, L.: Generalized Levinson theorem: applications to electron-atom scattering. Phys. Rev. A **54**, 4985–4991 (1996)
285. Portnoi, M.E., Galbraith, I.: Variable-phase method and Levinson's theorem in two dimensions: application to a screened Coulomb potential. Solid State Commun. **103**, 325–329 (1997)
286. Portnoi, M.E., Galbraith, I.: Levinson's theorem and scattering phase-shift contributions to the partition function of interacting gases in two dimensions. Phys. Rev. B **58**, 3963–3968 (1998)
287. Lin, Q.G.: Levinson theorem in two dimensions. Phys. Rev. A **56**, 1938–1944 (1997)
288. Lin, Q.G.: Levinson theorem for Dirac particles in two dimensions. Phys. Rev. A **57**, 3478–3488 (1998)
289. Lin, Q.G.: Levinson theorem for Dirac particles in one dimension. Eur. Phys. J. D **7**, 515 (1999)
290. Ma, Z.Q.: Proof of the Levinson theorem by the Sturm-Liouville theorem. J. Math. Phys. **26**(8), 1995 (1985)
291. Dong, S.H., Hou, X.W., Ma, Z.Q.: Levinson's theorem for the Schrödinger equation in two dimensions. Phys. Rev. A **58**, 2790–2796 (1998)
292. Dong, S.H., Ma, Z.Q.: Levinson's theorem for the Schrödinger equation in one dimension. Int. J. Theor. Phys. **39**, 469–481 (2000)
293. Dong, S.H.: Levinson's theorem for the nonlocal interaction in one dimension. Int. J. Theor. Phys. **39**, 1529–1541 (2000)
294. Dong, S.H.: Levinson's theorem for the Klein-Gordon equation in one dimension. Eur. Phys. J. D **11**, 159–165 (2000)
295. Blankenbecler, R., Boyanovsky, D.: Extensions of Levinson's theorem: application to indices and fractional charge. Physica D **18**, 367 (1986)
296. Niemi, A.J., Semenoff, G.W.: Anomalies, Levinson's theorem, and fermion determinants. Phys. Rev. D **32**, 471–475 (1985)
297. Yang, C.N.: In: Craigie, N.S., Goddard, P., Nahm, W. (eds.) Monopoles in Quantum Field Theory. Proceedings of the Monopole Meeting, Trieste, Italy, p. 237. World Scientific, Singapore (1982)

298. Waller, I.: Der Starkeffekt zweiter Ordnung bei Wasserstoff und die Rydbergkorrektion der Spektra von He und Li^+. Z. Phys. **38**, 635–646 (1926)
299. Van Vleck, J.H.: A new method of calculating the mean value of $1/r^s$ for Keplerian systems in quantum mechanics. Proc. R. Soc. Lond. A **143**, 679–681 (1934)
300. Pasternack, S.: On the mean value of r^S for Keplerian systems. Proc. Natl. Acad. Sci. USA **23**, 91–94 (1937). Erratum in Proc. Natl. Acad. Sci. USA **23**, 250
301. Pasternack, S., Sternheimer, R.M.: An orthogonality property of hydrogenlike radial functions. J. Math. Phys. **3**, 1280 (1962)
302. Armstrong Jr., L.: Group properties of hydrogenic radial functions. Phys. Rev. A **3**, 1546–1550 (1971)
303. Ding, Y.B.: On the Schrödinger radial ladder operator. J. Phys. A, Math. Gen. **20**, 6293 (1987)
304. Bethe, H.A., Salpeter, E.E.: Quantum Mechanics of One- and Two-Electron Atoms. Academic Press, New York (1957)
305. Blanchard, P.: A new recurrence relation for hydrogenic radial matrix elements. J. Phys. B, At. Mol. Phys. **7**, 993 (1974)
306. Bockasten, K.: Mean values of powers of the radius for hydrogenic electron orbits. Phys. Rev. A **9**, 1087–1089 (1974)
307. Hughes, D.E.: Recurrence relations for radial matrix elements obtained from hypervirial relations. J. Phys. B, At. Mol. Phys. **10**, 3167 (1977)
308. Drake, G.W.F., Swainson, R.A.: Expectation values of r^P for arbitrary hydrogenic states. Phys. Rev. A **42**, 1123 (1990)
309. Drake, G.W.F., Swainson, R.A.: Erratum: Expectation values of r^P for arbitrary hydrogenic states. Phys. Rev. A **43**, 6432–6432 (1991)
310. Swainson, R.A., Drake, G.W.F.: An alternative proof of some relations between hydrogenic matrix elements. J. Phys. B, At. Mol. Opt. Phys. **23**, 1079 (1990)
311. Ojha, P.C., Crothers, D.S.: On a simple relation between hydrogenic radial matrix elements. J. Phys. B, At. Mol. Phys.: **17**, 4797 (1984)
312. Shertzer, J.: Evaluation of matrix elements $\langle n, l\|r^{\beta}\|n, l'\rangle$ for arbitrary β. Phys. Rev. A **44**, 2832–2835 (1991)
313. Moreno, B., Piñeiro, A.L., Tipping, R.H.: Algebraic solution for the hydrogenic radial Schrödinger equation: matrix elements for arbitrary powers of several r-dependent operators. J. Phys. A, Math. Gen. **24**, 385 (1991)
314. Qiang, W.C., Dong, S.H.: An alternative approach to calculating the mean values of $\bar{r^k}$ for hydrogen-like atoms. Phys. Scr. **70**, 276–279 (2004)
315. Kramers, H.A.: Quantum Mechanics. North-Holland, Amsterdam (1957). Sect. 59
316. Sánchea, M.L., Moreno, B., López Piñeiro, A.: Matrix-element calculations for hydrogenlike atoms. Phys. Rev. A **46**, 6908–6913 (1992)
317. Morales, J., Peña, J.J., Portillo, P., Ovando, G., Gaftoi, V.: Generalization of the Blanchard's rule. Int. J. Quant. Chem. **65**, 205–211 (1997)
318. López-Bonilla, J.L., Morales, J., Rosales, M.A.: Hypervirial theorem and matrix elements for the Coulomb potential. Int. J. Quant. Chem. **53**(1), 3–7 (1995)
319. Núñez-Yépez, H.N., López-Bonilla, J.L., Salas-Brito, A.L.: Generalized hypervirial and recurrence relations for hydrogenic matrix elements. J. Phys. B, At. Mol. Opt. Phys. **28**, L525 (1995)
320. Martínez-y-Romero, R.P., Núñez-Yépez, H.N., Salas-Brito, A.L.: Relativistically extended Blanchard recurrence relation for hydrogenic matrix elements. J. Phys. B, At. Mol. Opt. Phys. **34**, 1261 (2001)
321. Martínez-y-Romero, R.P., Núñez-Yépez, H.N., Salas-Brito, A.L.: A useful form of the recurrence relation between relativistic atomic matrix elements of radial powers. J. Phys. B, At. Mol. Opt. Phys. **35**, L71 (2002)
322. Dong, S.H., Chen, C.Y., Lozada-Cassou, M.: Some recurrence relations among the radial matrix elements for the relativistic hydrogenic atoms. Phys. Lett. A **333**, 193–203 (2004)
323. Basida, A., Zúñiga, J., Alacid, A., Requena, A., Hidalgo, A.: Two-center matrix elements for Kratzer oscillators. J. Chem. Phys. **93**, 3408 (1990)

324. Morales, J.: Generalized recurrence relation for the calculation of two-center matrix elements. Phys. Rev. A **36**, 4101–4103 (1987)
325. Alves, N.A., Drigo Filho, E.: The factorisation method and supersymmetry. J. Phys. A, Math. Gen. **21**, 3215 (1988)
326. Yang, X.L., Guo, S.H., Chan, F.T., Wong, K.W., Ching, W.Y.: Analytic solution of a two-dimensional hydrogen atom. I. Nonrelativistic theory. Phys. Rev. A **43**, 1186–1196 (1991)
327. Guo, S.H., Yang, X.L., Chan, F.T., Wong, K.W., Ching, W.Y.: Analytic solution of a two-dimensional hydrogen atom. II. Relativistic theory. Phys. Rev. A **43**, 1197–1205 (1991)
328. Müller-Kirsten, H.J.W., Bose, S.K.: Solution of the wave equation for the logarithmic potential with application to particle spectroscopy. J. Math. Phys. **20**, 2471 (1979)
329. Huffaker, J.N., Dwivedi, P.H.: Diatomic molecules as perturbed Morse oscillators. IV. Franck-Condon factors for very high J. J. Chem. Phys. **68**, 1303 (1978)
330. Chakrabarti, B., Das, T.K.: Quality of the supersymmetric WKB quantization condition for non-shape-invariant potentials. Phys. Rev. A **60**, 104–111 (1999)
331. Dunham, J.L.: The Wentzel-Brillouin-Kramers method of solving the wave equation. Phys. Rev. **41**, 713–720 (1932)
332. Krieger, J.B., Rosenzweig, C.: Application of a higher-order WKB approximation to radial problems. Phys. Rev. **164**, 171–173 (1967)
333. Yi, H.S., Lee, H.R., Sohn, K.S.: Semiclassical quantum theory and its applications in two dimensions by conformal mapping. Phys. Rev. A **49**, 3277–3282 (1994)
334. Morehead, J.J.: Asymptotics of radial wave equations. J. Math. Phys. **36**, 5431 (1995)
335. Young, L.A., Uhlenbeck, G.E.: On the Wentzel-Brillouin-Kramers approximate solution of the wave equation. Phys. Rev. **36**, 1154–1167 (1930)
336. Langer, R.E.: On the connection formulas and the solutions of the wave equation. Phys. Rev. **51**, 669–676 (1937)
337. Hur, J., Lee, C.: Semiclassical theory for two-anyon system. Ann. Phys. **305**, 28–44 (2003)
338. Beckel, C., Nakhleh, J.: Application of the second-order WBK approximation to radial problems. J. Chem. Phys. **39**, 94 (1963)
339. Adhikari, R., Dutt, R., Khare, A., Sukhatme, U.P.: Higher-order WKB approximations in supersymmetric quantum mechanics. Phys. Rev. A **38**, 1679–1686 (1988)
340. Moritz, M.J., Eltschka, C., Friedrich, H.: Threshold properties of attractive and repulsive $1/r^2$ potentials. Phys. Rev. A **63**, 042102 (2001)
341. Friedrich, H., Trost, J.: Accurate WKB wavefunctions for weakly attractive inverse-square potentials. Phys. Rev. A **59**, 1683–1686 (1999)
342. Friedrich, H., Trost, J.: Phase loss in WKB waves due to reflection by a potential. Phys. Rev. Lett. **76**, 4869–4873 (1996)
343. Friedrich, H., Trost, J.: Nonintegral Maslov indices. Phys. Rev. A **54**, 1136–1145 (1996)
344. Hainz, J., Grabert, H.: Centrifugal terms in the WKB approximation and semiclassical quantization of hydrogen. Phys. Rev. A **60**, 1698–1701 (1999)
345. Pack, R.T.: On improved WKB (uniform asymptotic) quantum conditions, Dunham corrections, the Langer modification, and RKR potentials. J. Chem. Phys. **57**, 4612 (1972)
346. Howard, R.A.: Effects of the Langer transformation on the calculation of internuclear potential curves. J. Chem. Phys. **54**, 4252 (1971)
347. Fröman, N., Fröman, P.O.: JWKB Approximation. North-Holland, Amsterdam (1965)
348. Ma, Z.Q., Xu, B.W.: Quantization rules for bound states of the Schrödinger equation. Int. J. Mod. Phys. E **14**, 599 (2005)
349. Ma, Z.Q., Xu, B.W.: Quantum correction in exact improved quantization rules. Europhys. Lett. **69**, 685 (2005)
350. Ma, Z.Q., Gonzalez-Cisneros, A., Xu, B.W., Dong, S.H.: Energy spectrum of the trigonometric Rosen-Morse potential using an improved quantization rule. Phys. Lett. A **371**, 180–184 (2007)
351. Qiang, W.C., Dong, S.H.: Arbitrary l-state solutions of the rotating Morse potential through the exact quantization rule method. Phys. Lett. A **363**, 169–176 (2007)
352. Cao, Z.Q., Liu, Q., Shen, Q.S., Dou, X.M., Chen, Y.L.: Quantization scheme for arbitrary one-dimensional potential wells. Phys. Rev. A **63**, 054103 (2001)

353. He, Y., Cao, Z.Q., Shen, Q.S.: Bound-state spectra for supersymmetric quantum mechanics. Phys. Lett. A **326**, 315–321 (2004)
354. Ou, Y.C., Cao, Z.Q., Shen, Q.S.: Exact energy eigenvalues for spherically symmetrical three-dimensional potential. Phys. Lett. A **318**, 36–39 (2003)
355. Ou, Y.C., Cao, Z.Q., Shen, Q.S.: Formally exact quantization condition for nonrelativistic quantum systems. J. Chem. Phys. **121**, 8175 (2004)
356. Qiang, W.C., Dong, S.H.: Proper quantization rule. Europhys. Lett. **89**, 10003 (2010)
357. Serrano, F.A., Gu, X.Y., Dong, S.H.: Qiang-Dong proper quantization rule and its applications to exactly solvable quantum systems. J. Math. Phys. **51**, 082103 (2010)
358. Einstein, A.: Zum Quantensatz von Sommerfeld und Epstein. Verhand. Deut. Phys. Ges. **19**, 82–92 (1917)
359. Brillouin, L.: Remarques sur la mécanique ondulatoire. J. Phys. Radium **7**, 353–368 (1926)
360. Keller, J.B.: Corrected Bohr-Sommerfeld quantum conditions for nonseparable systems. Ann. Phys. **4**, 180–188 (1958)
361. Wentzel, G.: Eine Verallgemeinerung der Quantenbedingungen für die Zwecke der Wellenmechanik. Z. Phys. **38**, 518–529 (1926)
362. Kramers, H.A.: Wellenmechanik und halbzählige Quantisierung. Z. Phys. **39**, 828–840 (1926)
363. Brillouin, L.: La mécanique ondulatoire de Schrödinger: une méthode générale de resolution par approximations successives. C. R. Acad. Sci. **183**, 24–26 (1926)
364. Maslov, V.P.: Théorie des perturbations et methods asymptotiques. Dunod/Gauthier-Villars, Paris (1972)
365. Curtis, L.J., Ellis, D.G.: Use of the Einstein-Brillouin-Keller action quantization. Am. J. Phys. **72**, 1521 (2004)
366. Brack, M., Bhaduri, R.K.: Semiclassical Physics. Addison-Wesley, Reading (1977)
367. Berry, M.V., Mount, K.E.: Semiclassical approximations in wave mechanics. Rep. Prog. Phys. **35**, 315 (1972)
368. Seetharaman, M., Vasan, S.S.: Higher-order JWKB approximations for radial problems. I. Modification of the effective potential. J. Phys. A, Math. Gen. **17**, 2485 (1984)
369. Hruška, M., Keung, W.Y., Sukhatme, U.: Accuracy of semiclassical methods for shape-invariant potentials. Phys. Rev. A **55**, 3345–3350 (1997)
370. Bhaduri, R.K., Sakhr, J., Sprung, D.W.L., Dutt, R., Suzuki, A.: Shape invariant potentials in SUSY quantum mechanics and periodic orbit theory. J. Phys. A, Math. Gen. **38**, L183 (2005)
371. Bhaduri, R.K., Sprung, D.W.L., Suzuki, A.: When is the lowest order WKB quantization exact? Can. J. Phys. **84**, 573–581 (2006)
372. Leacock, R.A., Padgett, M.J.: Hamilton-Jacobi/action-angle quantum mechanics. Phys. Rev. D **28**, 2491–2502 (1983)
373. Ranjani, S.S., Geojo, K.G., Kapoor, A.K., Panigrahi, P.K.: Bound state wavefunctions through the quantum Hamilton-Jacobi formalism. Mod. Phys. Lett. A **19**, 1457–1468 (2004)
374. Cooper, F., Khare, A., Sukhatme, U.: Supersymmetry in quantum mechanics. Phys. Rep. **251**, 267–385 (1995)
375. de Lange, O.L., Raab, R.E.: Operator Methods in Quantum Mechanics. Clarendon, Oxford (1991)
376. Dong, S.H., Gonzalez-Cisneros, A.: Energy spectra of the hyperbolic and second Pöschl-Teller like potentials solved by new exact quantization rule. Ann. Phys. **323**, 1136–1149 (2008)
377. Shao, M.X., Mao, Y.Z., Zhang, D.T.: Geometric proof of exact quantization rules in one dimensional quantum mechanics. Int. J. Theor. Phys. **48**, 36–42 (2009)
378. Kasri, Y., Chetouani, L.: Application of the exact quantization rule for some noncentral separable potentials. Can. J. Phys. **86**, 1083–1089 (2008)
379. Qiang, W.C., Zhou, R.S., Gao, Y.: Application of the exact quantization rule to the relativistic solution of the rotational Morse potential with pseudospin symmetry. J. Phys. A, Math. Theor. **40**, 1677 (2007)
380. Gu, X.Y., Dong, S.H.: The improved quantization rule and the Langer modification. Phys. Lett. A **372**, 1972–1977 (2008)

381. Bagrov, V.G., Gitman, D.M.: Exact Solutions of Relativistic Wave Equations. Kluwer Academic, Dordrecht (1990)
382. Gu, X.Y., Dong, S.H., Ma, Z.Q.: Energy spectra for modified Rosen-Morse potential solved by the exact quantization rule. J. Phys. A, Math. Theor. **42**, 035303 (2009)
383. Hartmann, H.: Die Bewegung eines Korpers in einem ringformigen Potentialfeld. Theor. Chim. Acta **24**, 201–206 (1972)
384. Hartmann, H., Schuck, R., Radtke, J.: Die diamagnetische Suszeptibilität eines nicht kugelsymmetrischen Systems. Theor. Chim. Acta **46**, 1–3 (1976)
385. Mandal, B.P.: Path integral solution to non-central potential. Int. J. Mod. Phys. A **15**, 1225–1234 (2000)
386. Gönül, B., Zorba, İ.: Supersymmetric solutions of non-central potentials. Phys. Lett. A **269**, 83–88 (2000)
387. Rosen, N., Morse, P.M.: On the vibrations of polyatomic molecules. Phys. Rev. **42**, 210–217 (1932)
388. Manning, M.F., Rosen, N.: Minutes of the Middletown meeting, October 14. Phys. Rev. **44**, 951–954 (1933)
389. Qiang, W.C., Dong, S.H.: Analytical approximations to the solutions of the Manning-Rosen potential with centrifugal term. Phys. Lett. A **368**, 13–17 (2007)
390. Ikhdair, S.M., Sever, R.: Approximate l-state solutions of the D-dimensional Schrödinger equation for Manning-Rosen potential. Ann. Phys. (Berlin) **17**, 897–910 (2009)
391. Comtet, A., Bandrauk, A.D., Campbell, D.K.: Exactness of semiclassical bound state energies for supersymmetric quantum mechanics. Phys. Lett. B **150**, 159–162 (1985)
392. Popov, V.S., Karnakov, B.M., Mur, V.D.: On matching conditions in the WKB method. Phys. Lett. A **210**, 402–408 (1996)
393. Watson, J.K.G.: Semiclassical quantization and the Langer modification. J. Chem. Phys. **90**, 6443 (1989)
394. Curtis, L.J.: Classical mnemonic approach for obtaining hydrogenic expectation values of r^P. Phys. Rev. A **43**, 568–569 (1991)
395. Curtis, L.J.: Atomic Spectra and Lifetimes: A Conceptual Approach. Cambridge University Press, Cambridge (2003)
396. Ranjani, S.S., Kapoor, A.K., Panigrahi, P.K.: Bound states and band structure-unified treatment through the quantum Hamilton-Jacobi approach. Ann. Phys. **320**, 164–174 (2005)
397. Geojo, K.G., Sree Ranjani, S., Kapoor, A.K.: A study of quasi-exactly solvable models within the quantum Hamilton-Jacobi formalism. J. Phys. A, Math. Gen. **36**, 4591 (2003)
398. Ikhdair, S.M., Sever, R.: Any l-state solutions of the Woods-Saxon potential in arbitrary dimensions within the new improved quantization rule. Int. J. Mod. Phys. A **25**, 3941–3952 (2010)
399. Yin, C., Cao, Z.Q., Shen, Q.S.: Why SWKB approximation is exact for all SIPs. Ann. Phys. (N.Y.) **325**, 528–534 (2010)
400. Comtet, A., Bandrauk, A.D., Campbell, D.K.: Exactness of semiclassical bound state energies for supersymmetric quantum mechanics. Phys. Lett. B **150**, 159–162 (1985)
401. Barclaya, D.T., Maxwell, C.J.: Shape invariance and the SWKB series. Phys. Lett. A **157**, 357–360 (1991)
402. Barclaya, D.T.: Convergent WKB series. Phys. Lett. A **185**, 169–173 (1994)
403. Grandatia, Y., Bérard, A.: Ma-Xu quantization rule and exact JWKB condition for translationally shape invariant potentials. Phys. Lett. A **375**, 390–395 (2011)
404. Bhaduri, R.K., Sakhr, J., Sprung, D.W.L., Dutt, R., Suzuki, A.: Shape invariant potentials in SUSY quantum mechanics and periodic orbit theory. J. Phys. A, Math. Gen. **38**, L183 (2005)
405. Lozada-Cassou, M., Dong, S.H., Yu, J.: Quantum features of semiconductor quantum dots. Phys. Lett. A **331**, 45–52 (2004)
406. Dong, S.H., Lozada-Cassou, M.: Exact solutions of the Schrödinger equation with the position-dependent mass for a hard-core potential. Phys. Lett. A **337**, 313–320 (2005)
407. Yu, J., Dong, S.H., Sun, G.H.: Series solutions of the Schrödinger equation with position-dependent mass for the Morse potential. Phys. Lett. A **322**, 290–297 (2004)

408. Alhaidari, A.D.: Solutions of the nonrelativistic wave equation with position-dependent effective mass. Phys. Rev. A **66**, 042116 (2002)
409. Alhaidari, A.D.: Nonrelativistic Green's function for systems with position-dependent mass. Int. J. Theor. Phys. **42**, 2999–3009 (2003)
410. Alhaidari, A.D.: Solution of the Dirac equation with position-dependent mass in the Coulomb field. Phys. Lett. A **322**, 72–77 (2004)
411. Chen, G.: Approximate series solutions of the N-dimensional position-dependent mass Schrödinger equation. Phys. Lett. A **329**(1–2), 22–27 (2004)
412. Bastard, G.: Wave Mechanics Applied to Semiconductor Heterostructure. Les Editions de Physique, Les Ulis (1988)
413. Barranco, M., Pi, M., Gatica, S.M., Hernandez, E.S., Navarro, J.: Structure and energetics of mixed ^{4}He-^{3}He drops. Phys. Rev. B **56**, 8997–9003 (1997)
414. Harrison, P.: Quantum Wells, Wires and Dots. Wiley, New York (2000)
415. Serra, L., Lipparini, E.: Spin response of unpolarized quantum dots. Europhys. Lett. **40**, 667–672 (1997)
416. Arias de Saavedra, F., Boronat, J., Polls, A., Fabrocini, A.: Effective mass of one ^{4}He atom in liquid ^{3}He. Phys. Rev. B **50**, 4248–4251 (1994)
417. Gora, T., Williams, F.: Theory of electronic states and transport in graded mixed semiconductors. Phys. Rev. **177**, 1179–1182 (1969)
418. Von Roos, O.: Position-dependent effective masses in semiconductor theory. Phys. Rev. B **27**, 7547–7552 (1983)
419. Von Roos, O., Mavromatis, H.: Position-dependent effective masses in semiconductor theory. II. Phys. Rev. B **31**, 2294–2298 (1985)
420. Morrow, R.A.: Establishment of an effective-mass Hamiltonian for abrupt heterojunctions. Phys. Rev. B **35**, 8074–8079 (1987)
421. Trzeciakowski, W.: Boundary conditions and interface states in heterostructures. Phys. Rev. B **38**, 4322–4325 (1988)
422. Galbraith, I., Duggan, G.: Envelope-function matching conditions for GaAs/(Al, Ga)As heterojunctions. Phys. Rev. B **38**, 10057–10059 (1988)
423. Young, K.: Position-dependent effective mass for inhomogeneous semiconductors. Phys. Rev. B **39**, 13434–13441 (1989)
424. Einevoll, G.T., Hemmer, P.C., Thomsen, J.: Operator ordering in effective-mass theory for heterostructures. I. Comparison with exact results for superlattices, quantum wells, and localized potentials. Phys. Rev. B **42**, 3485–3496 (1990)
425. Weisbuch, C., Vinter, B.: Quantum Semiconductor Heterostructures. Academic Press, New York (1993)
426. Lévy-Leblond, J.M.: Position-dependent effective mass and Galilean invariance. Phys. Rev. A **52**, 1845–1849 (1995)
427. Gönül, B., Özer, O., Gönül, B., Üzgün, F.: Exact solutions of effective-mass Schrödinger equations. Mod. Phys. Lett. A **17**(37), 2453–2465 (2002)
428. Gönül, B., Gönül, B., Tutcu, D., Özer, O.: Supersymmetric approach to exactly solvable systems with position-dependent effective masses. Mod. Phys. Lett. A **17**(31), 2057–2066 (2002)
429. Plastino, A.R., Casas, M., Plastino, V.: Bohmian quantum theory of motion for particles with position-dependent effective mass. Phys. Lett. A **281**, 297–304 (2001)
430. Yu, J., Dong, S.H.: Exactly solvable potentials for the Schrödinger equation with spatially dependent mass. Phys. Lett. A **325**, 194–198 (2004)
431. Pérez-Alvarez, R., Rodriguez-Coppola, H.: Transfer matrix in 1D Schrödinger problems with constant and position-dependent mass. Phys. Status Solidi (b) **145**, 493–500 (1988)
432. Pérez-Alvarez, R., Parra-Santiago, J.L., Pajón-Suárez, P.: Cylindrical quantum well with position-dependent mass. Phys. Status Solidi (b) **147**, 127–133 (1988)
433. Milanović, V., Ikonić, Z.: Generation of isospectral combinations of the potential and the effective-mass variations by supersymmetric quantum mechanics. J. Phys. A, Math. Gen. **32**, 7001–7015 (1999)

434. Renan, R., Pacheco, M.H., Almeida, C.A.S.: Treating some solid state problems with the Dirac equation. J. Phys. A, Math. Gen. **33**, L509–L514 (2000)
435. Milanović, V., Ikonić, Z.: Intraband absorption of infrared radiation in a semiconductor quantum dot. Phys. Rev. B **39**, 7982–7984 (1989)
436. Koç, R., Koca, M.: A systematic study on the exact solution of the position dependent mass Schrödinger equation. J. Phys. A, Math. Gen. **36**, 8105–8112 (2003)
437. Roy, B.: Lie algebraic approach to singular oscillator with a position-dependent mass. Europhys. Lett. **72**, 1–6 (2005)
438. Dong, S.H., Peña, J.J., García-Ravelo, J., Pacheco-García, C.: Algebraic approach to the position-dependent mass Schrödinger equation for a singular oscillator. Mod. Phys. Lett. A **22**, 1039–1045 (2007)
439. Schmidt, A.G.M.: Wave-packet revival for the Schrödinger equation with position-dependent mass. Phys. Lett. A **353**, 459–462 (2006)
440. Chen, G.: Approximate series solutions of the N-dimensional position-dependent mass Schrödinger equation. Phys. Lett. A **329**, 22–27 (2004)
441. Chen, G.: Exact solutions of the position-dependent mass Schrödinger equation in D dimensions. Phys. Lett. A **331**, 312–315 (2004)
442. Ballesteros, A., Enciso, A., Herranz, F.J., Ragnisco, O., Riglioni, D.: A new exactly solvable quantum model in N dimensions. Phys. Lett. A **375**, 1431–1435 (2011)
443. Ikhdair, S.M., Sever, R.: Exact solvable effective mass d-dimensional Schrödinger equation for pseudoharmonic and modified Kratzer problems. Int. J. Mod. Phys. C **20**, 361–372 (2009)
444. Dirac, P.A.: The Principles of Quantum Mechanics, 4th edn. Oxford University Press, London (1958)
445. Greiner, W.: Relativistic Quantum Mechanics-Wave Equations. Springer, Berlin (1990)
446. Vaidya, A.N., Souza, L.E.S.: Algebraic calculation of the S-matrix for the Dirac-Coulomb problem in $2+1$ dimensions. Phys. Lett. A **293**, 129–132 (2002)
447. Greiner, W.: Quantum Mechanics—An Introduction. Springer, Berlin (1989)
448. Ikhadir, S.M., Mustafa, O., Sever, R.: Solution of the Dirac equation for vector and scalar potentials and some applications. Hadron. J. **16**(1), 57–74 (1993)
449. Dong, S.H., Gu, X.Y., Ma, Z.Q.: Exact solutions of the Dirac equation with a Coulomb plus scalar potential in $2+1$ dimensions. Int. J. Mod. Phys. E **11**, 483–489 (2002)
450. Dong, S.H., Lozada-Cassou, M.: On the analysis of the eigenvalues of the Dirac equation with a $1/r$ potential in D dimensions. Int. J. Mod. Phys. Lett. E **13**, 917–931 (2004)
451. Chen, C.Y., Liu, C.L., Lu, F.L., Sun, D.S.: Bound states of the Klein-Gordon equation with n-dimensional scalar and vector hydrogen atom-type potentials. Acta Phys. Sin. **52**(7), 1579–1584 (2003)
452. Friedel, J.: XIV. The distribution of electrons round impurities in monovalent metals. Philos. Mag. Ser. 7 **43**, 153–189 (1952)
453. Mahan, G.D.: Many-Particle Physics. Plenum, New York (2000)
454. Lin, D.H.: Friedel sum rule for relativistic spin-1/2 systems. Phys. Rev. A **72**, 012701 (2005)
455. Lin, D.H.: Friedel theorem for two-dimensional relativistic spin-1/2 systems. Phys. Rev. A **73**, 044701 (2006)
456. Lin, D.H.: Friedel theorem for Dirac fermions in D dimensions. Phys. Rev. A **74**, 032109 (2006)
457. Chen, G.: Relativistic spectral comparison theorem in two dimensions. Phys. Rev. **71**, 024102 (2005)
458. Hall, R.L.: Spectral comparison theorem for the Dirac equation. Phys. Rev. Lett. **83**, 468–471 (1999)
459. Chen, G.: Spectral comparison theorem for the N-dimensional Dirac equation. Phys. Rev. **72**, 044102 (2005)
460. Hall, R.L.: Special comparison theorem for the Dirac equation. Phys. Rev. Lett. **101**, 090401 (2008)
461. Hall, R.L.: Relativistic comparison theorems. Phys. Rev. A **81**, 052101 (2010)
462. Davydov, A.S.: Quantum Mechanics. Pergamon, New York (1965)

463. Einstein, A., Bergmann, P.: On a generalisation of Kaluza's theory of electricity. Ann. Math. **34**, 683–701 (1938)
464. Einstein, A.: Demonstration of the non-existence of gravitational fields with a non-vanishing total mass free of singularities. Rev. Univ. Nac. Tucumán Ser. A **2**, 11–16 (1941)
465. Einstein, A., Bargmann, V., Bergmann, P.: On the five-dimensional representation of gravitation and electricity. In: Theodore von Karman Anniversary Volume, pp. 212–225. California Institute of Technology, Pasadena (1941)
466. Einstein, A., Pauli, W.: On the non-existence of regular stationary solutions of relativistic field equations. Ann. Math. **44**, 131–137 (1943)
467. Green, M.B., Schwarz, J.H., Witten, E.: Superstring Theory. Cambridge University Press, Cambridge (1988)
468. Delgado, A.: Xtra-Dimensional world(s). J. Phys. Conf. Ser. **53**, 359 (2006)
469. Pérez-Lorenzana, A.: An introduction to extra dimensions. J. Phys. Conf. Ser. **18**, 224 (2005)
470. Randall, L., Sundrum, R.: Large mass hierarchy from a small extra dimension. Phys. Rev. Lett. **83**, 3370–3373 (1999)
471. Randall, L., Sundrum, R.: An alternative to compactification. Phys. Rev. Lett. **83**, 4960–4963 (1999)
472. Arkani-Hamed, N., Dimopoulos, S., Dvali, G., Kaloper, N.: Infinitely large new dimensions. Phys. Rev. Lett. **84**, 586–589 (2000)
473. Long, J.C., Chan, H.W., Price, J.C.: Experimental status of gravitational-strength forces in the sub-centimeter regime. Nucl. Phys. B **539**, 23–34 (1999)
474. Hoyle, C.D., Kapner, D.J., Heckel, B.R., Adelberger, E.G., Gundlach, J.H., Schmidt, U., Swanson, H.E.: Sub-mm tests of the gravitational inverse-square law. Phys. Rev. D **70**, 042004 (2004)
475. Hoyle, C.D., Schmidt, U., Heckel, B.R., Adelberger, E.G., Gundlach, J.H., Kapner, D.J., Swanson, H.E.: Submillimeter test of the gravitational inverse-square law: a search for "large" extra dimensions. Phys. Rev. Lett. **86**, 1418–1421 (2001)
476. Long, J.C., Chan, H.W., Churnside, A.B., Gulbis, E.A., Varney, M.C.M., Price, J.C.: 1. Upper limits to submillimetre-range forces from extra space-time dimensions. Nature **421**, 922–925 (2003)
477. Hoyle, C.D., Kapner, D.J., Heckel, B.R., Adelberger, E.G., Gundlach, J.H., Schmidt, U., Swanson, H.E.: Submillimeter tests of the gravitational inverse-square law. Phys. Rev. D **70**, 042004 (2004)
478. Wesson, P.S.: The effective properties of matter of Kaluza-Klein solitons. Phys. Lett. B **276**, 299–302 (1992)
479. Wesson, P.S., de Leon, J.P.: Kaluza-Klein equations, Einstein's equations, and an effective energy-momentum tensor. J. Math. Phys. **33**, 3883 (1992)
480. Wesson, P.S.: A physical interpretation of Kaluza-Klein cosmology. Astrophys. J. **394**, 19–24 (1992)
481. Wesson, P.S.: The geometrical unification of gravity with its source. Gen. Relativ. Gravit. **40**, 1353–1365 (2008)
482. Randall, L., Sundrum, R.: Fermion-Boson duality in integrable quantum field theory. Mod. Phys. Lett. A **13**, 2807–2818 (1998)
483. Arkani-Hamed, N., Dimopoulos, S., Dvali, G.: The hierarchy problem and new dimensions at a millimeter. Phys. Lett. B **429**, 263 (1998)
484. Maartens, R.: Brane-world gravity. Living Rev. Relativ. **7** (2004). irr-2004-7
485. Bailin, D., Love, A.: Kaluza-Klein theories. Rep. Prog. Phys. **50**, 1087 (1987)
486. van Dongen, J.: Einstein and the Kaluza-Klein particle. Stud. Hist. Philos. Sci. Part B, Stud. Hist. Philos. Mod. Phys. **33**, 185–210 (2002)
487. DeWitt, B.: In: DeWitt, B., DeWitt, C. (eds.) Relativity, Groups and Topology Lectures at 1963 Les Houches School. Gordon & Breach, New York (1964)
488. Kerner, R.: Generalization of the Kaluza-Klein theory for an arbitrary non-abelian gauge group. Ann. Inst. Henri Poincaré A **9**, 143–152 (1968)
489. Trautman, A.: Fibre bundles associated with space-time. Rep. Math. Phys. **1**, 29–62 (1970)

490. Cho, Y.M.: Higher-dimensional unifications of gravitation and gauge theories. J. Math. Phys. **16**, 2029 (1975)
491. Cho, Y.M., Freund, P.G.: Non-Abelian gauge fields as Nambu-Goldstone fields. Phys. Rev. D **12**, 1711–1720 (1975)
492. Scherk, J., Schwarz, J.H.: Dual field theory of quarks and gluons. Phys. Lett. B **57**, 463–466 (1975)
493. Cremmer, E., Scherk, J.: Spontaneous compactification of extra space dimensions. Nucl. Phys. B **118**, 61–75 (1977)
494. Cremmer, E., Julia, B., Scherk, J.: Supergravity in theory in 11 dimensions. Phys. Lett. B **76**, 409–412 (1978)
495. Cremmer, E., Julia, B.: The $N = 8$ supergravity theory. I. The lagrangian. Phys. Lett. B **80**, 48–51 (1978)
496. Antoniadis, I.: A possible new dimension at a few TeV. Phys. Lett. B **246**, 377 (1990)
497. Antoniadis, I., Arkani-Hamed, N., Dimopoulos, S., Dvali, G.: New dimensions at a millimeter to a Fermi and superstrings at a TeV. Phys. Lett. B **436**, 257–263 (1998)
498. Rizzo, T.: Introduction to extra dimensions. arXiv:1003.1698 [hep-ph] (2010)
499. Weinberg, S.: Gravitation and Cosmology. Wiley, New York (1972)
500. Witten, E.: Fermion quantum numbers in Kaluza-Klein theories. Princeton Preprint (1983). [Also see: Shelter Island, II. In: Jackiw, R., Khuri, N.N., Weinberg, S., Witten, E. (eds.) Proceedings of the 1983 Shelter Island Conference on Quantum Field Theory and the Fundamental Problems of Physics, p. 227. MIT Press, Cambridge (1985)]
501. Schrödinger, E.: Diracsches Elektron im Schwerefeld I. Sitzungsber. Preuss. Akad. Wiss. Phys.-Math. Kl. **11**, 105–128 (1932)
502. Lichnerowicz, A.: Spineurs harmoniques. C. R. Acad. Sci. Paris **257**, 7–9 (1963)
503. Destri, C., Orzalesi, C.A., Rossi, P.: Matter fields and metric deformations in multidimensional unified theories. Ann. Phys. (N.Y.) **147**, 321–364 (1983)
504. Wu, Y.S., Zee, A.: Chiral anomalies, higher dimensions, and differential geometry. Nucl. Phys. B **239**, 477–507 (1984)
505. Atiyah, M., Hirzebruch, F.: In: Haeflinger, A., Narasiman, R. (eds.) Essays on Topology and Related Topics, p. 18. Springer, Berlin (1970)
506. Schellekens, A.N.: Fermion mass spectrum in a generalized instanton background. Nucl. Phys. B **250**, 252–278 (1985)
507. Schellekens, A.N.: Compactification of Einstein-Yang-Mills theories on CP_n. Nucl. Phys. B **262**, 661–688 (1985)
508. Randjbar-Daemi, S., Salam, A., Strathdee, J.: Instanton induced compactification and fermion chirality. Phys. Lett. B **132**, 56–60 (1983)
509. Watamura, S.: Spontaneous compactification of $D = 10$ Maxwell-Einstein theory leads to $\mathrm{SU}(3) \times \mathrm{SU}(2) \times U(1)$ gauge symmetry. Phys. Lett. B **129**, 188–192 (1983)
510. Watamura, S.: Spontaneous compactification and CPN: $\mathrm{SU}(3) \times \mathrm{SU}(2) \times \mathrm{U}(1)$, $\sin^2\theta_W$, g_3/g_2 and SU(3)-triplet chiral fermions in four dimensions. Phys. Lett. B **136**, 245–250 (1984)
511. Rubakov, V.A., Shaposhnikov, M.E.: Do we live inside a domain wall? Phys. Lett. B **125**, 136–138 (1983)
512. Visser, M.: An exotic class of Kaluza-Klein models. Phys. Lett. B **159**, 22–25 (1985)
513. Squires, E.J.: Dimensional reduction caused by a cosmological constant. Phys. Lett. B **167**, 286–288 (1986)
514. Gibbons, G.W., Wiltshire, D.L.: Spacetime as a membrane in higher dimensions. Nucl. Phys. B **287**, 717–742 (1987)
515. Gogberashvili, G.: Hierarchy problem in the shell-Universe model. Int. J. Mod. Phys. D **11**, 1635–1638 (2002)
516. Rose, M.E.: Elementary Theory of Angular Momentum. Wiley, New York (1957)

Index

DOI 10.1007/978-94-007-1917-0, © Springer Science+Business Media B.V. 2011

MIX
Papier aus verantwortungsvollen Quellen
Paper from responsible sources
FSC® C105338

If you have any concerns about our products,
you can contact us on
ProductSafety@springernature.com

In case Publisher is established outside the EU,
the EU authorized representative is:
Springer Nature Customer Service Center GmbH
Europaplatz 3, 69115 Heidelberg, Germany

Printed by Libri Plureos GmbH
in Hamburg, Germany